ANIMAL CELL TECHNOLOGY: BASIC & APPLIED ASPECTS

JAACT'96 YOKOHAMA

JAACT'96 ORGANIZING COMMITTEE

Chairperson:	Kazuo Nagai	(Department of Bioengineering, Tokyo Institute of Technology)
Secretary General:	Masaaki Wachi	(Department of Bioengineering, Tokyo Institute of Technology)
Organizing Boards:	Akio Ametani	(The University of Tokyo)
	Shuichi Hashizume	(Morinaga Institute of Biological Science)
	Kazuki Shinohara	(National Research Institute of Fisheries Science)
	Eiji Suzuki	(The University of Tokyo)
	Takamoto Suzuki	(Kirin Brewery Co., Ltd.)

Program Committee

Masuo Aizawa	Tokyo Institute of Technology
Toshihiro Akaike	Tokyo Institute of Technology
John Barford	University of Sydney
Shun'ichi Dosako	Snow Brand Milk Products Co.,Ltd.
Hiroyoshi Hoshi	Research Institute of Functional Peptides
Wei-Shou Hu	University of Minnesota
Koji Ikura	Kyoto Institute of Technology
David Barnes	Oregon State University
David Jayme	Life Technologies Inc.
Shuichi Kaminogawa	The University of Tokyo
Kin'ichi Kawamura	Komatsugawa Chemical Engineering Co.,Ltd.
Yasuo Kitagawa	Nagoya University
Yoji Mitsui	Nat. Inst. Biosci. Human-Tech.
Alan Moore	Microbiological Associates Inc.
Tadao Ohno	Riken Gene Bank
Ralf Pörtner	Technische Universitat Hamburg-Harburg
Ryuzo Sasaki	Kyoto University
Kazuo Shin	Meiji Milk Products Co.,Ltd.
Hajime Unno	Tokyo Institute of Technology
John Wérenne	Universite Libre de Bruxelles

Animal Cell Technology: Basic & Applied Aspects

Volume 9

Proceedings of the Ninth Annual Meeting of the Japanese Association for Animal Cell Technology, Yokohama, Japan, September 1-4, 1996

Edited by

K. NAGAI
Department of Bioengineering,
Tokyo Institute of Technology, Yokohama, Japan

and

M. WACHI
Department of Bioengineering,
Tokyo Institute of Technology, Yokohama, Japan

SPRINGER SCIENCE+BUSINESS MEDIA, B.V.

A C.I.P. Catalogue record for this book is available from the Library of Congress.

ISBN 978-94-010-6170-4 ISBN 978-94-011-5161-0 (eBook)
DOI 10.1007/978-94-011-5161-0

Printed on acid-free paper

Contents

I. Establishing new cell lines and characterizing molecular biological properties of cultured cells

II. Effect of various substances on animal cell response

III. Basics and applications in immunology

IV. Technology and engineering for cell culture

V. Animal cell culture for transplantation and developmentof artificial organs

VI. Efficient production of protein and virus by cultured animal cells

*** Papers related to the subjects will be published in *Cytotechnology* specially issued for JAACT'96 in the more extended form.**

Preface

The plant cells, if not all, are generally believed to have totipotency which enables single cells to form mature plants when provided an optimal culture condition. On the other hand, in case of animal cells, the totipotency had been questioned as part of the genetic elements in the cells other than fertilized eggs or those derived from an early stage of embryo were suspected to have irreversible modification. Wilmut et al., however, recently reported that they succeeded in producing a cloned animal by fusing a nucleus derived from a cloned cell of mammary gland with an enucleated egg. This result revealed that, at least some of the somatic cells retain whole genetic informations necessary to support the production of an adult animal. This fact encourages us because animal cells have been shown to be able to express any of their native functions if supplied suitable enviroments. In this respect, animal cell technology is a powerful field of dicipline which enables establishment and modification of cloned cells with specific functions, application of the modulated cells to gene therapy, cultivation of the cells at an industrial scale to produce biologically active materials, reconstitution of an artificial tissues and organs and even production of cloned animals.

At the Ninth Annual Meeting of Japanese Association for Animal Cell Tecnology (JAACT) which was held in Yokohama from 1st through 4th of September, 1996, we welcomed more than 230 participants including 45 from abroad and accepted 133 presentations. In the present proceedings, the reports presented at the Meeting have been compiled. The papers will show the recent progress in the field of animal cell technology in the application as well as with basic aspects. Some of the presentations including the plenary lecture given by Dr. Edgar G. Engleman will be published in *Cytotechnology* specially issued for JAACT'96 in the more extended form. Those papers are indicated by asterisks in the table of contents.

The editors express their sincere thanks to all participants at the Meeting, to the organizers of the Symposium sessions, to the members of the organizing committee and program committee and the graduate and undergraduate students of Tokyo Institute of Technology for their dedication in assuring the success of the Meeting. We also appreciate Japanese Bioindustry Association, City of Yokohama and Nestle Science Promotion Foundation for their finacial support of the Meeting.

The Editors

STRESS-INDUCIBLE GENE EXPRESSION IN FISH CELLS

M. YAMASHITA
National Research Institute of Fisheries Science
Fukuura, Kanazawa, Yokohama 236, Japan

1. Introduction

Fish cells can generally grow over a wide range of temperatures, and the culture temperature is correlated with the body temperature in the fish's natural habitat. For example, rainbow trout can adapt to a temperature range of 0-22 °C; therefore, cultured trout cells usually grow in the range of 0 to 24 °C with an optimum at 22 °C. On the other hand, in the case of mammalian cells, the permissive growth temperature range is known to be 36-39 °C, with an optimum at 37 °C.

When cells are exposed to heat shock and other stresses, such as heavy metals and UV radiation, several stress proteins are induced [1]. This induction is mediated by a stress-activated heat shock factor, which can recognize specific heat shock promoters. In the case of rainbow trout cells, when cell cultures are transferred from 22 °C to 28 °C, the heat shock proteins have been reported to transiently appear [2].

On the other hand, the phenomenon connected with cold acclimation has been well studied in rainbow trout. Changes in the isozymes were reported for the cold-acclimated fish [3-8]. These phenomena were found to occur within 1-4 weeks of acclimation.

Temperature regulation of gene expression is critical for the growth and survival of cold-blooded animals living in variable thermal environments. In order to characterize the temperature adaptation in fish, the heat shock response upon exposure to stress conditions and the cold response and acclimation were examined in rainbow trout fibroblast RTG-2 cells. The major heat shock gene, the HSP70 gene, was cloned from the rainbow trout cells, and its stress-inducible expression was characterized. Furthermore, it was found that both induction and repression of protein synthesis took place in the trout cells during the cold acclimation [9, 10].

2. Heat Shock and Stress Responses

One of the major stress proteins is the HSP70 family, ranging from 66 to 78 kDa.

HSP70 is a heat-inducible member of the HSP70 family, and the heat shock cognate protein, HSC70, is constitutively expressed in the cells. Glucose-regulated protein GRP78 is induced by inhibition of glycosylation of protein synthesis. HSP70 recognizes unstructured polypeptides such as nascent polypeptides and heat-denatured proteins, and they release them only in the presence of hydrolyzable ATP. HSP70 carries out chaperone functions, such as binding to other polypeptides and allowing them to reach their correct intracellular destination in the cell, transporting other proteins across membranes, or preventing other proteins from misfolding.

Figure 1 shows the transactivation mechanism of the heat shock gene. Inactive heat shock transcription factor in cytoplasm is activated by cellular stress. Activated HSF forms a trimer and is translocated into the nucleus. The activated HSF binds to the heat shock element, HSE, of the heat shock gene promoter, and it transactivates the heat shock gene. The stress-inducible HSP70 gene has been cloned from rainbow trout recently [Yamashita, M. *et al.*, unpublished]. Its promoter region was found to contain three heat shock elements, HSE1, HSE2, and HSE3. The heat shock element corresponds to inverted multimers of the sequence nTTCn or nGAAn which binds with the heat shock factor. Each heat shock element of the rainbow trout HSP70 gene also consists of four or three 5-bp units. The CAT assay with expression CAT vectors containing deletion mutant of rainbow trout HSP70 promoter showed that only HSE1 has high heat-inducible transcriptional activity. In addition, the transient expression of the *lacZ* constructs of the heat shock promoter were observed in the zebrafish embryo at the 24 h-stage after the heat-shock treatment. These findings show that the rainbow trout HSP70 promoter is a useful tool for assay of a stress-inducible gene expression in fish.

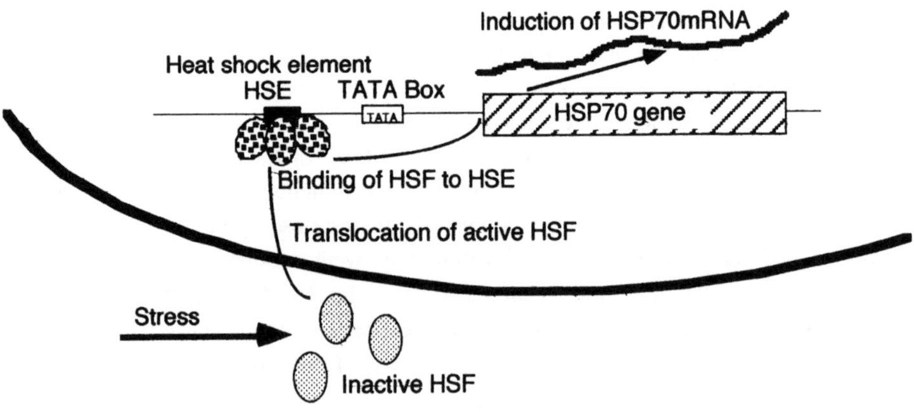

Figure 1. Transactivation mechanism of the HSP70 gene.

3. Molecular and Cellular Basis of Cold Acclimation

Cold induction of various proteins and enzymes has been reported in fish, bacteria and plants previously. One of the most drastic examples of cold acclimation is increased frost tolerance. Low molecular mass hydrophilic proteins were found at high concentrations in the serum of polar-dwelling marine fishes [3]. Other changes that occur during cold acclimation were found to be the appearance of new isozymes, such as trout acetylcholine esterase [4], lactate dehydrogenase [5], malate dehydrogenase [6] and myosin [7]. Change in lipid composition is mediated by desaturase in response to increase in cellular membrane fluidity [8]. However, the molecular mechanisms and exact roles of such cold acclimation processes are almost completely unknown. These cold-specific proteins in fish were examined with fish *in vivo*, and most cases of the cold induction are considered to be regulated hormonally in response to seasonal changes.

Figure 2 shows the hypothesis for the cellular regulation of cold response. Various kinds of genes are induced by the cold, and cold-inducible proteins and enzymes as the gene product play some roles in 1) metabolic compensation to enhance enzyme activity by changing to the more active isozymes at the cold temperature, 2) cold-inducible transcription and translation, 3) cold and freeze tolerance.

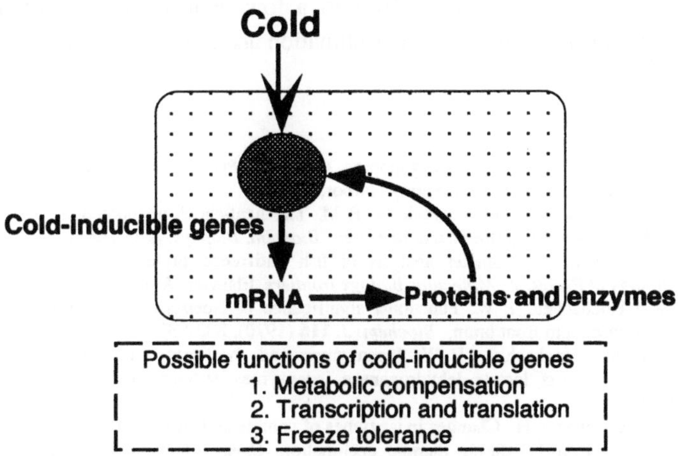

Figure 2. Mechanism of cold acclimation of fish cells.

To identify the cold-inducible proteins in the trout cells, cellular proteins were also subjected to two-dimensional gel electrophoresis [9]. A spot of a 70 kDa protein "cold70" gradually appeared during the cold treatment. From the analysis of the N-terminal amino acid sequence of cold70 using a blotted sample, the sequence exhibits extensive homologies with the yeast cell division cycle gene, CDC48p protein, and mammalian valosin-containing protein (VCP). The expression of cold70 was highly responsive to a temperature change; when the cells were returned to at 22℃, the amounts of the proteins markedly decreased within 1 day. According to studies on cold-sensitive yeast mutants, CDC48p is essential for the yeast to survive at a cold temperature. VCP acts as a substrate for a protein kinase related to mitosis and regulating protein interaction. From this fact that the homologous proteins are associated with cell division and mitosis, the accumulation of the trout cold70 might participate in the metabolic compensation for the delay in cell cycling due to cold temperature.

Because a limited number of changes in gene expression occurs during the cold acclimation of rainbow trout cells, a subtracted cDNA library of the cold-treated cells was constructed to isolate the cold-regulated cDNA clones [10]. A set of cDNA clones encoding three members of ferritin H isoforms was isolated as cold-inducible genes. Northern hybridization and nuclear run-on transcription assay indicated that the ferritin H genes were regulated by the cold-treatment at the transcriptional level.

The findings that exposure of trout cells to 4 °C alters the relative abundance of specific mRNAs and proteins suggest that vertebrates have an unknown cellular mechanism responsible for adaptation to cold temperature. The changes in the mRNA and protein contents during the cold-treatment may result from the expression of genes involved in the adjustment of the cellular metabolism to a low temperature, or else the induced proteins may be directly involved in the acclimation and growth of the trout cells.

4. References

1. Morimoto, R. I., Tissieres, A. and Georgopoulos, C.: *Stress Proteins in Biology and Medicine*, Cold Spring Harbor Laboratory Press, Cold Spring Harbor, 1990.
2. Kothary, R. K., Burgess, E. A. and Candido, E. P. M.: The heat-shock phenomenon in cultured cells of rainbow trout HSP70 mRNA synthesis and turnover, *Biochim. Biophys. Acta* **783** (1984), 137-143.
3. Gong, Z. and Hew, C. L.: Promoter analysis of fish antifreeze protein genes, *Biochemistry and Molecular Biology of Fishes, 2. Molecular biology frontiers*, Elsevier, Amsterdam, 1993, 307-324.
4. Boldwin, J. and Hochachka, P. W.: Functional significance of isoenzymes in thermal acclimatization. Acetylcholinesterase from trout brain, *Biochem. J.* **116** (1970), 883-887.
5. Somero, G. M. and Hochachka, P. W.: Isoenzymes and short-term temperature compensation in poikilotherms: activation of lactate dehydrogenase isoenzymes by temperature decreases, *Nature* **223** (1969), 194-195.
6. Kurokawa, T. and Nakano, H.: Changes in isozymes of soluble and mitchondrial malate dehydrogenase of carp (*Cyprinus carpio* L.) during thermal acclimation, *Comp. Biochem. Physiol.* **99B** (1991), 911-915.
7. Watabe, S., Hwang, G. C., Nakaya, M., Guo, X. F., and Okamoto, Y.: Fast skeletal myosin isoforms in thermally acclimated carp, *J. Biochem.*, **111** (1992), 113-122.
8. Tiku, P. E., Gracey, A. Y., Macartney, A. I., Beynon, R. J., and Cossins, A. R.: Cold-induced expression of delta 9-desaturase in carp by transcriptional and posttranslational mechanisms, *Science* **271** (1996), 815-818.
9. Yamashita, M., Ojima, N., and Sakamoto, T.: Induction of proteins in response to cold acclimation of rainbow trout cells, *FEBS Lett.* **382** (1996), 261-264.
10. Yamashita, M., Ojima, N., and Sakamoto, T.: Molecular Cloning and Cold-Inducible Gene Expression of Ferritin H Subunit Isoforms in Rainbow Trout Cells, *J. Biol. Chem.* **271** (1996), 26908-26913.

CELL CULTURE AND SOME PROPERTIES OF FISH IMMUNOCYTES

Tasuku Watanabe & Tadaaki Moritomo
Laboratory of Fish Pathology, Department of Veterinary Sciences, Nihon University.
1866 Kameino, Fujisawa 252, Japan

1. Introduction

The importance of the *in vitro* cultures of animal cells with their proper functions has been well recognized. There have been many trials on the cell culture maintaining their proper and sustainable functions. There are also many studies using such cell systems. The recent studies on immunocyte cultures reveal that cultured immunocytes provide adequate research systems for the investigations both of cytokine characterizations and their roles in immunobiology and host defense (Papadimitriou 1989, van Furth 1992). Especially, the intensive studies on chemoatractant cytokines (chemokines), such as interleukin 8 and macrophage inflammatory protein, provided tools for the definition of inflammation in molecular level (Baggiolini 1994). Most of such studies were performed using macrophage cell lines in mammals. There have also been several attempts on the cell culture of fish immunocytes. This overview describes the recent studies and their characteristics upon the trials of fish immunocyte cultures including our works.

2. Fish macrophage cell lines

There are also several studies on the cell cultures of fish immunocompetent cells. However, there are few markers of fish immunocytes. Macrophages, similar to mammalian macrophages, have some distinct and specific markers, such as phagocytosis, glass adherence, and peculiar morphology. Some cell lines of fish macrophages or monocytes have been established representing some properties of mononuclear phagocytes, as shown in Table 1. Most of them were derived from normal and healthy freshwater fish, differing from the mammalian macrophage lines. Human tumor necrotic factor and interleukin-2 enhanced the growth of a cell line derived from the peripheral blood monocytes of carp, whereas γ-interferon decreased the growth (Faisal 1990). The channel catfish monocyte-like cell line have the several properties of macrophages, such as nonspecific esterase and peroxidase positive cytoplasm, phagocytic for latex beads, and interleukin-1 production active on channel catfish and mouse T lymphocytes. The cell lines are also efficient antigen-presenting cells to autologous peripheral blood leukocytes (Vallejo 1991). A long-term macrophage cell line from goldfish kidney monocytes or macrophages stained positive for non-specific esterase, were phagocytic to sheep erythrocytes, and produced reactive oxygen and nitrogen intermediates when stimulated phorbol esters or lipopolysaccharide (Wang 1995).

TABLE 1. Cell lines derived from fish monocytes/macrophages

Species	Name	Origin	Properties	References
Carp *Cyprinus carpio*	CLC	Peripheral blood monocytes	Stimulated by tumor necrosis factor	Faisal & Ahne 1990
Channel catfish *Ictalurus punctatus*	M22	Peripheral blood monocytes	Antigen presentation, interleukin-1 production	Vallejo et al 1991
Goldfish *Carassius auratus*		kidney macrophages	Chemotaxis, respiratory burst	Wang et al 1995
Red sea bream *Pagrus major*	PMM	Resident perito-neal macrophages	IgM receptor, chemotactic factor	Watanabe et al 1997

We established cell lines derived from the resident peritoneal macrophages of red sea bream (RSB) *Pagrus major*, a popular marine teleost in Japan. These cell lines had some characteristics of macrophages, particularly IgM receptors on cell surface and the production and secretion of chemotactic factor-like substances.

3. Establishment of the cell lines from red sea bream macrophages

Resident peritoneal cells of marine fish are divided into four cell types (Watanabe 1997). Most of fresh water fish are one-cell type. Their peritoneal cells are macrophages. RSB have three kinds of peritoneal cells, macrophages, heterophilic granulocytes, and unclassified large cells which cannot observed in blood stream nor in head kidney hematopoietic tissue (Watanabe 1995). *Figure 1* shows the cytospined RSB peritoneal cells stained with May-Gruenward Giemsa, consisting of macrophages (m), heterophilic granulocytes (h) with eosinophilic granules, and large cells (l).

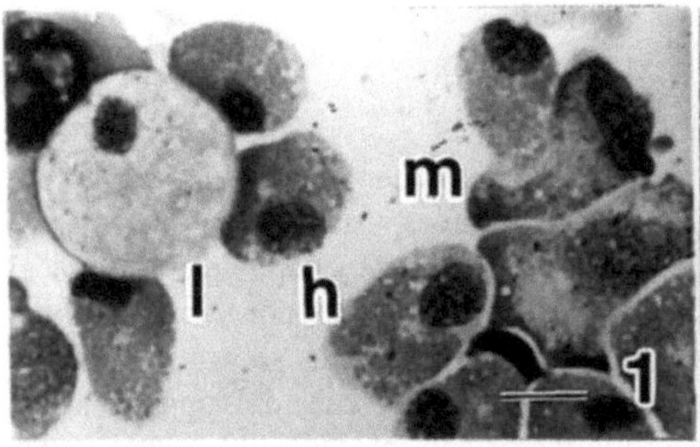

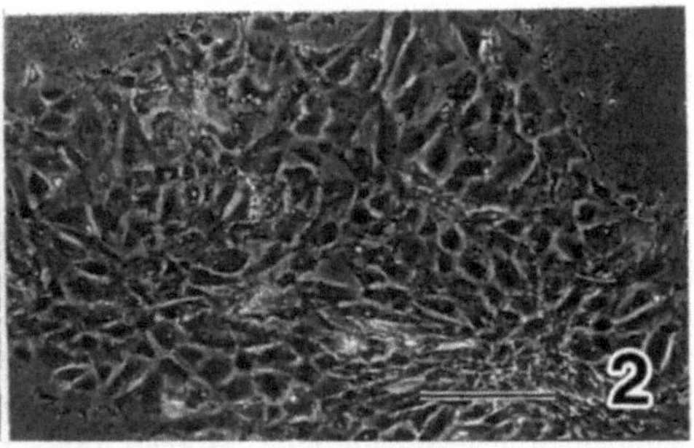

Figure 1. Three kinds of red sea bream peritoneal cells, showing macrophages (m), heterophilic granulocytes (h), and a large cell (l). Bar=10μm.
Figure 2. Phase contrast photomicrograph of the primary culture of red sea bream resident peritoneal macrophages. Bar=200μm.

We tried to culture the RSB resident peritoneal macrophages, as we often observed the mitosis in resident peritoneal macrophage populations. The media for cultivation of macrophages and procuring peritoneal cells were RPMI1640 medium supplemented with 0.4% NaCl and 20% of fetal bovine serum (FBS) (PM-RPMI20). Two-year-old RSB (1.2-1,7 kg body weight), cultured at the Fisheries Research Laboratory, the University of Tokyo, were anesthetized, injected with 20-40 ml of PM-RPMI20 into the abdominal cavity, and harvested peritoneal cells. About 10^7 cells/flask of the peritoneal cells were adhered to tissue culture flasks, washing with PM-RPMI20 three times, and the remaining macrophage (about 2.5 x 106 cells/flask) were cultured at 25 and 30℃ in 5% CO_2/95% air. When large cell colonies or confluent cell sheets were formed in flasks, cells were trypsinized and transferred to new flasks.

The cultured cells, designated PMM5 and PMM6, associated with *Pagrus major macrophage*, have been transferred more than 30 times from the initiation in May 1995. These cultured cells were mononuclear, adhered spreading to the flask (*Figure 2*), and proliferated slowly (doubling time: 4.8-5.0 days). These cultured cells were positive in non-specific esterase and acid-phosphatase, whereas they were negative in alkali-phosphatase and peroxidase, similar to the resident peritoneal macrophages. These morphological and cytochemical properties indicate that the cultured cells, PMM5 and PMM6, are derived from the RSB resident peritoneal macrophages.

4. Properties of cultured red sea bream resident peritoneal macrophages

The cultured cells in slide flasks were incubated with one ml of Latex beads (2.0μm in diameter) at 25°C for one hour. The supernatants were removed by decantation and the reactants were washed three times with PM-PBS, followed by staining with May Gruenward-Giemsa. The preparations were observed by applying both ultraviolet ray and visible ray. Most of these cultured cells had phagocytotic activity to Latex beads. Some cells engulfed more than 20 beads. Cultured cells and RSB RPM were incubated with RSB IgM in PM-PBS, RSB serum, or PM-RPMI20, followed by stained with biotin-anti-IgM in PB-PBS at 25°C for 30 minutes. After washed with PM-PBS three times, the cells were stained with FITC-conjugated avidin. The cultured macrophages, PMM5 and PMM6, had the receptors to RSB IgM on the cell surface. The IgM receptors and phagocytosis are the principal characteristic of phagocytes.

In vitro chemotaxis was assayed using a modification of previously described (Wang 1995). Bottom wells of blind well chambers were filled with 70μl of chemoattractant (10^{-8}M N-formylmethionyl leucocyl phenylalanine (FMLP) in PM-RPMI20, the supernatant of PMM5 and PMM6 cells, RSB serum, and PM-RPMI20) and top wells filled with 200μl of PM-RPMI containing 2.5×10^6 cultured cells. The two wells were separated by a polycarbonate membrane with a 5-μm pore size. Chambers were incubated at 25°C for three hours. Membranes were removed and stained with hematoxyline. The cultured cells that migrated through and adhered onto the underside of the membranes were enumerated by counting the total cell numbers in ten x200 fields using a microscope. These cultured resident peritoneal macrophages, PMM5 and PMM6, produce and release chemotactic factor-like substances into culture medium.

PMM cells migrated not only to FMLP, but also to the supernatant of the cultured PMM cells. Especially, the supernatant of PMM6 cells showed higher chemotaxic activity, as shown in Table 2. These data suggest that the various peritoneal cells of marine teleosts migrated from blood stream and/or peritoneum induced by chemotactic factor-like substances secreted by resident peritoneal macrophages. Such macrophage cell cultures provide good tools for the investigation of the relationships between inflammation and cytokines in fish, such as these chemotactic factors.

TABLE 2. Chemotaxis of red sea bream cultured resident peritoneal macrophages.

Cell	FMLP	Supernatant of PMM5	Supernatant of PMM6	RSB serum	PM-RPMI
PMM5	124.6±44.8†	179.8±51.9†	262.2±122.0†	3.6±3.6	10.0±3.5
PMM6	296.6±201.5	72.4±33.8	11.8±11.5	NT	78.0±16.1

RSB: Red sea bream †:P<0.01 *:P<0.05 NT: not tested

5. Carp hematopoietic cell culture

In vertebrates, the most dominant phagocytes have been called neutrophils or heterophilic granulocytes. These granulocytes have morphological similarities in fish. However, nucleus morphology varied from round to lobular and polymorphic shapes like mammalian neutrophils (Rowley 1988, Hine 1992).

Fish principal hematopoietic tissues are kidneys (mainly head kidney or pronephros). There are abundant granulocytes in carp *Cyprinus carpio* kidney hematopoietic tissue, heterophilic granulocytes type 1or neutrophils and heterophilic granulocytes type 2 or basophils (*Figure 3*). Carp hematopoietic cells can proliferate and form colonies in semi-solid agar in the presence of homologous carp serum (Moritomo 1993, 1996). Head kidney and body kidney was dissected and macerated on a stainless steel mesh with two forceps in RPMI-1640 medium to obtain separated hematopoietic cells. Then, erythrocytes were lysed by distilled water. The cells were dispersed in the RPMI1640 medium by pipetting and adjusted to a concentration of 1.5x106 cells/ml. The culture procedure modified from the Metcalf's method (Metcalf 1970) was used. The cells were cultured in semi-solid agar E-RDF medium supplemented with 20% FBS, 5% LPS treated carp serum. The cultured cells were consisted of undifferentiated blastic cells. Mitosis are often observed (*Figure 4*).

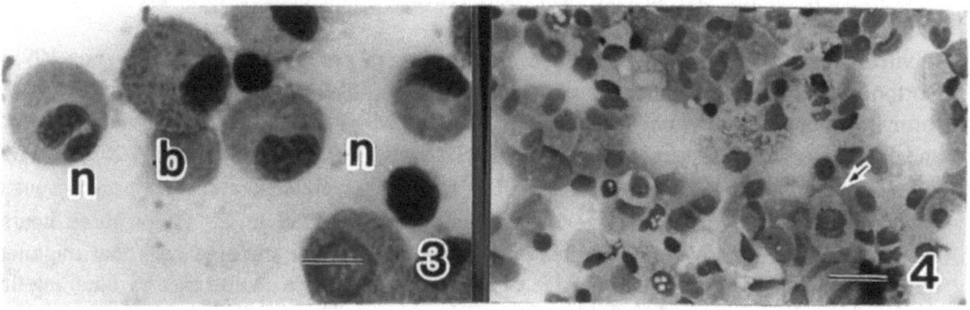

Figure 3. Head kidney hematopoietic cells of carp, showing neutrophils (n) and basophils (b). Bar=10μm.
Figure 4. Cultured carp hematopoietic cells, showing several kinds of blastic cells and mitosis (arrow). Bar=25μm.

When cultured in 10^{-7} M hydrocortisone added medium, formed colonies consisted of peroxidase positive cells. This enzyme is a proper characteristic of the heterophilic granulocytes or neutrophils in carp. They have similar morphology to the heterophilic granulocytes or neutrophils observed in carp peripheral blood and kidney hematopoietic tissues. This steroid hormone is a predominant adrenal hormone in fish, which is well known to control the osmotic pressure of fish body fluid.

Acknowledgments: We are grateful to Dr. Y. Suzuki, Dr. K. Furukawa, and Dr. M. Kono, the University of Tokyo, for providing us RSB IgM, biotin-conjugated anti-RSB IgM rabbit IgG, and red sea bream. This work was supported in part by Grant-in-Aid for Scientific Research from the Ministry of Education, Science and Culture.

6. References

Baggiolini M, Dewald B & Moser B (1994) Interleukin-8 and related chemotactic cytokines-CXC and CC chemokines. Adv. Immunol. 55: 97-179.

Chung S and Secombes CJ (1987) Activation of rainbow trout macrophages. J. Fish Biol. 31, 51-56.

Clem LW, Sizemore RC, Ellsaesser CF & Miller NW (1985) Monocytes as accessory cells in fish immune responses. Dev. Comp. Immunol. 9: 803-809.

Faisal M & W. Ahne W (1990) A cell line (CLC) of adherent peripheral blood mononuclear leukocytes of normal common carp Cyprinus carpio. Dev. Comp. Immunol. 14: 255-260.

Faisal M, Sami S & Rutan BJ (1992) Fish cell lines of leukocytic origin: maintenance and characterization. In Stolen JS, Fletcher TC, Anderson DP, Kaattari SL & Rowley AF (ed.) Techniques in Fish Immunology. SOS Publications, Fair Haven, pp.35-58.

Hine PM (1992) The granulocytes of fish. Fish & Shellfish Immunol. 2: 79-98.

Kanlis G, Suzuki Y, Tauchi M, Numata T, Shirojo Y & Takashima F (1995) Immunoglobulin in oocytes, fertilized eggs, and yolk sac larvae of red sea bream. Fisheries Sci. 61: 787-790.

Metcalf D (1970) Studies on colony formation in vitro by mouse bone marrow cells. II. Action of colony stimulating factor. J. Cell. Physiol. 76: 89-100.

Moritomo T, Itou A, Yamaguchi I & Watanabe T (1993) In vitro colony formation of carp granulocytes in a semi-solid agar culture system. Gyobyo Kenkyu 28: 59-63.

Moritomo T, Kihara R, Nakayama S, Itou T, Imagawa T & Watanabe T (1996) Culture of carp hematopoietic cells. Fish Pathol. 31: 59-63.

Papadimitriou JM & Ashman RB (1989) Macrophages: current views on their differentiation, structure, and function. Ultrastr. Pathol. 13: 343-372.

Rowley AF, Hunt TC, Page M, & Mainwaring G (1988) Fish. In: Rowley AF & Tatcliffe NA (ed.) Vertebrate Blood Cells (pp.19-127) Cambridge Univ. Press, Cambridge.

Vallejo AN, Ellsaesser CF, Miller NW & Clem LW (1991) Spontaneous development of functionally active long-term monocyte-like cell lines from channel catfish. In Vitro Cell. Dev. Biol. 27A: 279-286.

van Furth R (1992) Development and distribution of mononuclear phagocytes. In Gallin JI, Goldstein IM & Snyderman R (ed.) Inflammation: Basic Principles and Clinical Corretates 2ed. Ed., Raven Press Ltd., New York, pp.325-339.

Wang R, Neumann NF, Shen Q, & Belosevic M (1995) Establishment and characterization of a macrophage cell line from the goldfish. Fish & Shellfish Immunol. 5: 329-346.

Watanabe T, Kamijo A, Narita H, Ohta H, Kitayama K, Kubo N, Moritomo T, Kono M & Furukawa K (1995) Resident peritoneal cells of red sea bream *Pagrus major*. Fisheries Sci. 61: 937-941.

Watanabe T, Shoho T, Ohta H, Kubo N, Kono M & Furukawa K (1997) Cell culture of red sea bream Pagrus major resident peritoneal macrophages. Fisheries Sci. in press.

Watanabe T, Kitayama K, Takagi T, Murata J. Kono M, Takase T, & Furukawa K (1997) Heterogeneity of peritoneal cells in marine teleosts. Fisheries Sci. in press.

Telomerase regulation in differentiated serum-free mouse embryo cells

Y. KATAKURA[1], K. YAMAMOTO[1], O. MIYAKE[1], E. NAKATA[1], T. YASUDA[1], N. UEHARA[1], D. W. BARNES[2] AND S. SHIRAHATA[1]
[1]*Graduate School of Genetic Resources Technology, Kyushu University*
6-10-1 Hakozaki, Higashi-ku, Fukuoka 812-81, Japan
[2]*Dept of Biochemistry and Biophysics, Oregon State University*

ABSTRACT We developed a non-radioisotopic assay to detect telomerase activity in serum-free mouse embryo cells (SFME cells). Using this assay, telomerase regulation mechanism in differentiated SFME cells was investigated. SFME cells treated with 10% serum or with 10 ng/ml of TGF-β showed astrocytes-like phenotypes and lowered telomerase activity. These results indicate that telomerase activity is down-regulated with the differentiation of SFME cells into astrocytes.

1. Introduction

Telomeres are located at the ends of eukaryotic chromosomes and consist of an array of hexanucleotide tandem repeats, TTAGGG. One of the roles of these telomeres is to protect the ends of chromosomes against exonucleases and ligases. Segments of these telomeric repeats are lost with each cell division because DNA polymerases cannot replicate the terminal end of a linear DNA molecule (end replication problem). Due to this incomplete replication, telomeres progressively shorten as the number of cell division increases. The average reduction was estimated to be 40-200 bp per chromosome end per cell division. This telomere shortening has been proposed as a regulatory mechanism that controls cellular senescence, cellular immortalization and cellular carcinogenesis. Telomere shortening has been observed in normal somatic cells depending upon their cell divisions. However, immortal cells including cancer cells avoid this progressive loss of telomeric DNA by expressing the enzyme telomerase. Telomerase is known to be a RNA-dependent ribonucleoprotein enzyme, which is capable of re-constituting tracks of short repetitive telomeres to the ends of the chromosome. This addition of TTAGGG repeats compensates for the end replication problem and stabilizes the lengths of telomeres, allowing the cells to divide indefinitely. Telomerase inhibitors have been noted as potential anti-tumor agents, which can confer mortality to cancer cells without damaging normal somatic cells. Telomerase activity is also detected in germ cells and stem cells. When terminally differentiated, these cells are expected to lose or repress telomerase activity. Serum-free mouse embryo cells (SFME cells), when cultured in serum-free medium supplemented with insulin, transferrin, epidermal growth factor (EGF), and high-density lipoprotein (HDL), do not exhibit growth crisis, lack detectable chromosomal aberrations, are nontumorigenic *in vivo*, are dependent on EGF for survival, and are growth inhibited by serum (1, 8). SFME cells are thought to be neural stem cells with the ability to differentiate into glial cells upon addition of TGF-β or serum and can express glial fibrillary acidic protein (2) and is able to increase the cystatin C mRNA level (9). In this study, we attempted to clarify whether the differentiation of SFME cells correlates with the regulation of telomerase activity.

2. Materials and Methods

2.1. DETECTION OF TELOMERASE ACTIVITY

We adopted a highly sensitive assay for detecting telomerase activity, the telomeric repeat amplification protocol (TRAP) (6). Using this method, telomerase activity in a small number of cells can be easily and reproducibly detected. Furthermore, with a few modifications, telomerase activity can be detected without using radioisotope. To measure telomerase activity semi-quantitatively, cell lysates prepared from 10^6 cells were titrated between 10 and 10^4 cells equivalent per assay, which was used to determine a linear range of telomerase activity versus cell number to estimate the appropriate cell number to be used in each assay. Telomere elongation reaction was carried out in the presence of 2 μl of appropriately diluted lysate, 50 μM dNTP, 0.1 μg TS primer (AATCCGTCGAGCAGAGTT), 1.0 μg of T4 Gene 32 protein (Ambion, Austin, Texas, USA) and 2.0 U of Taq DNA polymerase (Boehringer Mannheim, Indianapolis, IN, USA) at 20°C for 30 min. Following incubation, 0.1 μg of CX primer (CCCTTACCCTTACCCTTACCCTAA) was added to this mixture and the tubes were transferred to a thermal cycler and subjected to 31 rounds at 94°C for 45 sec, 50°C for 45 sec and 72°C for 60 sec. A portion of this reaction mixture was analyzed by polyacrylamide gel electrophoresis. The TRAP assay products were visualized by SYBR Green I (TAKARA, Shiga, Japan) staining. Telomerase activity was evaluated using the densitometric sum of the 6-bp ladder bands corresponding to TTAGGG.

2.2. CELL LINE AND CULTURE CONDITION

SFME cells was established by D. Barnes *et al.* The basal nutrient medium consisted of an one-to-one mixture of Dulbecco-modified Eagle's medium (F12:DME medium) containing 4.5 g/l glucose and Ham's F12 supplemented with 15 mM HEPES, pH 7.4, 1.2 g/l sodium bicarbonate, 10 nM sodium selenite, 200 U/ml penicillin, 200 μg/ml streptomycin and 25 μg/ml ampicillin. SFME cells were cultured in F12:DME medium supplemented with insulin (10 μg/ml), transferrin (25 μg/ml), HDL (10 μg/ml), and EGF (50 ng/ml), in dishes precoated with bovine fibronectin (10 μg/ml). Details and general protocols for culturing SFME cells are described elsewhere (7).

3. Results

3.1. DIFFERENTIATION OF SFME CELLS INTO ASTROCYTES

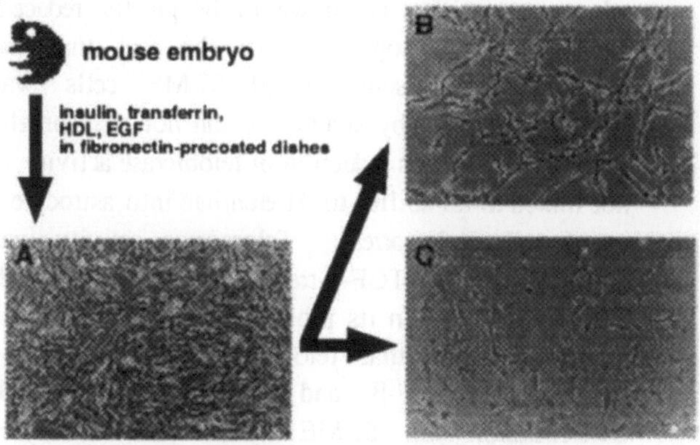

Fig. 1 Differentiation of SFME cells into astrocytes
SFME cells were cultured in serum-free medium (A), medium containing 10 ng/ml of TGF-β (B) and medium containing 10% serum (C)

SFME cells do not lose proliferative potential or show gross chromosomal aberrations when cultured for an extended period of time in serum free medium (Fig. 1A) (1). SFME cells can differentiate into astrocytes by treatment with serum or TGF-β (Figs. 1 C and B, respectively). This report has been confirmed by the induced expression of glial fibrillary acidic protein from SFME cells cultured in the presence of serum or TGF-β (2). We then used this differentiation event to attempt to clarify the telomerase regulation mechanism.

3.2. PROLIFERATION OF SFME CELLS IN THE ABSENCE OR PRESENCE OF TGF-β OR SERUM

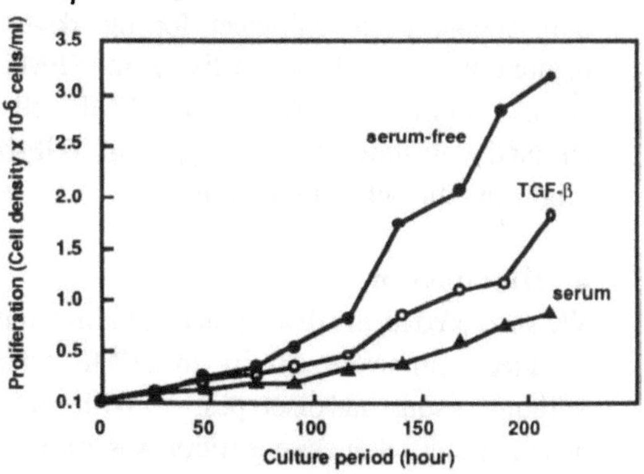

Fig. 2 Proliferative potential of SFME cells

First, we investigated the proliferative potential of SFME cells in the absence or presence of TGF-β or 10% serum (Fig. 2). SFME cells vigorously proliferated in serum-free medium. However, addition of serum to this culture medium inhibited the proliferation of SFME cells. SFME cells treated with 10 ng/ml of TGF-β showed moderate proliferative potential, although these TGF-β-treated SFME cells is thought to differentiate into astrocytes (2).

3.3. TELOMERASE ACTIVITY OF DIFFERENTIATED SFME CELLS

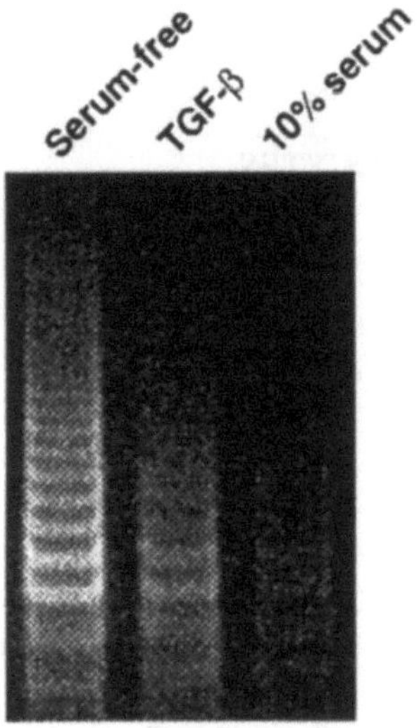

Fig. 3 Telomerase activities of differentiated SFME cells

Figure 3 shows the telomerase activities of SFME cells cultured in serum-free medium, and of SFME cells treated with 10 ng/ml of TGF-β or with 10% serum. In SFME cells treated with 10% serum, telomerase activity is shown to be greatly reduced. However, taken together with the fact that the proliferation of serum-treated SFME cells was strongly inhibited by serum, we can not abandon the possibility that this reduction of telomerase activity is not linked to its ability to differentiate into astrocytes, or to cell cycle arrest. Telomerase activity was down-regulated in TGF-β-treated SFME cells, which was shown to retain its proliferative potential. This clearly indicates that telomerase activity can be regulated by TGF-β, and is down-regulated with differentiation of SFME cells into astrocytes independently of cell proliferation.

3.4 DOWN-REGULATION OF TELOMERASE ACTIVITY IN SFME CELLS TREATED WITH VARYING CONCENTRATIONS OF TGF-β.

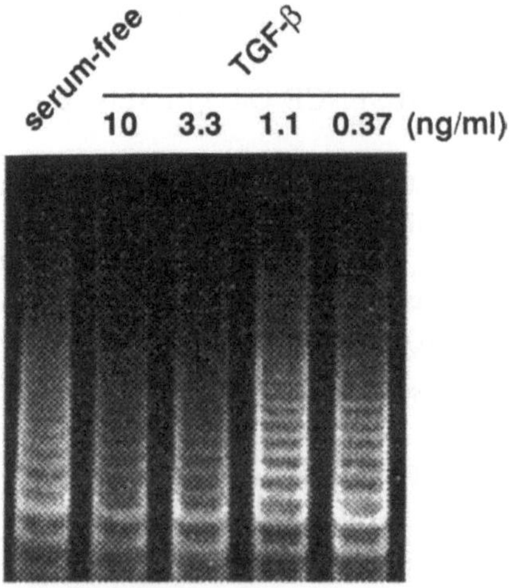

Fig. 4 Down-regulation of telomerase activity in SFME cells treated with varying concentrations of TGF-β

Figure 4 shows the TGF-β-concentration dependent down-regulation of telomerase activity in SFME cells. 3.3 ng/ml of TGF-β is shown to be sufficient for the down-regulation of telomerase activity in SFME cells. This result suggests that SFME cells cultured with more than 3.3 ng/ml of TGF-β might initiate cellular senescence.

4. Discussion

We succeeded in the development of a method to detect telomerase activity in SFME cells without using radioisotopes. With this assay, strong telomerase activity was detected in SFME cells cultured in serum-free medium. This result suggests the immortality of

SFME cells when cultured in serum-free medium, which is consistent with the observation that SFME cell cultures have been passaged continuously without evidence of growth crisis or senescence for over 200 doublings (1). Furthermore, taken together with the fact that treatment of SFME cells with serum or TGF-β led to the appearance of glial fibrillary acidic protein, a specific marker for astrocytes (2), telomerase activity was down-regulated in differentiated SFME cells. Several papers reported that telomerase activity is down-regulated in the terminally differentiated cells (3, 4). However, these reports could not clarify whether this down-regulation of telomerase activity is linked to either to cellular differentiation or to cell-cycle arrest, because terminally differentiated cells completely lost the proliferative potentials. In fact, some researchers have reported that telomerase activity is repressed in quiescent cells (5). Here, eventhough TGF-β-treated SFME cells retains its proliferative potential, telomerase activity was greatly down-regulated. This result clearly demonstrates that telomerase activity could be regulated by TGF-β in SFME cells and is down-regulated upon differentiation of SFME cells into astrocytes independently of cell proliferation.

5. References

1. Loo, D. T., Fuquay, J. I., Raqson, C. L., and Barnes, D. W. (1987) "Extended culture of mouse embryo cells without senescence: inhibition by serum." *Science*, **236**, 200-202
2. Sakai, Y., Rawson, C., Lindburg, K., and Barnes, D. W. (1990) "Serum and transforming growth factor b regulate glial fibrillary acidic protein in serum-free-derived mouse embryo cells." *Proc. Natl. Acad. Sci. USA*. **87**, 8378-8382.
3. Sharma, H. W., Sokoloski, J. A., Perez, J. R., Maltese, J. Y., Sartorelli, A. C., Stein, C. A., Nichols, G., Khaled, Z., Telang, N. T., and Narayanan, R. (1995) "Differentiation of immortal cells inhibits telomerase activity." *Proc. Natl. Acad. Sci. USA*. **92**, 12343-12346.
4. Albanell, J., Han, W., Mallado, B., Gunawardane, R., Scher, H. I., Dmitrovsky, E., and Moore, M. A. S. (1996) "Telomerase activity is repressed during differentiation of maturation-sensitive but not resistant human tumor cell line." *Cancer Res*. **56**, 1503-1508.
5. Holt, S. E., Wright, W. E., and Shay, J. W. (1996) "Regulation of telomerase activity in immortal cell line." *Mol. Cell. Biol*. **16**, 2932-2939.
6. Kim, N. W., Piatyszek, M. A., Prowse, K. R., Harley, C. B., West, M. D., Ho, P. L. C., Coviello, G. M., Wright, W. E., Weinrich, S. L., and Shay, J. W. (1994) "Specific association of human telomerase activity with immortal cells and cancer." *Science* **266**, 2011-2015.
7. Loo, D., Rawson, C., Helmrich, A., and Barnes, D. (1989) "Serum-free mouse embryo cell: growth responses in vitro." *J. Cell. Physiol*. **139**, 484-491
8. Ernst, T., Jackson, C., Barnes, D. (1991) "Karyotypic stabiity of serum-free mouse embryo (SFME) cells" *Cytotechnology*, **5**, 211-222.
9. Solem, M., Rawson, C., Lindburg, K., and Barnes, D. (1990) "Transforming growth factor b regulates cystatin C in serum-free mouse embryo (SFME) cells." *Biochem. Biophys. Res. Commun*., **172**, 945-951.

IMMORTALIZATION OF MAMMALIAN CENTRAL NERVOUS SYSTEM STEM CELLS FROM p53 KNOCK-OUT FOREBRAIN

T. Hisatsune[#*], K. Yamada[#], R.D.G. McKay[*], and S. Kaminogawa[#]
#Dept. Appl. Biol. Chem., The University of Tokyo
***Lab. Mol. Biol., NINDS, NIH**

Summary

Neuronal and glial cells making the central nervous system are generated from central nervous system (CNS) stem cells during the development of CNS. CNS stem cells are potentially applicable for regeneration of damaged brain, such as in Parkinson's Disease, or Alzheimer's Disease. Recent studies have shown that mitogen and/or oncogene keep CNS stem cells proliferating. However, the long-term expansion of CNS stem cells needs a period of crisis, when almost all cells stop dividing and presumably wait for another intrinsic genomic change. The p53 tumor suppressor gene encodes a protein that is important in the regulation of cell proliferation and apoptosis. In this study, we used p53 knock-out embryos to analyze the effect of the p53 gene on apoptosis of the CNS stem cells. In a primary culture assay, p53 knock-out CNS stem cells are continuously growing in the presence of basic Fibroblast Growth Factor (bFGF), whereas expansion of wild-type CNS stem cells stopped shortly after cultivation due to substantial apoptosis. This indicates the critical role of the p53 locus on apoptosis of CNS stem cells. In this study, we have established immortalized CNS stem cells from p53 knock-out forebrain, MSP-1. To inhibit spontaneous differentiation during cell expansion, MSP-1 had been introduced by c-Myc-Estrogen Receptor (ER)-fusion gene, in which c-Myc can be conditionally activated by adding beta-estradiol. The integration of introducing c-Myc-ER has been confirmed in MSP-1 cells by Southern blot, Northern blot, and Western blot. The MSP-1 precursor cells express nestin and proliferate exponentially under the growing media, supplemented with bFGF, beta-estradiol and 10% FCS. The vigorous growth of MSP-1 may be supported both by c-Myc and the P53 null mutation. MSP-1 cells differentiate into neuronal and glial cell types after switching to the differentiating media (N2 serum-free media without growth factor) for 6 days. MSP-1 can be expanded from just one cell. Clonally expanded MSP-1 cells differentiated into neuronal and glial cell types after withdrawal of growth factors. The availability of immortal stem cells capable of neuronal differentiation offers an unusually powerful way of identifying the molecular basis of neuronal commitment.

1. Introduction

The central nervous system performs complex tasks depending on the activity of a large number of neurons and glial cells that are organized into numerous regions specialized for distinct functions. All of the cells of the CNS arise from dividing apparently homogeneous progenitor cells, CNS stem cells, in appropriate numbers, times, and positions during the development of CNS (McKay, 1989). Recent studies have shown that several signaling cascades affect termination of the cell-cycle of CNS stem cells. In an in vitro system, the period of mitosis of CNS stem cells can be prolonged by extrinsic signals driven by mitogen such as bFGF or EGF (Cattaneo and McKay, 1990; Reynolds and Weiss, 1992). In addition, several oncogenes such as SV40 or Myc can be used for eternal expansion of CNS stem cells (Frederiksen et al. 1988; Renfranz et al. 1991). However, long-term expansion of CNS stem cells needs a period of crisis, when almost all cells stop dividing and presumably wait for another intrinsic genomic change. This implies intrinsic key molecules positively affecting termination of the cell-cycle in CNS stem cells.

The P53 tumor suppressor gene encodes a protein that is important in the regulation of cell proliferation, apoptosis, and differentiation (Haffner and Oren, 1995). Metz et al. have shown that hematopoietic stem cells can be immortalized from P53 knockout mice by the myc and raf oncogenes without any cell crisis (Metz et al. 1995). Profound expression of p53 mRNA in developing cortex especially at proliferating zone has been reported (Schmidt et al, 1991); however, the role of P53 on development of CNS is still unknown. In this communication, we have shown that p53 is involved in apoptosis of CNS stem cells during in vitro expansion, and established immortalized CNS stem cells from p53 knock-out forebrain, MSP-1, capable of differentiating neuronal and glial cell types.

2. Materials and Method

2.1 Cultivation of CNS stem cells

Transgenic mice carrying a null allele in the P53 locus were obtained from Taconic, Inc. (Germantown, NY). The forebrain proliferating cells of individual 14.5 day embryos from $P53^{+/-}$ male x $P53^{+/-}$ female breeding were cultured in 10 cm dishes, which were precoated with poly-ornithine and fibronectin, in 5 ml of N2-DMEM/F12 serum free media supplemented with 10 ng/ml of bFGF. Four days later, the passed cells were collected as P53 knock-out, $P53^{+/-}$, or $P53^{+/+}$ (wild-type) cells. The passed cells are a homogenous population for CNS stem cells, stained with CNS stem cell marker nestin not with neuronal nor glial marker (Vicario-Abejon et al., 1995).

2.2 Establishment of immortalized CNS stem cells

p53 knock-out forebrain cells were infected with a retroviral vector harboring Myc-ER fusion gene and neomycin resistance gene (Eiler et al., 1989). At 24 hr after infection, cells were placed under neomycin selection in N2-DMEM/F12 medium in the presence of bFGF and beta-estradiol. Neomycin-resistant clones were isolated and expanded. MycER integration of clonal cell lines was assessed by Southern, Northern, and Western blotting. Clonal cell line MSP-1 (MyeER immortalized p53 knock-out CNS stem cells) was used in this study. Expanded clonal cells were occasionally cultured in 10% FCS supplemented with bFGF plus beta-estradiol for maximal proliferation.

2.3 Differentiation of CNS stem cells

Cells expanded in N2-SFM were seeded in a precoated 8 well chamber slide at a concentration of 4 x $10^4/cm^2$. The cells were expanded with bFGF for 2-4 days, then were cultured in the absence of bFGF for a further 9-12 days. For clonally expansion, MSP-1 cells were seeded at a low density in a pre-coated 10 cm dish and grown for 7-10 days, then were changed to differentiation media. Differentiation to neuronal and glial phenotypes were evaluated by immunostaining with MAP-2 and GFAP, respectively. Cells were fixed with 4% paraformaldehyde for 10 min. After permilazation with 0.1% Triton X-100/3% normal goat serum/PBS, the cells were incubated overnight with the primary antibodies against nestin (1:500), MAP2a-c(1:500), GFAP(1:100), GalC(1:100). Then, the cells were incubated with corresponding fluorscein-conjugated secondary antibody for 2 h. After washing, the cells were mounted in DABCO/immunoblot.

3. Result and Discussion

3.1 Role of p53 on apoptosis of CNS stem cells

The effect of the P53 gene on expansion of CNS stem cells is tested in a short-term culture from E14.5 telencephalon. Fig. 1A shows that the cell number of $P53^{-/-}$ culture is continuously increasing to 50000 cells/cm2, whereas the number of $P53^{+/-}$ or $P53^{+/+}$ culture has dropped 3 days after the passage. The number of dead cells was evaluated as dead/live cell ratio during this culture period. Fig. 1B shows dead/live ratio of $P53^{-/-}$ culture constantly decreasing to 0.1 on 4 days of this culture, whereas the ratio of $P53^{+/-}$ or $P53^{+/+}$ culture is increasing. At the end of this culture, there were many live cells and few dead cells in $P53^{-/-}$ culture, whereas there are many dead cells and few live cells in P53$^{+/+}$ culture (Fig. 1C&D). These results clearly show that the P53$^{-/-}$ CNS stem cells would escape from a certain type of cell death, which appears in the CNS stem cells with P53 gene. With regard to wild type culture, the P53 locus could act as an inducer of cell death on CNS stem cells.

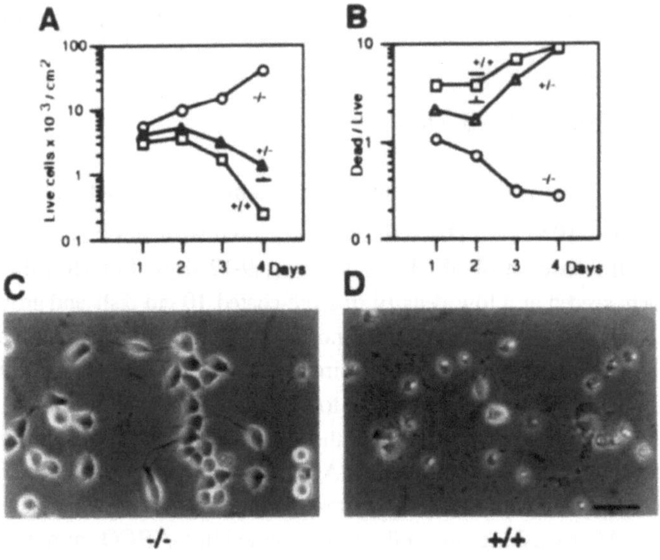

Figure 1. Involvement of p53 locus in apoptosis of CNS stem cells

A. Time-course of cell proliferation of P53$^{-/-}$, P53$^{+/-}$, or P53$^{+/+}$ CNS stem cells.
B. Time-course of dead/live cell ratio of individual CNS stem cell culture.
C&D. P53$^{-/-}$ (C), and P53$^{+/+}$ (D) CNS stem cells 4 days after replating.

3.2 Establishment of immortalized p53 knock-out CNS stem cells

The potency of progressive expansion of p53 knock-out CNS stem cells in culture is a good chance to make CNS stem cell lines, which offers an unusually powerful way of identifying the molecular basis of CNS development. After 6 weeks of clonal expansion, several CNS stem cell clones have been isolated from p53 knock-out embryos. All of these cell lines differentiate into neuronal and glial cells after switching into differentiation media. Using a simple clonal expansion method, we can isolate CNS stem cell lines. Nonetheless, we also observed spontaneous differentiation of all the clones, implying that P53 null-mutation has the potential to expand CNS stem cells, but this single mutation is not enough to inhibit spontaneous differentiation after long-term culture. To inhibit spontaneous differentiation during expansion of clonal cells, a culture of p53 knock-out CNS stem cells had been infected with regulatable Myc-ER gene (Eilers et al, 1989). After cloning and long-term expansion, we have succeeded in establishing a CNS stem cell clone MSP (MycER integrated Stem cell clone from P53 null-brain)-1. The MSP-1 cells express nestin and proliferate exponentially under the growing media, where no spontaneous differentiation was observed. After switching to differentiation media, MSP-1 cells gave rise to neuronal (MAP-2) and glial cell (GFAP and Gal-C) types (Fig. 2). MSP-1 can be expanded from clonally density culture. Clonally expanded MSP-1 cells differentiated into neuronal and glial cell types after withdrawal of growth factors (not shown). P53 null-mutation along with MycER fusion genes is quite useful to immortalize CNS stem cells.

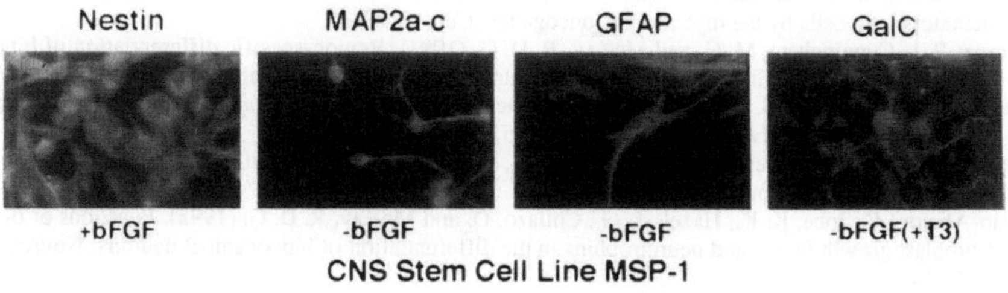

Figure 2. Characterization of immortalized CNS stem cell line MSP-1

MSP-1 cells cultured with bFGF were fixed and were immunostained with anti-nestin antibody. MSP-1 cells were differentiated in the absence of bFGF for 10 days, were fixed and were stained with anti-MAP-2 or anti-GFAP. Gal-C expression from differentiating MSP-1 cells required the presence of T3.

3.3 Concluding remarks

We plan to do several lines of research using MSP-1 cells. In vivo aspect, we have implanted MSP-1 cells into normal or damaged brain, and checked neuronal integration. In vitro aspect, we tried to fully characterize differentiated neuronal and glial cell types derived from MSP-1. By RT-PCR analysis, differentiated MSP-1 cells express glutamate receptor (NMDA-type, AMPA-type, etc.), and GABA-receptor. Now, we are in progress to identify receptor expressing cells (neuronal or glial) by the Ca-imaging method. In regard to the neurotransmitter, GAD (enzyme for GABA synthesis) was immunostained in differentiated MSP-1 neurons. We will check the release of glutamates or neuropeptides, such as substance P, enkephalin, somatostatin, colecystokinin, or VIP. MSP-1 cells will be a very important cell line for analyzing the molecular basis of neuronal development and function of the mammalian forebrain, which is responsible for our intelligence.

4. References

Cattaneo, E. and McKay, R. D. G. (1990). Proliferation and differentiation of neuronal stem cells regulated by nerve growth factor. Nature *347*, 762-765.

Eilers, M., Picard, D., Yamamoto, K. and Bishop, J. M. (1989). Chimaeras of Myc oncoprotein and stroid receptors cause hormone-dependent transformation of cells. Nature *340*, 66-68.

Frederiksen, K., Jat, P. S., Valtz, N., Levy, D. and McKay, R. D. G. (1988). Immortalization of precursor cells from the mammalian CNS. Neuron *1*, 439-448.

Haffner, R. and Oren, M. (1995). Biochemical properties and biological effects of P53. Current Opinion in Genetics and Development *5*, 84-90.

McKay, R. D. G. (1989). The origins of cellular diversity in the mammalian central nervous system. Cell *58*, 815-821.

Metz, T., Harris, A. W. and Adams, J. M. (1995). Absence of p53 allows direct immortalization of hematopoietic cells by the myc and raf oncogenes. Cell *82*, 29-36.

Renfranz, P. J., Cunningham, M. G. and McKay, R. D. G. (1991). Region-specific differentiation of hippocampal stem cell line HiB5 upon implantation into the developing mammalian brain. Cell *66*, 713-729.

Reynolds, B. A. and Weiss, S. (1992). Generation of neurons and astrocytes from isolated cells of the adult mammalian central nervous system. Science *255*, 1707-1710.

Schmid, P., Lorenz, A., Hameister, H. and Montenarh, M. (1991). Expression of p53 during mouse embryogenesis. Development *113*, 857-865.

Vicario-Abejon, C., Johe, K. K., Hazel, T. G., Collazo, D. and McKay, R. D. G. (1995). Functions of basic fibroblast growth factor and neurotrophins in the differentiation of hippocampal neurons. Neuron *15*, 105-114.

NEUROPROTECTIVE FUNCTION OF ERYTHROPOIETIN IN THE CENTRAL NERVOUS SYSTEM

SEIJI MASUDA, MARIKO CHIKUMA, EMI MORISHITA,
MASAYA NAGAO and RYUZO SASAKI
Department of Food Science and Technology, Faculty of Agriculture, Kyoto University, Kyoto 606-01, Japan

SUMMARY

It has been thought that erythropoietin exclusively acts on erythroid precursor cells *in vivo*. However, we describe that erythropoietin is produced by astrocytes and its receptor is expressed in neurons. Erythropoietin production by astrocytes is dependent on oxygen tension and activated by insulin and insulin-like growth factors. Furthermore, erythropoietin protected the cultured hippocampal and cerebral cortical neurons from glutamate neurotoxicity. These results indicate that, in the central nervous system, erythropoietin acts on neurons in a paracrine fashion, and erythropoietin and its receptor system is independent of that for erythropoiesis.

1. INTRODUCTION

Erythropoietin (EPO) regulates erythropoiesis by supporting survival of erythroid precursor cells, and by stimulating their proliferation and differentiation. In adults, the kidney is a major production site of EPO [1,2]. Oxygen is a primary regulator of EPO production [3-5]; hypoxia increases EPO production by activating transcription of the EPO gene and also stabilizing EPO mRNA.

Recently, several findings indicate that EPO and EPO receptor (EPOR) are expressed in the central nervous system (CNS). We showed that neuronal cell lines, brain tissues and primary cultured neurons from rat embryos express EPOR protein as well as its mRNA [6-8]. Using radiolabeled EPO, EPO binding sites have been detected in some parts of the adult mouse brain, including the hippocampus and cerebral cortex, where there are neurons vulnerable to ischemia [9]. These results implicates that EPO and EPOR are involved in hypoxia-induced neuronal death. The blood-brain barrier formed in the late embryonic and early postnatal period [10] insulates the CNS from the circulation, preventing the access of blood Epo to the neurons. Therefore, if the EPO action on neurons is physiological, a novel site for Epo production should be found in the brain. Tan *et al.* reported EPO mRNA is expressed in the adult rat brain and its expression is hypoxia-inducible [11]. Here, we describe biosynthesis [12,13] and biological function [7] of EPO in the CNS.

2. MATERIALS AND METHODS

Astrocyte Culture and EPO Assay with Enzyme-linked Immunoassay - The brain was removed from Wistar fetal rats on day 18. Cerebral tissue was dissociated by passing it through a nylon mesh. The cell suspension was washed twice with Dulbecco's modified Eagle medium supplemented with 10% fetal calf serum. The cells were plated and incubated at 21%, 5%, or 1% O_2/5% CO_2/ balance N_2. To examine the effect of insulin and IGFs to stimulate EPO production, experiments were started by adding various amounts of IGFs or insulin. EPO in the spent medium was measured by enzyme-linked immunoassay [14]. The amounts of cellular proteins were determined with protein assay kit using bovine immunoglobulins as a standard.

Hippocampal and Cerebral Cortex Neuron Culture - Primary cultures of dissociated hippocampal and cerebral cortical neurons were prepared from the brains of 19-day Wistar fetal rats. Cells were plated in poly-L-lysine-coated wells. After one day of culture, the medium was changed with the serum-free B-27/Neurobasal medium. The cells after culture for 7-10 days were used for experiments.

Assay of Glutamate Neurotoxicity. - For exposure to glutamate, cultured neurons were incubated in the medium containing glutamate for 15 min in a CO_2 incubator. The medium was replaced with fresh medium without glutamate and the cells were cultured for a further 24 h. EPO was present in neuron cultures for 24 h before the onset of glutamate challenge and removed just before exposure of neurons to glutamate. EPO was absent in the further 24-h culture. Neuronal cell death was assessed by the trypan blue exclusion assay.

Immunochemical Detection of the Cellular EPOR - Neurons were fixed by the method previously described[7] and incubated at 60°C for 1 h to ensure the denaturation of EPOR. The cells were incubated with the anti-EPOR antibody and biotinylated goat anti-chicken IgG antibody, and stained with diaminobenzidine.

3. RESULTS

Oxygen-dependent Production of EPO by Cultured Rat Astrocytes - Astrocytes were cultured in an oxygen tension of 21% and 5% (Fig. 1A). In 5% oxygen, the high production was observed after 12 days of culture. To determine whether the regulation of the EPO production by oxygen tension was reversible, the cells cultured for 18 days in 5% oxygen were placed in different oxygen tensions for three days (Fig. 1B and C). The EPO production in 1% oxygen was approximately 20 times higher than that in 21% oxygen. Thus, Epo production of astrocytes was also regulated by oxygen as well as that of kidney and liver.

Dose-dependent, and Oxygen-independent Stimulation of EPO Production by IGFs and Insulin - Astrocytes were cultured under 5% oxygen in the presence or absence of insulin and

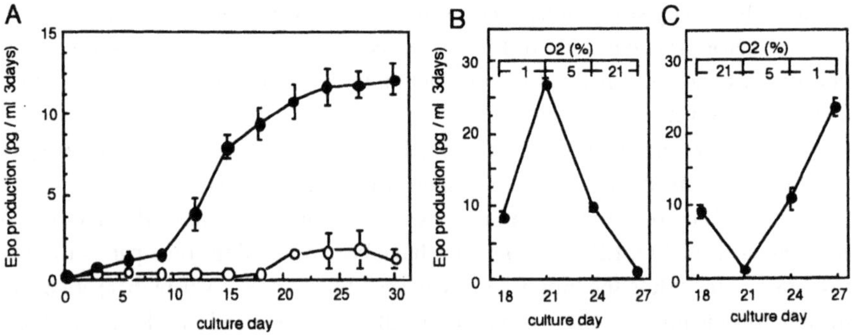

Figure 1. Oxygen-dependent production of EPO by primary cultures of rat astrocytes. - The cells were cultured with changes of the medium every 3 days and EPO in the spent medium was measured. Each point is the mean ± s.d. of sextuplicate cultures. A, the cells were cultured in 21% (○) and 5% (●) of O_2. B, oxygen tension was changed to 1, 5, and 21% of 3-days intervals. C, oxygen tension was changed to 21, 5, and 1% in this order.

IGFs. EPO production was stimulated by IGFs or insulin in a dose-dependent manner (Fig. 2A). Half maximal stimulation of EPO production was achieved by 3 nM IGF-I, 10 nM IGF-II and 100 nM insulin. To determine whether or not the stimulation by these factors was oxygen-dependent, astrocytes were cultured with IGFs or insulin under various oxygen concentrations (Fig. 2B-D). Insulin and IGFs increased EPO production under all oxygen concentrations tested. Furthermore, the dose-response curves of individual growth factor under different oxygen concentrations were essentially similar. These results suggest that there are different mechanisms of EPO induction by these two stimuli.

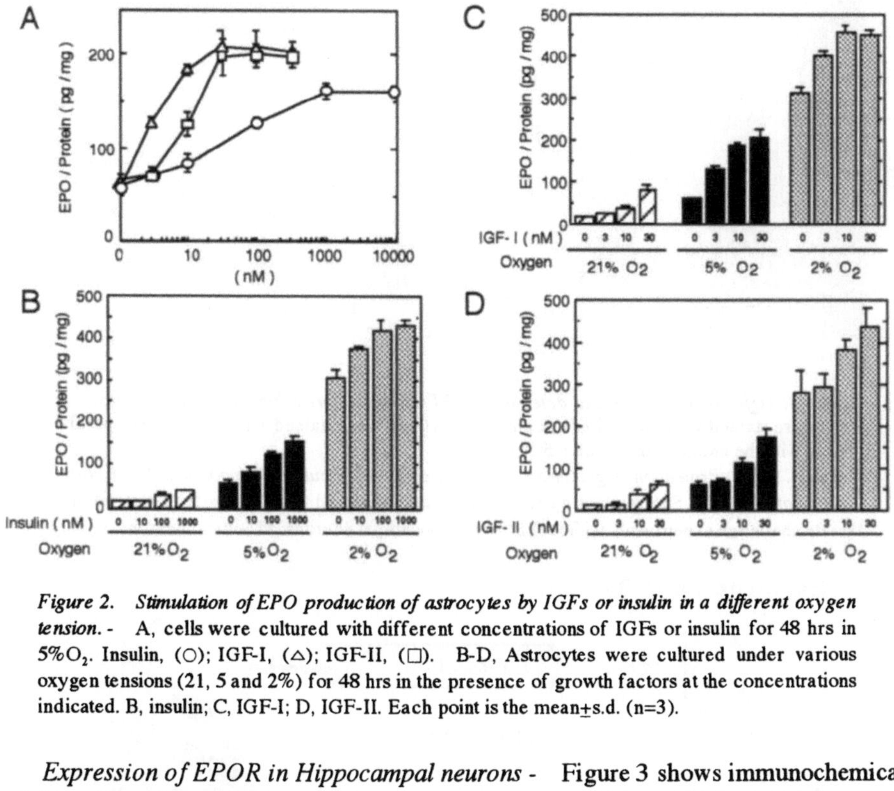

Figure 2. Stimulation of EPO production of astrocytes by IGFs or insulin in a different oxygen tension. - A, cells were cultured with different concentrations of IGFs or insulin for 48 hrs in 5%O_2. Insulin, (○); IGF-I, (△); IGF-II, (□). B-D, Astrocytes were cultured under various oxygen tensions (21, 5 and 2%) for 48 hrs in the presence of growth factors at the concentrations indicated. B, insulin; C, IGF-I; D, IGF-II. Each point is the mean±s.d. (n=3).

Expression of EPOR in Hippocampal neurons - Figure 3 shows immunochemical staining of the cultured cells using an anti-EPOR antibody. Cultured hippocampal neurons were reactive with the anti-EPOR antibody and the reactivity was abolished by preincubation of the antibody with the antigen.

Glutamate-induced Neuronal Death of Cultured Neurons and Protective Effect of EPO - Glutamate neurotoxicity can be most directly studied using the neurons prepared from the brain in the late embryonal stage and subsequently cultured for maturation [15,16]. To assess glutamate-induced cell death and the protective effect of EPO in a quantitative manner, we determined the ratios of the number of trypan blue-incorporating cells to the total number of neurons. Figure 4 shows the results. Approx. 80-85% of the neurons were viable in the control cultures. Exposure of the cells to 1 mM glutamate for 15 min reduced the viable cells to 30% in hippocampal neurons and 40% in cortical neurons. EPO prevented the glutamate-induced cell death in a dose-dependent manner. Culture with 3 pM EPO (approx. 100 pg/ml ≈ 10 munits/ml) increased the number of viable cells significantly, and culture with 30 pM EPO almost completely protected neurons from glutamate toxicity. The coexistence of the soluble EPOR over EPO abrogated almost completely the protective effect of EPO.

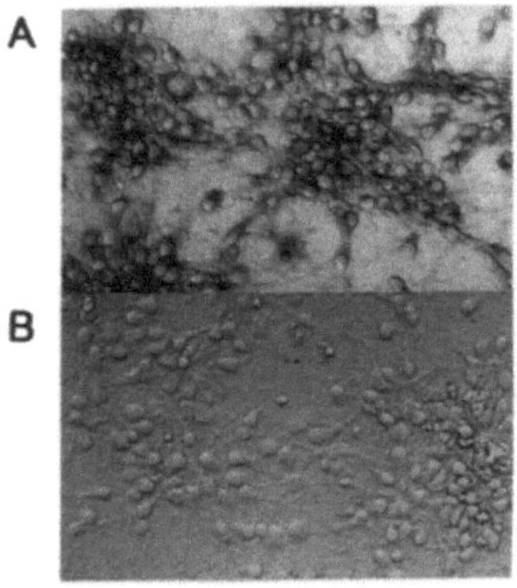

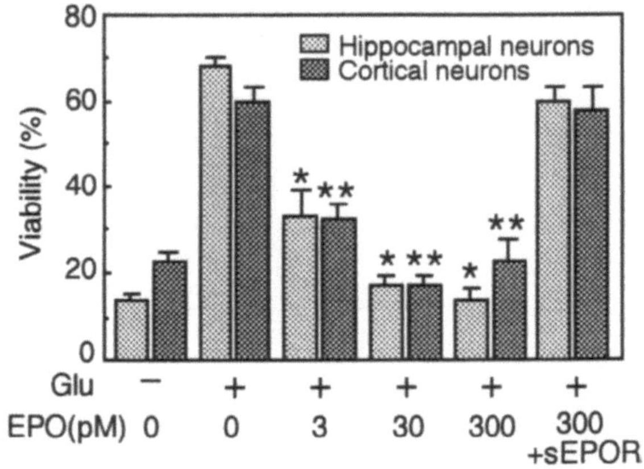

Figure 3. (left) Immunochemical detection of EPOR in cultured hippocampal neurons. A Cells were stained with the anti-EPOR antibody. B Cells were stained with the antibody preabsorbed with the antigen. Scale bar = 50 μm.

Figure 4. (right) Protection of glutamate-induced cell death of cultured hippocampal and cortical neurons by EPO. Star symbols mean p <0.001, significantly different from the values in glutamate challenge experiments without EPO. * Hippocampal neurons; **cortical neurons.

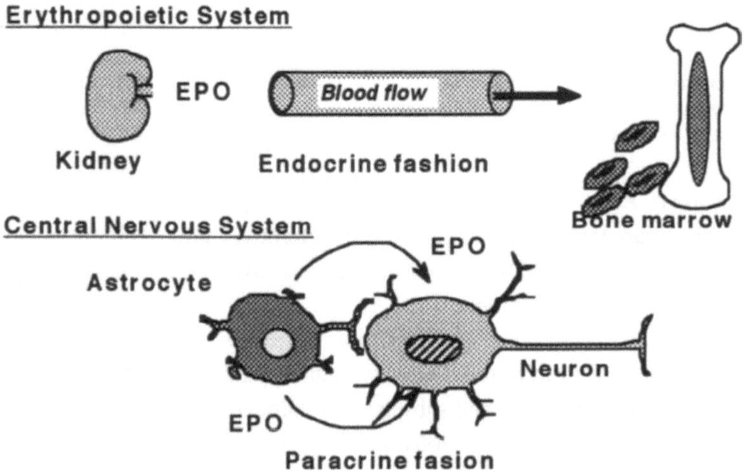

Figure 5. EPO in the peripheral and the CNS

4. DISCUSSION

In fetal stage, EPO is produced in the liver where erythropoiesis takes place. In adult, EPO produced by the kidney travels in the circulation to reach erythropietic tissues, bone marrow and spleen in the peripheral. So far the erythroid lineage has been thought to be a sole physiological target of EPO. Several lines of evidence indicate that the EPO/EPOR system independent of that in erythropoiesis is present in the CNS where astrocytes produce EPO and neurons express EPOR, and that EPO protected glutamate neurotoxicity (Fig. 5). Since astrocytes are located closely to or interact directly with neurons in the CNS. EPO acts on neurons through a paracrine mechanism.

The occurrence of hypoxia-induced brain EPO production[9, 11, 12] suggests that an increase in EPO concentration upon hypoxia reduces the neurotoxicity owing to an extracellular increase of glutamate. EPO may not meet the requirement for neuroprotection in submaximal but severe hypoxia or ischemia, both of which cause degeneration of the neurons vulnerable to these insults. A relatively long period (several hours) is required for manifestation of the neuroprotective effect of EPO. Therefore, EPO may play a neuroprotective role in mild and long-lasting hypoxic or ischemic conditions, under which neurons can survive.

REFERENCES

1. Jacobson L. O., Goldwasser E., Fried W. and Plzak L., Role of the kidney in erythropoiesis, *Nature* **179**, (1957) 633-634.
2. Krantz, S. B., Erythropoietin, *Blood*, **77** (1991) 419-434
3. Beru, N., McDonald, J., Lacombe, C., and Goldwasser, E., Expression of the erythropoietin gene, *Mol. Cell.Biol.*, **6** (1986) 2571-2575
4. Goldberg, M.A., Dunning, S.P., and Bunn, H. F., Regulation of the erythropoietin gene : evidence that the oxygen sensor is a heme protein, *Science* **242** (1988) 1412-1415
5. Koury; S. T., Koury; M. J., Bondurant. M.C., Caro, J., and Graber, S. E., Quantitation of erythropoietin-producing cells in kidney of mice by in situ hybridization : correlation with hematocrit, renal erythropoietin mRNA, and serum erythropoietin concentration, *Blood*, **74** (1989) 645-651
6. Masuda, S., Nagao, M., Takahata, K., Konishi, Y., Gallyas, F., Jr., Tabira, T. and Sasaki, R., Functional erythropoietin receptor of the cells with neural characteristics: comparison with receptor properties of erythroid cells, *J. Biol.Chem.*, **268** (1993) 11208-11216
7. Morishita, E., Masuda, S., Nagao, M., Yasuda, Y. and Sasaki, R., Erythropoietin receptor is expressed in rat hippocampal and cerebral cortical neurons, and erythropoietin prevents in vitro glutamate-induced neuronal death, *Neuroscience*, (1996) *in press*
8. Morishita, E., Narita, H., Nishida, M., Kawashima, N., Yamagishi, K., Masuda, S., Nagao, M. and Sasaki, R., Anti-erythropoietin receptor monoclonal antibody : epitope mapping, quantification of the soluble receptor, and detectionof the solubilized transmembrane receptor and the receptor-expressing cells, *Blood*, **88** (1996) 465-471
9. Digicaylioglu, M., Bichet, S., Marti, H. H., Wegner, R.H., Rivas, L. A., Bauer, C. and Gassmann, M., Localization of specific erythropoietin binding sites in defined areas of the mouse brain, *Proc. Natl. Acad. Sci. USA*, **92** (1995) 3717-3720
10. Perry, V.H. and Gordon, S., Macrophages and microgria in the nervous system, TINS, **11**, (1988) 273-277
11. Tan, C.C., Eckardt, K.U., Firth. J.D. and Ratcliffe, P.J., Feedback modulation of renal and hepatic erythropoietin mRNA in response to graded anemia and hypoxia, *Am. J. Physiol.*, **263** (1992) F474-481
12. Masuda, S., Okano, M., Yamagishi, K., Nagao, M., Ueda, M. and Sasaki, R., A novel site of EPO production : oxygen-dependent production in cultured rat astrocytes, *J. Biol. Chem.*, **269** (1994) 19488-19493
13. Masuda, S., Chikuma, M. and Sasaki, R. Insulin-like growth factors and insulin stimulate erythropoietin produyction in primary cultured astrocytes, *Brain Res., in press*
14. Goto, M., Murakami, A., Akai, K., Kawanishi, G., Ueda, M., Chiba, H. and Sasaki, R., Characterization and use of monoclonal antibodies directed against human erythropoietin that recognize different antigenic determinants, *Blood*, **74**, (1989) 1415-1423
15. Choi D. W. Ionic dependence of glutamate neurotoxicity, *J. Neurosci.*, **7**, (1987) 369-379.
16. Rothman S. M., Thurston J. H. and Hauhart R. E., Delayed neurotoxicity of excitatory amino acids *in vitro*, *Neuroscience*, **22**, (1987)471-480.

SELECTION AND STABILITY FOR A RECOMBINANT CHO CELL LINE EXPRESSING HUMAN GM-CSF IN GENE AMPLIFICATION

Takeshi Omasa, Seima Itami, Daisuke Kameoka, Yoshio Katakura and Ken-ichi Suga
Biochemical Engineering Laboratory,
Department of Biotechnology, Graduate School of Engineering,
Osaka University, Suita 565, JAPAN
Telephone: +81-6-879-7437, Facsimile: +81-6-879-7439
e-mail: omasa@bce.bio.eng.osaka-u.ac.jp
URL: http://www.bio.eng.osaka-u.ac.jp/suga/home.html

ABSTRACT: We investigated the relationship between the specific production rate and the concentration of the toxic drug methotrexate (MTX), inhibitory to nucleic acid synthesis in the medium, during the construction of recombinant CHO cells. The results of selection of MTX tolerant cells when the MTX concentration was increased under 10 different stepwise conditions showed that the number of days required to obtain cells resistant to 1000 nM MTX depended on the selection conditions. Tolerant cells were obtained most rapidly at an MTX concentration of 10nM. With increasing MTX concentration, the specific production rate increased, while the specific growth rate gradually decreased. In the case of a gradual increase in the MTX concentration, the specific production rate reached a higher level. We also investigated the stability of the specific production rate during cultivation over a long period with or without MTX. In long-term cultivation, the specific production rate decreased, while the specific growth rate remained at the same level. The specific production rate in a culture without MTX decreased more rapidly than that in a culture with MTX.

1. Introduction

Gene amplification techniques are used for production of glycoproteins in recombinant mammalian cell cultures. Among these techniques, the dihydroforate reductase (dhfr) gene amplification system in the Chinese Hamster Ovary (CHO) cell line is the most widely employed. In the application of gene amplification systems to industrial processes, the key factors are the method of selecting the recombinant CHO cell line and the stability of recombinant protein production. In this study, we investigated the relationship between the specific production rate and the concentration of the toxic drug methotrexate (MTX) in the medium during the construction of recombinant CHO cells.

2. Materials and Methods

The gene amplification systems employed in the experiment were the dihydroforate reductase (dhfr) and glutamine synthetase (gs) systems. In the dhfr system, the CHO DG44 cell line ($dhfr^-$), which was kindly provided by Dr. Chasin of Columbia University was used as the host cell. The vector pSV2/hGM-CSF was constructed from pSV2-dhfr (ATCC 37146) and pcD-hGM-CSF (ATCC 57594). CHO DG44 cells were transformed with pSV2/hGM-CSF using the calcium phosphateco-precipitation method. For dhfr gene amplification, cells were cultivated in a selection medium (Iscove's modified DMEM with 10% dialyzed fetal calf serum (FCS)) with increasing MTX concentration under various patterns.

In the gs gene amplification system, the CHO K1 cell line (RCB 0285) was used as the host cell. The vector pSV2-GS/hGM-CSF was constructed from pSV2-dhfr, pcD-hGM-CSF, and pGSRK-1 (ATCC 63067). CHO cells were also transformed using calcium phosphateco-precipitation method. For gs gene amplification, cells were cultivated in a selection medium (RDF(HO) [1] with 10% FCS supplemented with glutamate (0.336g ℓ^{-1}) and NH_4Cl (0.122g ℓ^{-1}), with increasing methionine sulfoximine (MSX) concentration under various patterns. Flow karyotype analysis was performed using a flow cytometer (Coulter EPICS elite). Chromosomes were isolated using the Polyamine-Digitonin method [2].

3. Results and Discussion

3.1 EFFECT OF STEPWISE INCREASE OF MTX CONCENTRATION ON SPECIFIC GROWTH AND PRODUCTION RATES

Gene amplification is a useful way to increase the copy number of an objective gene. Recombinant cells are cultivated in a medium containing a toxic drug (e.g. MTX), which inhibits the enzyme expressed by an encoded marker gene. Although most cells can not survive, a few that contain a higher number of selectable marker genes can grow in this medium. The copy number of the objective genes also increases with the marker gene. Thus, by increasing the drug concentration, highly productive cell lines can be obtained. However, since a suitable pattern of drug concentration increase has not been clarified, we investigated the relationship between the specific production rate and the concentration of the toxic drug MTX in the medium during the construction of recombinant CHO cells.

Cell growth and production were investigated under 10 kinds of selection conditions. Figure 1 shows method of increasing the MTX concentration stepwise in the gene amplification medium. The initial MTX concentration was either 0 or 10 nM. At initial MTX concentrations of 20 mM or more, a tolerant cell line could not be obatined. Figure 2 shows the number of days required for selection under several MTX concentrations in the dhfr system. The length of time required to obtain the cells resistant to 1000 nM MTX differed depending on the selection conditions. At an initial

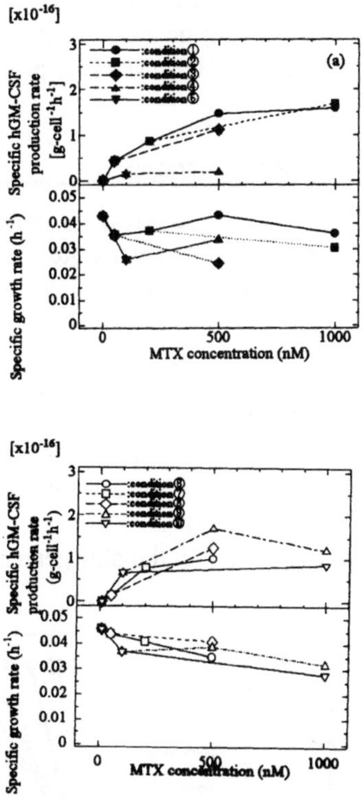

Figure 3 Comparison of specific growth and production rates under several selection conditions
Initial MTX concentration ...... (a) 0 nM, (b) 10 nM

3.2 STABILITY AND FLOW KARYOTYPE ANALYSIS IN GENE AMPLIFICATION

The stability of cell growth and production over a long period of cultivation is an important factor in industrial production. We therefore investigated the stability of the specific production rate during a long-term cultivation with or without MTX. The specific production rate decreased, while the specific growth rate remained at the same level (data not shown). The specific production rate decreased more rapidly in the culture with MTX than in that without the drug. It is well known that in gene amplification, stable recombination occurs in the intra chromosome segment, and not in the extra chromosome segment [2,3]. We thus investigated the difference in the chromosome patterns of recombinant cells. Figure 4 shows the results of flow karyotype analysis. The small chromosome segment decreased in recombinant CHO cells and the middle chromosome increased in wild-type CHO cells. The small chromosome might increase in size because of amplification of the dhfr and hGM-CSF genes.

MTX concentration of 10 nM, tolerant cells were obtained rapidly compared with an initial concentration of 0 nM. In the gs gene amplification system, the number of days required to obtain the cells resistant to 800 μM MTX also differed with the selection condition (data not shown).

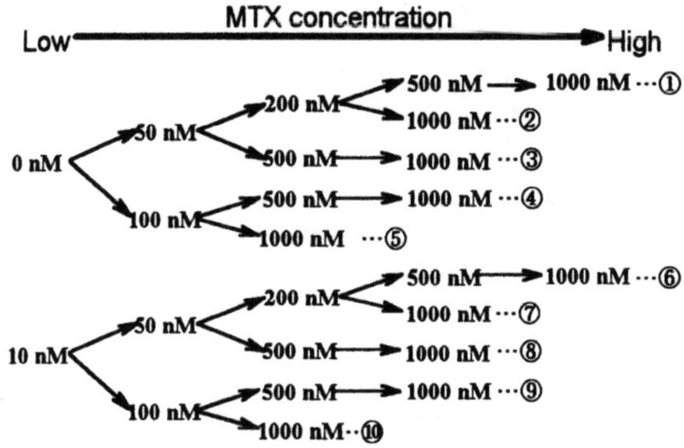

Figure 1 Stepwise selection in the dhfr gene amplification system

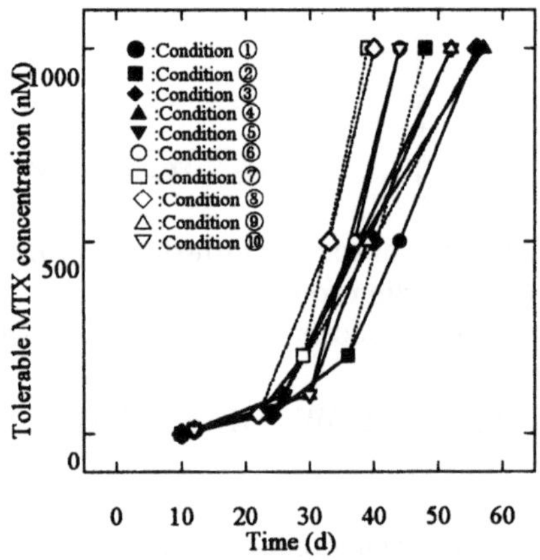

Figure 2 Days required for selection under several MTX concentrations in the dhfr gene amplification system

Figures 3 (a) and (b) show the specific growth and human granulocyto-macrophage colony stimulating factor (hGM-CSF) production rates at various MTX concentrations under conditions number ① to ⑤ and ⑥ to ⑩, respectively. The specific hGM-CSF production rate increased with increasing MTX concentration, while the specific growth rate gradually decreased. It can be seen that the hGM-CSF gene was amplified with amplification dhfr gene. When the different conditions are compared, the specific hGM-CSF production rate is seen to increase the higher the MTX level in the case of a gradual increase in the MTX concentration. However, this tendency was not so marked at an initial MTX concentration of 10 nM.

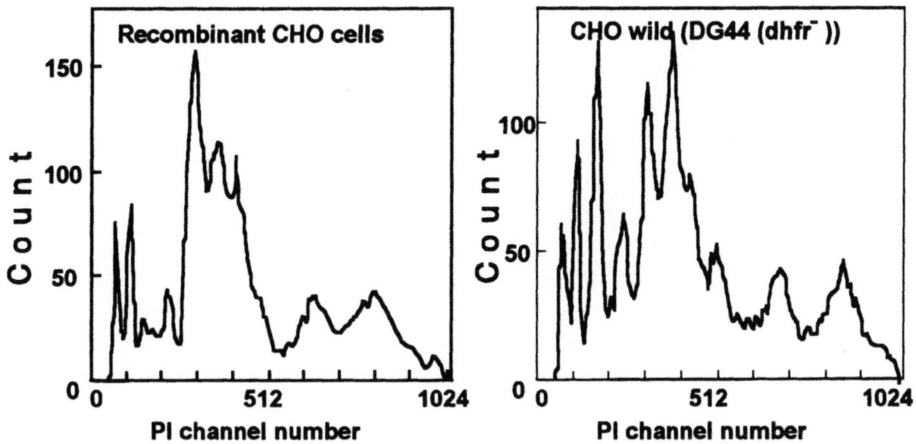

Figure 4 Flow karyotype analysis in wild and recombinant CHO cell line

4. Acknowledgments

We thank Dr. Chasin of Columbia University for providing the CHO DG44 cell lines and Dr. Kuriki of Kansai Medical University for directing us in the flow karyotyping.

5. References

[1] Omasa T., Ishimoto M., Higashiyama K., Shioya S., and Suga K. (1992) 'The enhancement of specific antibody production rate in glucose- and glutamine-controlled fed-batch culture' Cytotechnology, **8**, 75-84

[2] Kuriki K., Sonta S., and Murata K.(1993) 'Flow karyotype analysis and sorting of the Chinese hamster chromosomes: Comparing the effects of the isolation buffers' J. Clin. Lab. Anal., 7, 119-122.

[3] Schimke R. T. , Brown P. C., and Kaufman R. J. (1982) 'Gene amplification and drug resistance in mammalian cells' Natl. Cancer Inst. Monogr. **60**,79-86.

[4] Ruiz J. C. and Wahl G. M. (1990) 'Chromosomal destabilization during gene amplification' Mol. Cell. Biol. **10**,3056-3066

References

EFFECT OF CULTURE CONDITIONS ON GLYCOSYLATION IN HYBRIDOMA ANTIBODY PRODUCTION

Takeshi Omasa, Jyun-ya Tanaka, Kenji Amimoto, Yoshio Katakura and Ken-ichi Suga
Biochemical Engineering Laboratory,
Department of Biotechnology, Graduate School of Engineering,
Osaka University, Suita 565, JAPAN
Telephone: +81-6-879-7437, Facsimile: +81-6-879-7439
e-mail: omasa@bce.bio.eng.osaka-u.ac.jp
URL: http://www.bio.eng.osaka-u.ac.jp/suga/home.html

ABSTRACT: To investigate the effect of culture conditions on glycosylation patterns, we carried out the batch cultivation (using serum-containing or serum-free media) and glucose-limited continuous cultivation (serum-free medium). Con A lectin blotting showed that N-linked glycosylation occurred under all the culture conditions used. The sugar composition of the oligosaccharide chain in each culture was studied using gas-chromatography and was found to be different under each culture condition. The detailed structure of each oligosaccharide chain was then analyzed by two-dimensional mapping techniques using pyridylamination method. The oligosaccharide chain content in human IgG was higher in batch cultivation using a serum medium than that in serum-free cultures. However, oligosaccharide chain content with a galactose residue at the chain end, which plays an important role in the immune system, was shown to be high in glucose-limited continuous cultivation under a low specific growth rate.

1. Introduction

The oligosaccharide chain of a glycoprotein can affect features such as protein folding, solubility, biological activity, and antigenicity. Glycosylation variability is affected by culture conditions. In this study, we investigated the effect of culture conditions on glycosylation in hybridoma antibody production.

2. Materials and Methods

The cell line employed in the experiments was the mouse-mouse hybridoma 3A21 (RCB 1985), which produces an anti-RNase A monoclonal antibody (IgG)[1]. The serum-free culture medium was RDF-ITES with bovine serum albumin (BSA) [1]. A serum medium was prepared by supplementing the serum-free medium with 10% FCS. In glucose-limited continuous cultivation, the feed glucose concentration in RDF-ITES medium was 1.0 g ℓ^{-1}. In glutamine-limited continuous cultivation, the feed glutamine concentration was 0.16 g ℓ^{-1}. The specific growth rate was calculated as previously reported [2]. Antibody purified from each culture supernatant using RNase A affinity column chromatography was used to investigate N-linked glycosylation. The presence of N-linked glycosylation was determined by concanavalin A lectin blotting. The monosaccharide contents of each complex sugar chain were determined using gas chromatography with trimethylsilyl labeling. The detailed structure of each oligosaccharide chain was analyzed by the pyridylamination method [3].

3. Results and Discussion

Glucose-limited continuous cultivation (serum-free) was used to investigate the effect of culture conditions on glycosylation patterns. The antibody was purified by affinity chromatography from the cultures and its sugar chain was analyzed. The presence of N-linked glycosylation was checked by SDS-polyacrylamide gel electrophoresis (PAGE) with/without glycopeptidase F and concanavalin A lectin blotting. Figure 1 shows the results of SDS-PAGE and lectin blotting under various specific growth rates. Glycosylation is seen o have occurred under all the culture tested.

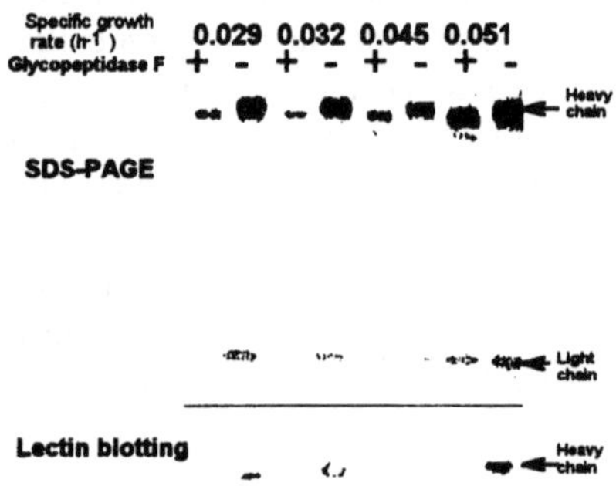

Figure 1 Results of lectin blotting under various specific growth rates in glucose-limited continuous cultivation.

The sugar composition of the olygosaccharide chain under different culture conditions was determined using gas-chromatography. Table 1 shows the monosaccharide contents of each complex sugar chain. The sugar composition is seen to be different under each culture condition. However, because the details of the oligosaccharide chains are not apparent from monosaccharide analysis alone, we also analyzed the chain structures by the pyridylamination method.

Table 1 Monosaccharide contents of each complex sugar chain

Monosaccharide	Fuc	Man	Gal	GlcNAc	Man /GlcNAc
Glucose-limited continuous cultivation μ=0.045 (h^{-1})	0.9	3	1.3	3.0	1.0
Glutamine-limited continuous cultivation μ=0.041 (h^{-1})	0.5	3	1.2	1.6	1.9
Batch cultivation (serum-free)	0.6	3	1.2	1.5	2.0
Batch cultivation (serum)	0.6	3	1.1	1.9	1.6

The values were normalized as mannose=1.0

The oligosaccharide chain was separated from purified antibody by hydrazinolysis. The chain was then pyridylaminated and analyzed using two-dimensional HPLC mapping. Figure 2 shows the analysis of the pyridylaminated oligosaccharide chain using reverse-phase chromatography and, Fig. 3 depicts the estimated structure of each of the pyridylaminated oligosaccharide chains shown in Fig. 2 by two-dimensional HPLC mapping.

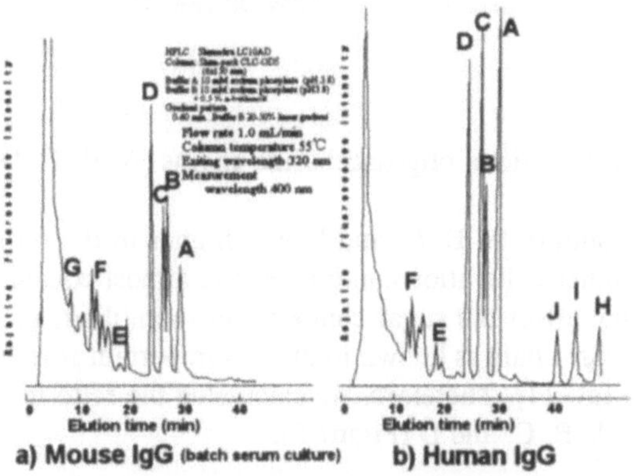

Figure 2 Analysis of pyridylaminated oligosaccharide chains using reverse-phase chromatography

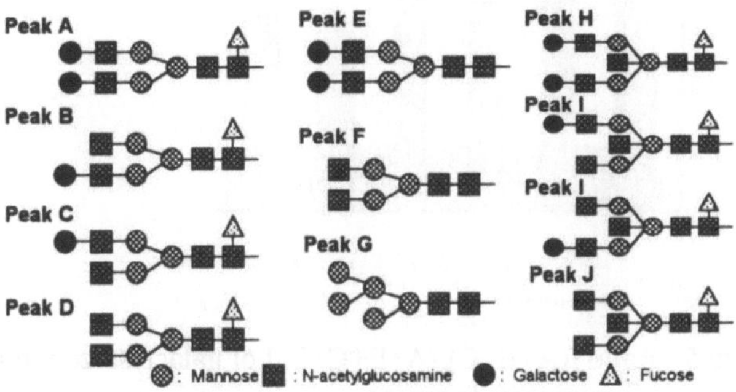

Figure 3 Estimated oligosaccharide structure

A comparison of the analyses of the two kinds of IgG depicted in Fig. 2 a) and b) shows the oligosaccharide structure of mouse IgG to be almost same as that of human IgG except for a lack of sugar chains H, I, and J. Additionally, a high-mannose type structure (structure G) was observed in mouse but not in human IgG. The structures A, B, C, and D were the main oligosaccharides in both mouse and human IgG. We therefore compared the ratio of the sum of A, B, C, and D to the whole sugar chain under each cultivation condition. The results are shown in Fig.4.

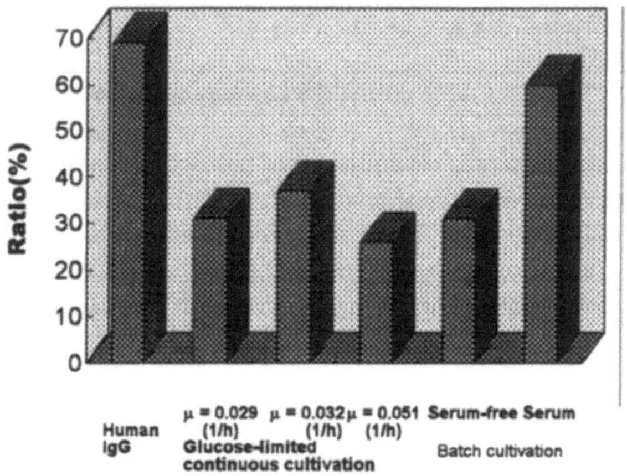

Figure 4 Ratio of oligosaccharide chains (A+B+C+D)

The ratio of the sum of A, B, C, and D was higher in the serum batch cultivation compared with the other cultivation conditions, and almost equaled that of human IgG. Unknown factors that affect the sugar composition in antibody production might exist in serum. The galactose chain is known to play an important role in the immunological activities of human IgG [4]. Therefore, we calculated the ratio of galactose-containing chains in structures A, B, C, and D (Figure 5).

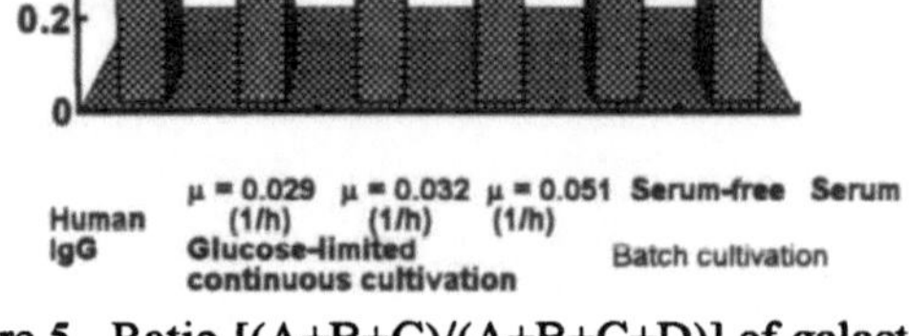

Figure 5 Ratio [(A+B+C)/(A+B+C+D)] of galactose-containing oligosaccharide chains

The ratio of galactose-containing oligosaccharide chains was greatest under continuous cultivation at a higher specific growth rate. The ratios were lower in batch cultivations than in continuous cultures, whereas, the specific growth rates were higher in batch cultivations. We therefore, we supposed that the specific growth rate affected the ratio of galactose-containing oligosaccharide chains.

4. Acknowledgments

We thank Dr. Natsuka and Prof. Hase of Osaka university for directing us in the oligosaccharide structure analysis.

5. References

[1] Omasa, T., Higashiyama, K., Shioya, S., and Suga, K (1992) 'Effect of lactate concentration on hybridoma culture in lactate-controlled fed-batch operation' Biotechnol. Bioeng. **39**, 556-564

[2] Omasa, T., Kobayashi, M., Nishikawa, T., Shioya, S., Suga, K., Uemura, S., Kitani, T., and Imamura, Y.(1995) 'Enhancement of antibody production by growth factor addition in perfusion and hollow-fiber culture systems' Biotechnol. Bioeng. **48**, 673-680

[3] Hase S., (1994) 'High-performance liquid chromatography of pyridylaminated saccharides' Methods Enzymol., **230**, 225-237

[4] Tsuchiya N., Endo T., Natsuta K., Yoshinoya S., Aikawa T., Kosuge E., Takeuchi F., Miyamoto T. and Kobata A. (1989) 'Effects of galactose depletion from oligosaccharide chains on immunological activities of human IgG' J. Reumatol. **16**, 285-290.

ANTISENSE RB RNA INHIBITS IN VITRO MYOGENIC DIFFERENTIATION IN A MYOBLAST CELL LINE, C2

MASAYUKI KOBAYASHI, YUKIKA YAMAUCHI, AND AKIKO TANAKA
Morinaga Milk Branch, Research Institute of Innovative Technology for the Earth, c/o Biochemical Research Institute, Morinaga Milk Ind. Co., Ltd.
5-1-83 Higashihara, Zama, Kanagawa 228, Japan

1. Introduction

Mammalian cell growth and differentiation are mutually exclusive. Accumulating evidence suggests the implication of retinoblastoma gene (*Rb-1*) product, pRb, in the exit from cell cycle and the entry into cell-differentiation (1). During progression of cell cycle and differentiation, the activity of pRb is modulated in a phosphorylation/dephosphorylation dependent-manner. The activity of pRb is also regulated at gene expression level, and the expression level increases with progression of differentiation in several lineages, including skeletal muscle differentiation in vitro (2).

An antisense RNA expression vector is a useful tool for clarifying the specific roles of pRb in differentiating myogenic cells, because it stably and selectively suppresses the synthesis of target proteins. We constructed several antisense RNA expression vectors directed to *Rb-1* mRNA, and showed the effective reduction of cellular pRb content by an antisense RNA complementary to 3'-untranslated region (3'UTR) of *Rb-1* mRNA (3). Recently, we have constructed a dicistronic mRNA expression vector, pMK10$_{59}$ (4). The vector expresses a single mRNA concomitantly encoding a gene of interest, internal ribosomal entry site (IRES) derived from encephalomyocarditis virus, and neomycin-resistance gene (Neo). In cells transfected with pMK10$_{59}$, a high degree of secondary structure of the IRES in the transcript is recognized by ribosomes, and the Neo placed downstream of the IRES is translated IRES-dependently.

In the present study, we showed directly the essential role of pRb in myogenic differentiation of C2 cells using the antisense RNA strategy. We adopted the pMK10$_{59}$ vector to express the antisense RNA directed to 3'UTR of *Rb-1* mRNA as the first sequence in a dicistronic mRNA.

2. Materials and Methods

2.1. CONSTRUCTION OF AN ANTISENSE RB RNA EXPRESSION VECTOR

The full-length mouse *Rb-1* cDNA (~4.6-kbp) contains a single large open reading frame from nt 99 to a stop codon at nt 2862 (Fig. 1). An antisense Rb RNA expression vector directed to 3'UTR of *Rb-1* mRNA (3'UTR14) was constructed using a pMK10$_{59}$ vector. This vector contains a powerful and versatile chicken β-actin promoter with the CMV immediate early enhancer and rabbit β-globin intron II acceptor site (CAG promoter). The duplicated 3'UTR14 (3'UTR14x2) was inserted into the cloning site of pMK10$_{59}$ in the sense or antisense orientation.

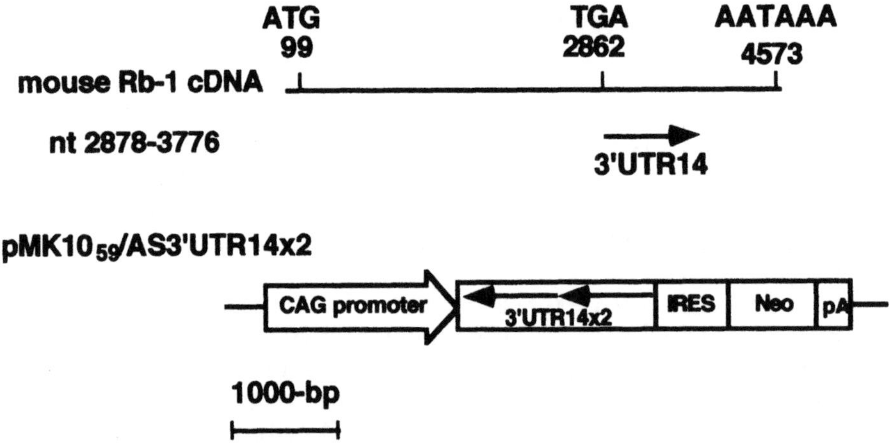

Figure 1. Construction of an antisense RNA expression vector directed to mouse *Rb-1* mRNA.

2.2. CELL CULTURES

Mouse myoblast C2 cells were routinely maintained in DMEM containing 15% fetal calf serum (FCS) and 4.5 g/l glucose (GM, growth medium) at 37 °C with 5% CO_2/air. To induce cell differentiation, 3-4 x

10^5 myoblast cells were seeded in 100-mm dishes with GM. After an 18 h-incubation at 37 °C with 5% CO_2/air, the medium was changed to DMEM supplemented with 3% horse serum and 1 g/l glucose (DM, differentiation medium). Extensive morphological and biochemical differentiation was obtained after 4-5 days in culture. C2 cells were transfected with pMK10$_{59}$-derived plasmids by lipofection using a LipofectAMINE reagent (GIBCO-BRL, Gaithersburg, MD). The vector expressing sense 3'UTR14x2 RNA was transfected as a control.

3. Results and Discussion

C2 clones expressing antisense RNA directed to 3'UTR of *Rb-1* mRNA were isolated. As shown in Fig. 2, mRNA and protein contents for pRb in the antisense Rb transfected myoblasts decreased to 25-35% of those in the control clone. These transfectants exhibited accelerated growth properties especially in culture medium with low concentrations of FCS.

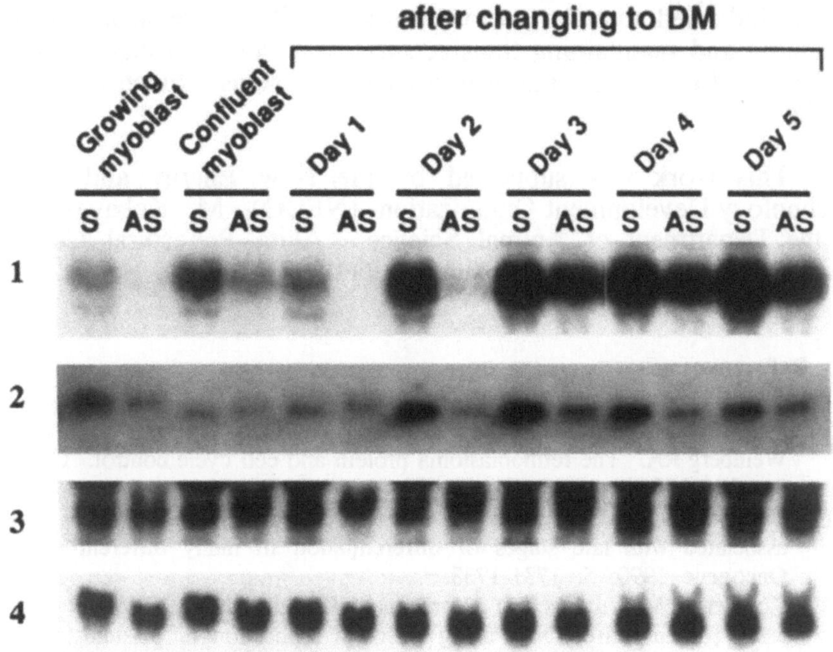

Figure 2. Expression of the antisense Rb RNA decreased cellular contents of *Rb-1* products in the C2 clone (S; a C2 clone expressing sense 3'UTR RNA, AS; a C2 clone expressing antisense 3'UTR RNA). 1. *Rb-1* mRNA, 2. pRb, 3. sense or antisense 3'UTR14x2 RNA, 4. G3PDH mRNA

When cultures were kept in DM, accumulations of *Rb-1* mRNA and pRb in the antisense Rb clones were only 30-50% of those in the control clone, but the expression levels of other Rb family proteins (p107 and p130) were affected little by the antisense RNA, when compared to the control clone. Under the condition, the control clones extensively formed multi-nucleated myotubes, whereas the clones expressing antisense Rb RNA continued to grow, remained single-nucleated cells and expressed no sarcometric myosin heavy chain. This inhibition of differentiation was abrogated by the re-introduction of expression vector for *Rb-1* mRNA without 3'UTR or sense 3'UTR of *Rb-1* mRNA to the C2 clone expressing antisense RNA. After the induction of myogenesis, the cellular contents of MyoD protein at days 2 and 3 severely decreased and the elevation of myogenin protein was not observed clearly in C2 clone expressing antisense Rb RNA.

These results demonstrate an indispensable role of pRb for completion of terminal differentiation in C2 cells. We consider that the antisense Rb expression strategy is a molecular scalpel to reveal directly the functions of pRb in cell-growth, death and differentiation. Furthermore, it was indicated that the vectors using the IRES are advantageous in obtaining and maintaining transfectants expressing antisense RNAs, by utilization of a coexpressing drug-resistance gene as a functional selection marker.

This work was supported by the New Energy and Industrial Technology Development Organization (NEDO). M. Kobayashi is now at the Department of Animal Science, Akita Prefectural College of Agriculture, 2-2 Minami, Ohgata, Akita 010-04, Japan.

4. References

(1) Weinberg RA. The retinoblastoma protein and cell cycle control. Cell 1995; 81: 323-330.

(2) Coppola JA, Lewis BA, Cole MD. Increased retinoblastoma gene expression is associated with late stages of differentiation in many different cell types. Oncogene 1990; 5: 1731-1733.

(3) Kobayashi M, Yamauchi Y, Yamaguchi K, Tanaka A. Transient expression assay for antisense RNAs using episomal replication of plasmids: effective reduction of retinoblastoma gene (*Rb-1*) product by its antisense RNA complementary to 3'-untranslated region. Antisense Res Dev 1995; 5: 141-148.

(4) Kobayashi M, Yamauchi Y, Tanaka A, Shimamura S. Improved dicistronic mRNA expressing vectors for efficient selection of transfectants highly expressing foreign genes. Biotechniques 1996; 21: 398-402.

Establishment of a Stable hGH Producing Verots S3 Transformant

Ayuko Sato[1,2], Nakao Nomura[1,2], Kaoru Saijo[2], Tadao Ohno[2]
and Masatoshi Matsumura[1]
[1]*Institute of Applied Biochemistry, University of Tsukuba, Japan*
[2]*RIKEN Cell Bank, The Institute of Physical and Chemical Research (RIKEN), Japan*

ABSTRACT

It is well known that the large T antigen produced by Verots S3 is temperature sensitive. This implies that production of hGH by the transformants of Verots S3 can be controlled by shifting the temperature. In this study, a stable hGH producing Verots S3 transformant was established, and the growth of the stable transformant of Verots S3 immobilized in porous carrier, Cellsnow, monitored by the glucose consumption rate, was observed faster than that on solid microcarrier Cytodex3 in the later of the culture. However, the hGH production by the Verots S3 transformant with Cytodex3 was much higher than that with Cellsnow.

1. Introduction

Most of host cell lines used for biotechnological applications are cultured in serum-containing medium. However, the use of serum medium in animal cell culture has some disadvantages in terms of purification of products and cost of medium. The Verots S3 cell line established by Ohno et al. (1991) can be grown in protein-free, biotin-containing Eagle's MEM without supplements. Since the Verots S3 cell line constitutively expresses temperature-sensitive SV40 large T antigen, and the large T antigen produced by Verots S3 cells enhances replication of SV40 origin containing plasmids at 33 ℃ but not at 39.5℃. It may be able to regulate the expression of exogeneous genes in SV40 replication origin containing plasmids in serum-, protein-, and lipid-free conditions by shifting the temperature. The large scale culture of Verots S3 in protein free medium for the transient production of hGH was done by Nomura et al. (1996). They also found that porous microcarriers were more suitable for cell growth than solid microcarriers.

In this work, we tried to establish the stable transformants of Verots S3 producing hGH and culture them efficiently by shifting the temperature. Spinner culture of the Verots S3 transformant with different types of carriers was also performed.

2. Materials and methods

2.1. CELL LINE

Verots S3 (RCB0707) cells were obtained from RIKEN Cell Bank, Tsukuba Science City, Japan. Verots S3 cells were maintained at 37 ℃ in MEM containing 5% fetal bovine serum.

2.2. PLASMID

pXGH5ori containing human growth hormone (hGH) gene and SV40 DNA replication origin was digested with BamHI. pMDSG was digested with BamHI and the resulting 2.2kbp fragment containing eco-gpt was separated by agarose gel electrophoresis. This fragment was then inserted into the BamHI sites of Plasmid pXGH5ori and named pXGH5ori-gpt (Fig.1).

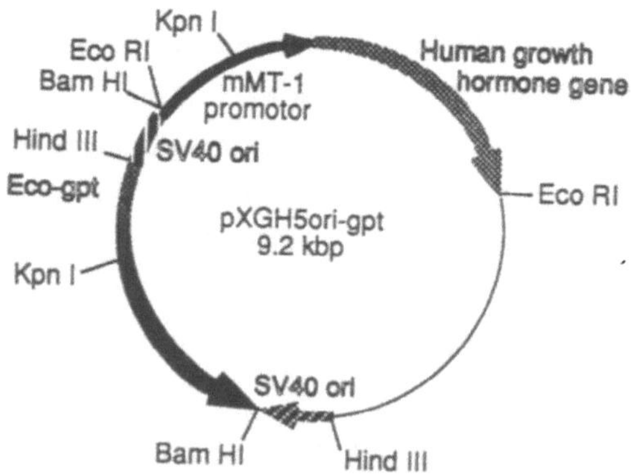

Fig.1 Restriction map of pXGH5ori-gpt

2.3. TRANSFECTION

Verots S3 cells were maintained in the culture medium of MEM added 5% fetal bovine serum. Two days after seeding approximately $2x10^5$ cells on 100 mm plates, the cells were transfected with 40 μ g of the plasmid pXGH5ori-gpt using Lipofectamine Regent (GIBCO). DNA-liposome complexs were added onto the cells rinsed with serum free medium.

2.4. SELECTION FOR TRANSFORMANTS

After 24 hours of transfection, the culture medium was replenished with fresh medium. Three days after transfection, the monolayers of the transfected cells were harvested with trypsin-EDTA, and $6x10^4$ cells were dispersed in MEM containing 5% fetal bovine serum on 35 mm plates. Twenty four hours later, the culture medium was replenished with MEM containing 5% fetal bovine serum, xantine (250 μ g/ml), hypoxantine (15 μ g/ml), tymidine (10 μ g/ml), aminopterin (2 μ g/ml) and mycophenolic acid (25 μ g/ml). Thereafter, medium was changed every 3 days. Clonies were visible in 14-20 days and isolated by using penicillin cup after 20 days.

2.5. REPEATED BATCH CULTURE AND CARRIER

Cellsnow P9-402 and Cytodex3 were employed as porous and solid carriers respectively. Carrier concentration and inoculum size was 3.0 g/l and $1.3x10^5$ cells/ml, respectively. To monitor the cell growth, glucose concentration was measured every day and medium was replenished when glucose was depleted.

3. Results and discussion

3.1. TRANSFORMATION OF Verots S3

Verots S3 cells were transfected with the plasmid pXGH5ori-gpt. The transfected cells were cultured in MEM containing aminopterin and mycophenolic acid. Twenty three colonies were obtained and isolated by using penicillin cup and cultured. One of these clones was named as 4-5(+) and then used for the following experiments.

3.2. EFFECT OF TEMPERATURE

The growth of 4-5(+) and the production of hGH by the 4-5(+) was compared at different temperatures. The cell growth at 33 ℃ was slower than that at 39.5 and 37 ℃. The highest specific production rates at 39.5 ℃ and 37 ℃ were 23.6 and 29.3ng/ 10^4 cells/day on day one and two, respectively, but thereafter the specific production rates were quickly decreased. This may be resulted from temperature-sensitive shut-down of hGH production at the nonpermissive temparature. The highest specific production rate at 33 ℃ was lower than those at 39.5 ℃ and 37 ℃. However, the specific production rate at 33 ℃ was maintained stably for 5 days (Fig.2). Hence, it is assumed that temperature shifting from 37 ℃ to 33 ℃ after a few days of inoculatiion may facilitate more effective production of hGH.

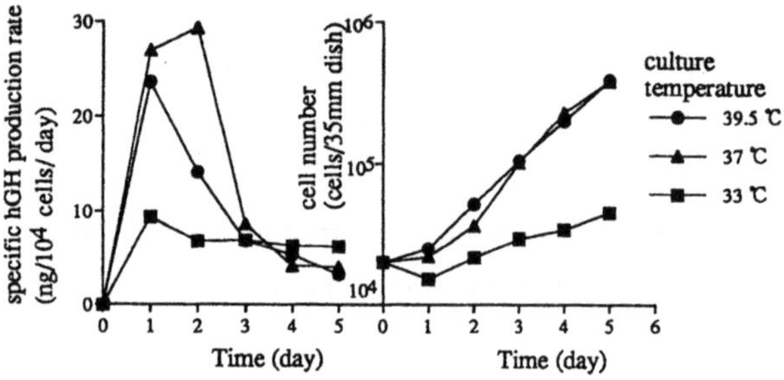

Fig.2 Growth of 4-5(+) and specific hGH production rate at different temperatures

The culture temperature was shifted from 37 ℃ to 33 ℃ after 2 days of inoculation. The specific production rate reached the highest on day one and thereafter that was decreased. However, as the cells were seeded at a concentration of $5x10^3$ cells/35mm dish, and specific production rate was increased again by shifting the temperature from 37 ℃ to 33 ℃ (Fig.3). The specific production rate at low inoculated cell number was higher than that at high inoculated cell number.

The accumulation of hGH productivity by 4-5(+) seeded at a concentration of $5x10^3$ cells/35mm dish with shifting the temperature from 37 ℃ to 33 ℃ was also the highest (data not shown).

3.3. SPINNER CULTURE

The spinner culture of 4-5(+) with different types of carriers was carried out. The growth of 4-5(+) immobilized in porous carrier, Cellsnow, monitored by the glucose consumption rate, was faster than that on solid microcarrier Cytodex3 in the later of the culture. However, the hGH production by 4-5(+) with Cytodex3 was much higher than that with Cellsnow (Fig.4).

The spinner culture of figure 4 was carried out at 37 ℃. The result showed that the specific production rate at low cell density was higher than that at high cell density. On the basis on the results of fig. 3, the production rate at 33 ℃ is higher than that at 37 ℃, but spinner culture was only carried out at 37 ℃ presently. Further experiments at 33 ℃ will be performed in the future.

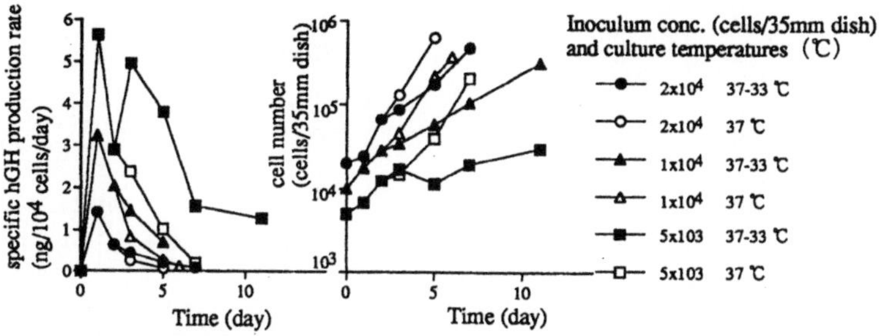

Fig.3 Growth of 4-5(+) and specific hGH production rate with or without temperature shift after 2 days of inoculation

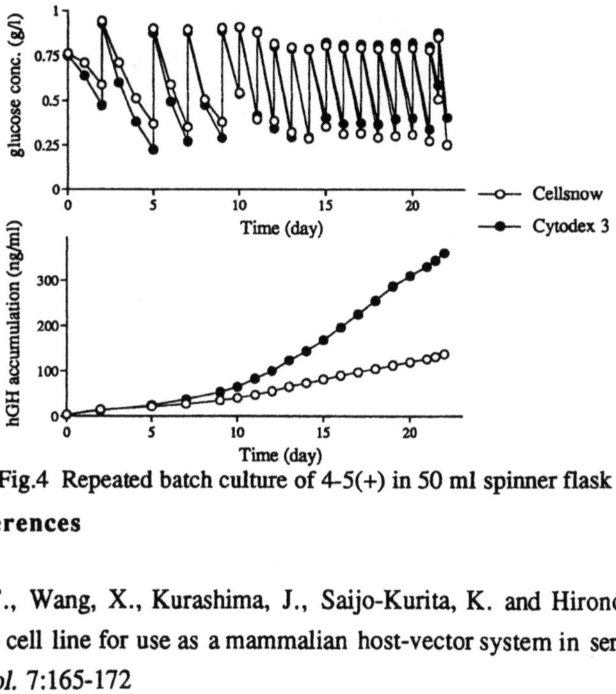

Fig.4 Repeated batch culture of 4-5(+) in 50 ml spinner flask

4. References

Ohno, T., Wang, X., Kurashima, J., Saijo-Kurita, K. and Hirono, M., (1991) A novel Vero cell line for use as a mammalian host-vector system in serum-free medium, *Cytotechnol.* 7:165-172

Nomura, N., Wang, P. C., Saijo, K., Ohno, T. and Matsumura, M., (1996) Production of human growth hormone by protein-free cultivation of anchorage-dependent cells with porous cellulose carrier, *J. Ferment. Bioeng.* (in press)

CHARACTERIZATION OF AN ANTIGEN-PRODUCING CELL LINE HUGK-14 AT THE MOLECULAR LEVEL

NOBORU NAKAMICHI, ASAO NODA, TOMONORI YONEZU, KATSURO KOIKE[1], TOSHIHARU MATSUMURA

Meiji Cell Technology Center, Meiji Milk Products Co., Odawara 250,

[1]The Cancer Institute, JFCR, Kami-Ikebukuro, Tokyo 170

Abstract

In order to evaluate its applicability as substrate for producing hepatitis B vaccine, a seed lot of human continuous cell line huGK-14 of hepatocellular carcinoma origin was established and investigated in its molecular genetic backgrounds. It was shown here that HBV DNA was integrated in eight different sites of the cellular DNA, in each of which HBV genome was rearranged, fragmented, and/or partly deleted. Complete HBV genome, as well as free HBV genome, was not detected in the cells nor in the culture medium. Clones of cDNA containing a complete coding frame for HBs antigen protein (type adr) were obtained. The cells were shown to be stable during more than 60 population doublings over the period of six months of cultivation in the mode of HBV integration and HBs mRNA expression. These results substantially fulfill the safety requirements for the seed lot to be used in vaccine production in terms of HBV genome integration and HBs gene expression.

Introduction

The possibility has long been discussed that a hepatitis B virus (HBV) vaccine could be produced using an HBV surface antigen (HBs Ag) protein-producing human cell line, and that such a vaccine might be advantageous compared to those obtained from the pooled serum of infected donors or from a non-human recombinant host cell line in several respects (McAleer et al., 1983).

The HBV genome consists of ca. 3.2kb partly double stranded circular DNA (Tiollais et al., 1985). The HBV genome transcribes 3.5kb, 2.1kb, and a few other poly A$^+$ RNAs. The 3.5kb poly A$^+$ RNA which contains the whole genomic sequence is synthesized for the replication of viral genome, while other poly A$^+$ RNAs code four viral protein species (Figure 1). The S, PreS1 and PreS2 regions of the genome contain coding frames for HBs antigen proteins of three types, i.e., the major protein (S), middle protein (PreS2+S) and large protein (PreS1+PreS2+S). For simplicity, PreS1 and PreS2 are abbreviated as S1 and S2 in the following. The C and X regions code HB core antigen and X protein respectively. The P gene codes viral reverse transcriptase, and is extended from a part of C region to a part of X region, overlapping the whole of S1, S2 and S regions. There are several subtypes in HBV, which are very similar in terms of their genomic structure.

Previously, we noted that clonal lines of a human hepatocellular carcinoma cell line, huH-1 (Huh et al., 1981), released HBs antigen in culture medium efficiently (Koike et al., 1987), and that HBV genomes integrated in these cells were deleted and/or rearranged (Mizusawa et al., 1985).

In an attempt to find the possibility of HBV vaccine production, HBs antigen was obtained from the culture fluid of a selected clone of huH-1, i.e., huGK-14, and characterized in some detail (Oda et al., 1986). In this paper, we report the characteristics of the seed lot and passaged cultures thereof in their HBV genome status and the expression of HBs gene.

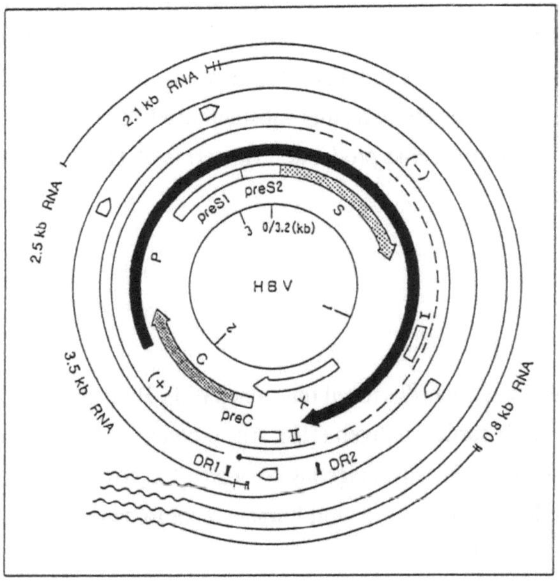

Figure.1 Genomic structure of HBV. Reproduced from Koike (1994).

Materials and Methods

The cell banks and cell cultures of huGK-14 were developed as shown in Figure 2. Briefly, cells of clone huGK-14 propagated in monolayer cultures were freeze-preserved in ampoules as seed cells (SC). From an SC ampoule, cells were reconstituted, propagated, and frozen again for a master cell bank (MCB), and from an MCB ampoule for a working cell bank (WCB) in the same manner. The number of cell population doublings (PDL) was 12 and 18 respectively for MCB and WCB when the PDL of SC was set at 0. Cells reconstituted from WCB ampoule were grown in monolayer, trypsinized and then introduced into a perfusion-type cell reactor in which they were immobilized on glass fiber cloth. The culture environment was kept at 37 ℃, with controlled the supply of oxygen gas, nitrogen gas, and carbon dioxide gas, and the culture medium was continuously supplied and harvested during cultivation. The cell reactor system will be described elsewhere in detail. Cells between 18 PDL and 33 PDL were introduced in a reactor and maintained for up to 6 months. Cells obtained from the reactor after 2, 4 and 6 months will be referred to as P2M, P4M and P6M respectively. Growth medium contains DM-160 supplemented with 1% or 5% fetal bovine serum (FBS). Maintenance medium used for maintaining the cells in the reactor contains William's E medium supplemented with 0.1% bovine serum albumin (BSA).

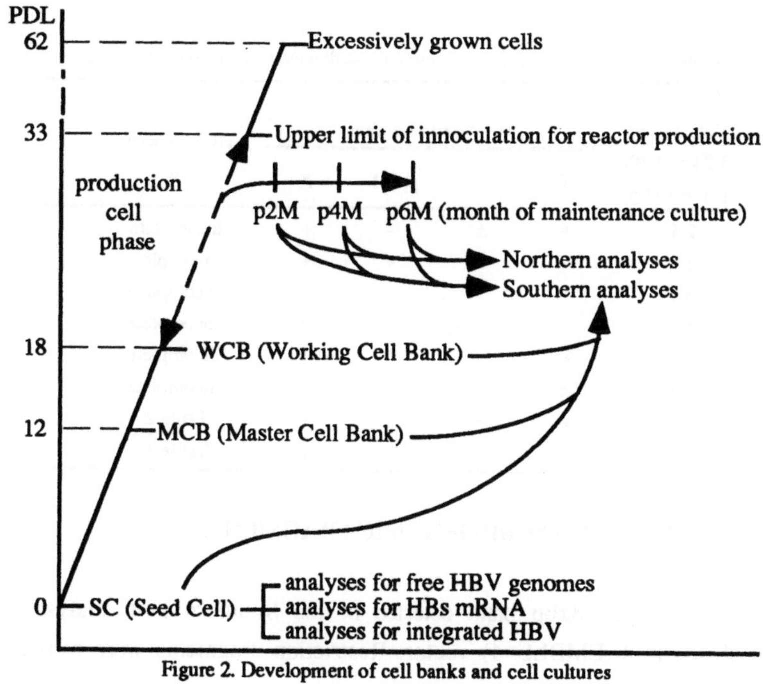

Figure 2. Development of cell banks and cell cultures

Results

ATTEMPTED DETECTION OF FREE HBV DNA IN THE CELLS

In a Southern hybridization analysis using ^{32}P-labeled HBV DNA as a probe, free HBV DNA was undetectable either in the cells or in the spent medium (<0.2 copy HBV per total DNA equivalent with one copy of cellular DNA). Here, an HBs particle-containing fraction for examination was separated from the spent culture medium by a sucrose density gradient centrifugation.

CELLULAR DNA FRAGMENTS CONTAINING HBV GENOMES

In Southern hybridization analyses huGK-14 cellular DNA completely digested with Pst I was examined for detection of HBV DNA-containing fragments using four partial probes, i.e., S-gene probe, C-gene probe, PreS1 probe, X-gene probe, and a whole HBV genome probe. The HBV whole genome probe detected eight bands in the Pst I-digested cellular DNA. Six of the eight bands lacked whole HBV genome, as indicated by the loss of one or more bands to be detected by the partial probes. Two other bands (17kbp and 8kbp) hybridized with all partial probes, which suggests that additional characterizations are necessary (Table 1).

Table 1. HBV DNA-containing fragments as detected by Southern analyses

Band sizes detected by 3.2 kbp HBV probe (Kbp)	HBV partial probes				HBV genome status
	C	S	preS1	X	
2.4	+	±	-	+	incomplete
3.0	-	+	+	+	incomplete
3.2	+	-	+	-	incomplete
3.6	-	+	±	±	incomplete
4.0	+	-	-	+	incomplete
6.0	±	±	+	-	incomplete
8.0	+	+	+	+	Table 2
17.0	+	+	+	+	Table 3

ANALYSES OF 17KBP AND 8KBP GENOMIC FRAGMENTS

The 17kbp band and the 8kbp band detected in the Southern blot analyses were respectively cloned in λ (EMBLE 4) vector. Restriction fragments were respectively obtained from the 17kbp fragment-containing and the 8kbp fragment-containing clones using BamHI (B) and EcoRI (E), of which restriction maps were constructed respectively using the whole and the partial HBV DNA probes. Based on the respective restriction maps, subclones were further constructed and their partial nucleotide sequences determined. In addition, the 8kbp-clone was analyzed by a PCR method using primers with their sequences at different regions of HBV DNA.

The results of these structural analyses showed that the 17kbp genomic fragment contains less than one unit of HBV genome each in two discontinuous locations (Table 3), and that the 8kbp genomic fragment contains less than one unit of HBV genome in one location (Table 2). Deletions, insertions, and inversions of HBV genome were noted and no complete set of HBV genome was recovered from any of the three locations.

Table 2. Structure of integrated HBV DNA in 8kbp genomic clone

HBV genes	C	S1	S2	S	X	HBV genome status
	17 and 99 bp Del	*	*	*	12 bp Del	incomplete

*:complete gene; Del: deletion; bp: base pairs

Table 3. Structure of integrated HBV DNA in 17kbp genomic clone (8.8 and 7.6kbp subfragments)

	HBV genes									HBV genome
	C	S1	S2	S	X	C	S2	S1	C	status
8.8 kbp fragment	Par	*	47 bp Del	*	*	Par	Par Inv	* Inv	Par Inv	incomplete
7.6 kbp fragment	*	*	47 bp Del	*	*	Par	Par Inv	Inv	Par Inv	incomplete

*:complete gene; Par: partial gene; Del: deletion; Inv: inverted

STRUCTURE AND EXPRESSION OF HBs mRNA

Poly A^+ RNA extracted from the seed cells showed an apparently single 2.1kbp band in a Northern blot using an HBs DNA fragment as a probe. From the Poly A^+RNA, two clones of cDNA were obtained using HBs DNA as a probe, and one PCR-amplified cDNA preparation. All three cDNA preparations were sequenced showing that they all contain the same coding frame of the HBs (type adr).

Variations were noted in the amino acid sequence of the HBs protein as deduced from the present sequence data and from the reported type adr mRNAs. The sequence of 226 amino acid residues as deduced from the cDNA sequence was the same as those of three reported type adr HBs proteins except that they vary in seven positions of amino acid residues. They are considered to be variations within type adr. These sequence data will be described in detail elsewhere.

STABILITY OF INTEGRATED HBV GENOMES AND HBs mRNA EXPRESSION

Preparations of PstI-digested DNA and poly A^+ RNA isolated from the cells at various stages during the course of seed lot construction and serial passages (Figure 2) were examined by Southern analyses using an S-gene probe and a whole HBV genome probe, and by Northern analyses using an S-gene probe. In these analyses, no apparent changes in the mode of HBV genome integration or of HBs mRNA expression were noted (data not shown).

Discussion and Conclusion

The idea that an HBs vaccine using a human continuous cell line may have advantages over those obtained using human blood, or other animal or microbial resources is not new: This idea has already been developed, particularly with the PCL/PRF/5 cell line of hapatocellular carcinoma origin, which is known as the Alexander cell line (Alexander et al., 1978, Skelly et al., 1979, McAleer, et al., 1983). The technology of HBs antigen production, as well as the mode of integration of HBV genome, has been studied in Alexander cell lines in some detail (McAleer et al. 1983, Zerial et al., 1986, Rivkina, et al., 1988).

Since the use of WI-38 human diploid fibroblasts for the production of polio vaccine, the advantages of using human diploid cell lines have long and widely been accepted for biopharmaceutical production. The advantages of using human continuous cell lines are becoming recognized, although still in limited cases. To the authors' knowledge, Namalwa cells of Burkit lymphoma origin, and BALL-1 cells of acute B cell leukemia origin are used for biopharmaceutical production of α-interferon in Japan.

With the present seed lot of the continuous human cell strain, huGK-14, we rejected in this study the possibility that it contains intact HBV genome either in the chromosome, in the cells, or in the culture medium, while HBs mRNAs are steadily produced during a time sufficiently long for the production of HBs antigen. We believe that these results substantially fulfill the safety requirements for the seed lot to be used in HBV vaccine production in terms of HBV genome integration and HBs gene expression.

References

Alexander J., Macnab., G., & Saunders, R. (1978) Studies on in vitro production of hepatitis B surface antigen by a human hepatoma cell line, *Perspectives in Virology*, **10**:103-120.

Huh, N., & Utakoji, T. (1981) Production of HBs-antigen by two new human hepatoma cell lines and its enhancement by dexamethasone, *Jpn. J. Canc. Res.* **72**:178-179.

Koike, K. (1994) *Kan-Tan-Sui* **29**:235-247 (in Japanese).

Koike, K., Yoshida, E., Katagiri, K., Kata-yanagi, M., Oda, M., Tsunoo, H., Yaginuma, K., & Kobayashi, M. (1987) Production of Hepatitis B virus surface antigen particles by the humanhepatoma cell line huGK-14 in a serum-free medium, *Jpn. J. Canc. Res.* **78**:1341-1346.

McAleer, WJ., Markus, HZ, Bailey, FJ, Herman, AC., Harder, BJ., Wampler, ED, Miller, WJ., Keller, PM., Buynak, EB., & Hilleman, MR. (1983) Production of purified hepatitis B surface antigen from Alexander hepatoma cells grown in artificial capillary units, *J. Virol. Methods* **7**:263-271.

Mizusawa, H., Taira, M., Yaginuma, K., Kobayashi, M., Yoshida, E., & Koike, K. (1985) Inversely repeating integrated hepatitis B virus DNA and cellular flanking sequences in the human hepatoma-derived cell line huSP, *Proc. Natl. Acad. Sci.* **82**:208-212.

Oda, M., Tsuno, H., & Koike, K. (1986) HB vaccine derived from a human cell line, *Kan-Tan-Sui* **13**:585-590 (in Japanese).

Rivkina, M.B., Lunin, V.G., Mahov, A.M., Tikchonenko, T.I., & Kukain, R.A. (1988) Nucleotide sequence of integrated hepatitis B virus DNA and human flanking regions in the genome of the PLC/PRF/5 cell line, *Gene* **64**:285-296.

Skelly, J., Copeland, J.A., Howard, C.R., & Zuckerman, A.J. (1979) Hepatitis B surface antigen produced by a human hepatoma cell line, *Nature* **282**:617-618.

Tiollais, P., Pourcel, C., & Dejean, A. (1985) The hepatitis B virus, *Nature* **317**:489-495.

References

VJ REARRANGEMENT AT PREVIOUSLY EXCLUDED Igλ ALLELE IN HUMAN PLASMA CELL LINE NAT-30

HIROTAKA HARUTA[1)], SANETAKA SHIRAHATA[1)] and HIROFUMI TACHIBANA[2)]

[1)] *Graduate School of Genetic Resources Technology, Kyushu University, Hakozaki 6-10-1 , Higashi-ku, Fukuoka City, Japan.*

[2)] *Department of Food Science and Technology, Faculty of Agriculture, Kyushu University, Hakozaki 6-10-1, Higashi-ku, Fukuoka City, Japan.*

Abstract

NAT-30 (μ,λ) is a human plasma cell line derived from Namalwa, a human Burkitt lymphoma cell line (1). It was previously demonstrated that concanavalin A (Con A) stimulation of the NAT-30 cells caused the loss of the original light chain of the NAT-30 cells and the expression of the new light chains in this cell line (2). This phenomenon was referred to as Light Chain Shifting (2). In this report, we showed that the new VJ rearrangement followed by the transcription spontaneously occurred in NAT-30 cells without Con A stimulation at low frequency and Con A stimulation may enhance this phenomenon in the magnitude of several hundreds fold.

Introduction

The Light Chain Shifting is a curious phenomenon that the human plasma cell line NAT-30 (μ,λ) changes its light chain to another one (2). Culture with concanavalin A (Con A) for one month, more than 10% of the NAT-30 cells carried out Light Chain Shifting (2). It was already shown that these new light chains were derived from another VJ rearrangement occurred at the previously excluded allele of the λ light chain locus (2). It has been considered that VJ rearrangement event is strictly regulated and never happens in plasma cell stage (3,4). The Light Chain Shifting shows the potential for the occurrence of another VJ rearrangement in plasma cells. Investigating this phenomenon will be profitable for understanding of the regulation mechanism of VJ rearrangement and allelic exclusion in human B lineage cells, which are not understood well.

In this paper, we investigated whether Light Chain Shifting occurs or not in the unstimulated NAT-30 cells by detecting mRNA and VJ coding joints for new light chains in the cDNA and genomic DNA of unstimulated NAT-30 cells.

Results

Detection of mRNA for new light chains in the unstimulated NAT-30 cells

Some of the new light chains, derived by Light Chain Shifting, were already cloned. We have tested whether two of these new light chains, λCA2 and λCA3 (Fig.1) were expressed in the unstimulated NAT-30 cells or not using RT-PCR. Total RNA from unstimulated NAT-30 cells were extracted using Trizol reagent (GIBCO, BRL, UK) and served for cDNA synthesis as described elsewhere (5). Sense direction primers specific for variable region of the λCA2 chain (P-VλCA2, 5'-ACT-CTg-AgC-AgT-ggg-CAC-Ag-3'), λCA3 chain (P-VλCA3, 5'-AgT-ggC-AgC-ATT-gCC-AgC-AAC-3') and original λ chain λC5 (P-VλC5, 5'-AAC-AgC-TCC-AAC-ATT-ggg-AA-3') and anti-sense direction primer specific for λ chain constant region (P-Cλ, 5'-gAA-gCT-CCT-CAg-Agg-Agg-g-3') were used for PCR amplification to detect the expression of the λCA2, λCA3 and original λ chain λC5 (Fig.2). We could detect not only the expression of the original λ chain but also the expression of the new light chains, suggesting the possibility that the Light Chain Shifting occurs spontaneously in unstimulated NAT-30 cells (Fig.2). However, it should be noted that unstimulated NAT-30 cells does not express any λ light chains as proteins except original λ chain λC5 (2).

λC5 Variable Region

```
  1  ATG GCC TGG TCT CCT CTC TTT CTG TTC CTT CTC ATT CAC TGC ACA GGG TCC TGG GCC CAG TCT GTG TTG ACG CAG   75
 76  CCG CCC TCA GTG TCT GCG GCC CCA GGA CAG AAA GTC ACC ATC TCC TGC TCT GGA AAC AGC TCC AAC ATT GGG AAT  150
151  AAT TAT GTA TCC TGG TAC CAG CAC CTC CCA GGA ACA GCC CCC AAA CTC CTC ATT TAT GAC AAT AAT AAG CGA CCC  225
226  TCA AGT ATC CCT GAC CGA TTC TCT GGC TCC AAG TCT GGC ACG TCT GCC ACC CTG GGC ATC ACC GGA CTC CAG ACT  300
301  GGG GAC GAG GCC GAT TAT TAC TGC GCA ACA TGG ACT AGC ATC CTC CGT GTC AAT TGG CTG TTC GGC GGA GGG ACT  375
376  AAA CTG ACC GTC CTA GGT
```

λCA2 Variable Region

```
  1  ATG GCC TGG TCT CCT CTC CTT CTG TTC CCT CTC CTC CTC CAC TGC ACA GGG TCT CTC TCC CAG CCT GTG CTG ACT   75
 76  CAA TCA TCC TCT GCC TCT GCT TCC CTG GGA TCC TCG GTC AAG CTC ACC TGC ACT CTG AGC AGT GGG CAC AGT AGC  150
151  TAC ATC ATC GCA TGG CAT CAG CAA CAG CCA GGG AAG GCC CCT CGG TAC TTG ATG AAG CTT GAA GGT AGT GGA AGC  225
226  TAC AAC AAG GGG AGC GGA GTT CCT GAT CGC TTC TCA GGC TCC AGC TCT GGG GCT GAC CGC TAC CTC ACC ATC TCC  300
301  AAC CTC CAG TCT GAG GAT GAG GCT GAT TAT TAC TGT GAG ACC TGG GAC AGT AAC ACT CAG GTG TTC GGC GGA GGG  375
376  ACC AAG CTG ACC GTC CTA
```

λCA3 Variable Region

```
  1  ATG GCC TGG TCT CCT CTC CTT CTG TTC CTC CTC GCT CAC TGC ACA GGT TCT TGG GCC AAT TTT ATG CTG ACT CAG   75
 76  CCC CAC TCT GTG TCG GAG TCT CCG GGG AAG ACG GTA ACC ATC TCC TGC ACC CGC AGC AGT GGC AGC ATT GCC AGC  150
151  AAC TAT GTG CAG TGG TAC CAG CAG CGC CCG GGC AGT GCC CCC ACC ACT GTG ATC TAT GAG GAT AAC CAA AGA CCC  225
226  TCT GGG GTC CCT GAT CGG TTC TCT GGC TCC ATC GAC AGC TCC TCC AAC TCT GCC TCC CTC ACC ATC TCT GGA CTG  300
301  AAG ACT GAG GAC GAG GCT GAC TAC TAC TGT CAG TCT TAT GAT AAC AGC AAT CAG GTG TTC GGC GGA GGG ACC AAG  375
376  CTG ACC GTC CTA
```

Fig.1 DNA sequences of light chain variable regions from original NAT-30 (λC5), new light chain producers CA2 (λCA2) and CA3 (λCA3).

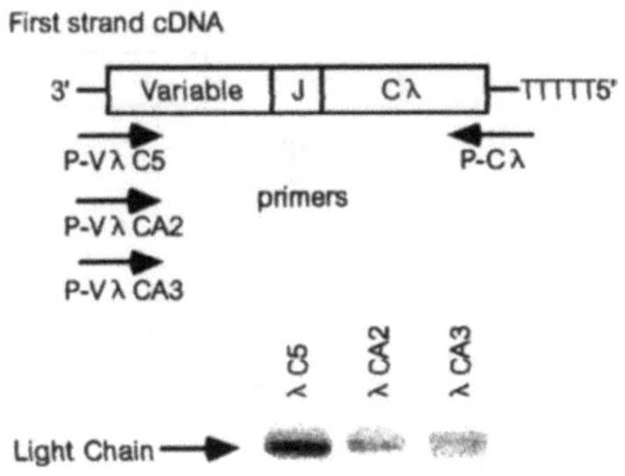

Quantification of the VJ-coding joints for new light chains in the genome of the unstimulated NAT-30 cells

The result of RT-PCR suggested that the new light chain expression occurs spontaneously in NAT-30 cells. Then we tried to quantify the percentage of the VJ coding joints for new light chains in unstimulated NAT-30 cells using PCR of the genomic DNA followed by Southern blot analysis. Genomic DNA of the unstimulated NAT-30, cloned new light chain producers CA2 and CA3 cells were extracted using DNAzol reagent (GIBCO, BRL, UK). Genomic DNA of CA2 and CA3 cell lines were employed as a standards for CA2 and CA3 VJ coding joints, respectively. To quantify the ratio of VJ coding joint for λCA2 chain, each genomic DNA of the unstimulated NAT-30 and CA2 cells were mixed with calf thymus DNA to make the final concentrations of 100, 10, 1, 0.1, 0.01 (w/w)%, respectively. To obtain higher sensitivity, we performed a nested PCR. As a first step, 200 ng of these samples were amplified in 20 μl reaction volumes using the sense direction primer P-λF (5'-ACT-AgA-ATT-CAT-g(Ag)C-CTg-(Cg)(AT)C-(CT)CC-TCT-C(CT)T-(CT)CT-(Cg)(AT)(CT)-CC-3') and anti-sense direction primer P-JIR (5'-ggg-CTg-ACC-ACA-AgT-TgA-gA-3'). These primers can amplify most of the human λ light chain genes. As a second step, 1 μl of reaction mixtures from the first step were amplified using the P-VλCA2 primer and anti-sense direction primer specific for Jλ region (P-Jλ, 5'-TCA-gTT-TAg-TCC-CTC-CgC-C-3') followed by electrophoresis on 2.0% agarose gels. After electrophoresis, the amplified products were transferred to the nylon membrane Hybond-N+ (Amersham) and Southern hybridization was performed using ^{32}P-labeled variable region fragment of the λCA2 gene as a probe. According to the λCA2 standard, about 0.1% of the unstimulated NAT-30 cells had VJ coding joint for λCA2 (Fig.3). The same experiment was demonstrated for λCA3 chain using the P-VλCA3 and P-Jλ primers, showing that less than 0.01% of the unstimulated NAT-30 cells had VJ coding joint for λCA3 (Fig.3). These results suggests that new VJ rearrangements occurs in the NAT-30 cells at low frequency.

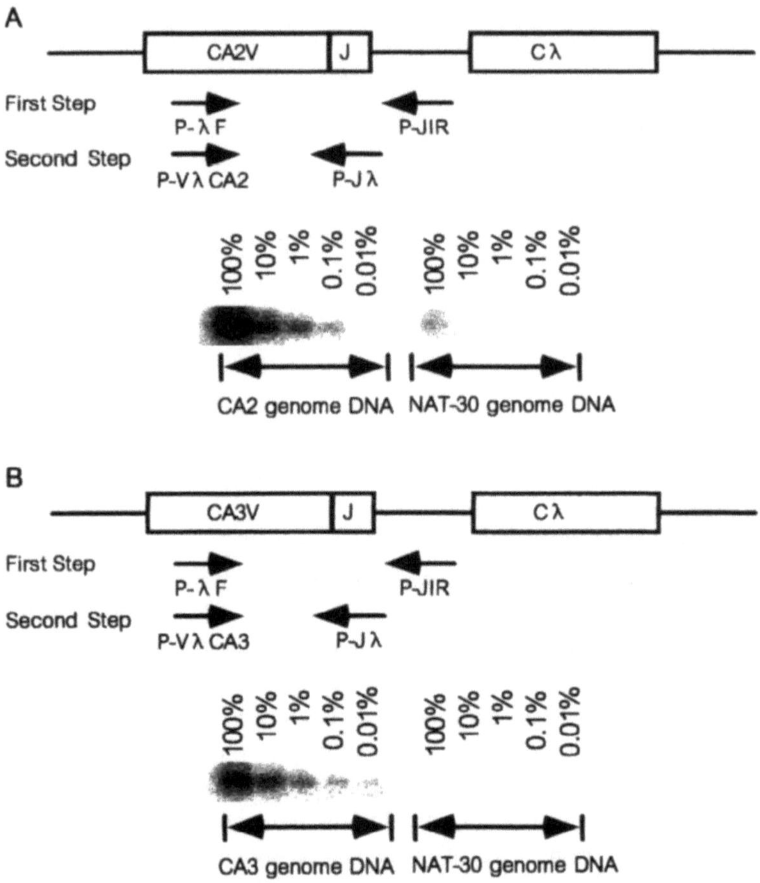

Fig.3 Quantification of VJ coding joints for λCA2 (A) and λCA3 (B) in NAT-30 genome.

Discussion

We showed here that NAT-30 cells spontaneously carried out the new light chain expression without Con A stimulation. The ratio of the new light chain producer in the unstimulated NAT-30 cells was about 0.1% whereas that of the Con A-stimulated NAT-30 cells was more than 10% (2). These results suggest that if meet its conditions, another VJ rearrangement occurs even in the plasma cells, and that Con A may enhance the Light Chain Shifting event. It was previously revealed that Namalwa and NAT-30 cells express RAG-1 and RAG-2 constitutively (2). RAG-1 and RAG-2 are required for VJ rearrangement and considered not to be expressed in the plasma cell stage (3,4). However, Namalwa cells did not carry out the Light Chain Shifting even stimulated with Con A, while NAT-30 cells carried it out (2). These findings suggest that the expression of

RAG-1 and RAG-2 are not enough for occurrence of the Light Chain Shifting and there are at least two more factors needed to carry out the Light Chain Shifting event efficiently. One is the difference between Namalwa and NAT-30, and the other is what effected by Con A stimulation. Elucidating the differences between Namalwa and NAT-30 may be an important trial for understanding of the Light Chain Shifting event. It is also interesting what happens after Con A stimulation of NAT-30 cells. Some candidate functions of these factors are 1) reducing the expression of original light chain, 2) inducing the regulatory factors of VJ rearrangement, 3) breaking the allelic exclusion mechanism at the light chain locus. So far tested by Western-blot analysis, the unstimulated NAT-30 cells did not express any λ light chain proteins except original λC5 (2). We are not sure that this was caused simply by the low level expression of new light chains in the unstimulated NAT-30 or by other mechanism related to allelic exclusion. The investigation of the Light Chain Shifting may contribute to the understanding of the regulation of the light chain expression.

Acknowledgments

This work was supported in part by grants from Program for Promotion of Basic Research Activities for Innovative Biosciences (PROBRAIN)

References

1) Murakami H, Hashizume S, et al. (1985) Human-human hybridomas secreting antibodies specific to human lung carcinoma, *In Vitro Cell. Dev. Biol.* 21, 593-596.
2) Tachibana H, Ushio Y, et al. (1996) Inducing the loss of Immunoglobulin λ light chain production and the rearrangement of a previously excluded allele in human plasma B cell lines with concanavalin A, *J. Biol. Chem.* 271, 17404-17410.
3) Ma A, Fisher P, Dildrop R, et al. (1992) Surface IgM mediated regulation of *RAG* gene expression in Eμ-N-*myc* B cell lines, *EMBO J.* 11, 2727-2734.
4) Oettinger MA, Schatz DG, Gorka C, and Baltimore D. (1990) RAG1 and RAG2, adjacent genes that synergistically activate V(D)J recombination, *Science* 248, 1517-1523.
5) Larrick JW, Danielsson L, Brenner CA, et al. (1989) Polymerase chain reaction using mixed primers: cloning of human monoclonal antibody variable region genes from single hybridoma cells, *BIO/TECHNOLOGY* 7, 934-938.

RATIONAL DRUG DESIGN USING MAMMALIAN CELL LINES EXPRESSING SITE-DIRECTED MUTANTS OF THE HUMAN H_1 HISTAMINIC RECEPTOR

N. MOGUILEVSKY, E. DIFFERDING*, M. GILLARD* and A. BOLLEN
Applied Genetics, University of Brussels, rue de l'Industrie 24, B-1400 Nivelles and
**UCB Pharma S.A., Chemin du Foriest, B-1420 Braine-l'Alleud, Belgium.*

Abstract

The binding cavity of histamine and histamine antagonists has been explored using site-directed mutagenesis of the human histamine H_1 receptor and the amino acids involved in ligand binding have been identified. Whereas Asp_{107} and Phe_{199} appeared important for both agonists and antagonists, two additional amino acids (Asn_{198} and Trp_{103}) were required for efficient histamine binding. The binding site of antagonists was best defined as resulting from a strong ionic bond to Asp_{107}, an orthogonal interaction between one of the aromatic rings with Phe_{199}, and probably a hydrophobic interaction between the second aromatic ring and the lipophilic amino acids of the upper part of TMIV and TMV. This is consistent with structure-activity data of most described antagonists.

1. Introduction

Release of histamine initiates a cascade of biological effects in response to the interaction with histamine receptors, of which three subtypes are known. The stimulation of the H_1 receptor leads to the activation of phospholipase C, resulting in Ca^{++} mobilization and neuronal excitation. As a result, the main symptoms of allergic rhinitis develop, that is dilatation of capillary vessels and secretion of mucus. Histamine antagonists, which inhibit the binding of histamine to the H_1 receptor, are widely used in the treatment of allergy (1), despite their unwanted side effects. The quest for improved products with higher specificity and activity remains therefore a challenge for the pharmaceutical industry.

Two complementary approaches have been followed in this work. We first build a topological model of a consensus anti H_1 pharmacophore and confronted it to a model of the H_1 histaminic receptor to derive hypothesis on amino acid residues involved in drug binding. We next challenged our predictions by mutating the human H_1 receptor, cloned and expressed in mammalian cells, and measuring drug binding parameters. These experiments led to the definition of agonist and antagonist interaction sites on the receptor and to proposals for designing new H_1 antagonists.

2. The consensus anti H_1 pharmacophore

Numerous H_1 antagonists have appeared in the literature. A detailed analysis of structure-activity data on 15 representative molecules led to the following pharmacophore (Fig. 1).

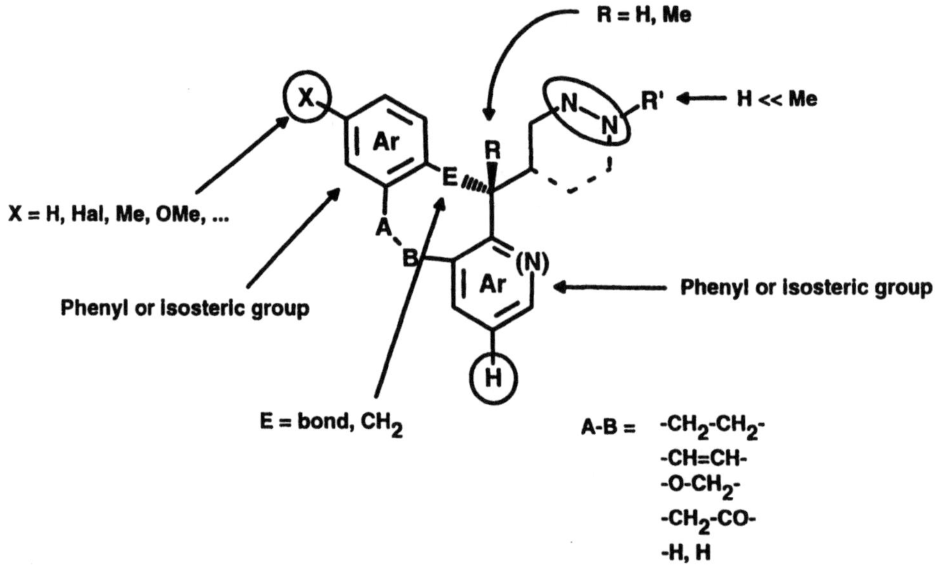

A detailed presentation of this pharmacophore has been recently published (2).

3. A qualitative model of the human H_1 receptor

The human H_1 receptor has been recently cloned (3). On the basis of the hydropathy plot (4) and an analysis of the sequence according to J. Baldwin (5), seven fragments corresponding each to 22 amino acids were selected as helical transmembrane (TM) domains of the receptor. The helices were built using the Biopolymer module within SYBYL and arranged according to the motif of bacteriorhodopsin (6). Helix IV was then manually rotated and adjusted to maximise the interaction between its hydrophobic surface and the lipid bilayer of the membrane. This qualitative model was used as the basis for the selection of amino acids potentially implicated in ligand binding. There is general agreement that the binding site of small biogenic amines is located inside the receptor and is defined by the helices III to V, the major anchor point being an aspartate of helix III (Asp_{107}) involved in an ionic interaction with the ammonium part of histamine. This ionic bond is supposedly stabilized by several aromatic amino acids among which Trp_{103}. The imidazole ring of drug compounds could then interact with several H-bond donors such as Ser_{155} in helix IV and Asn_{198} or Thr_{194} in helix V. In addition, several aromatic amino acids are well located for aromatic-aromatic (π) interactions, such as Trp_{158} (helix IV) or Phe_{199} (helix V) (Fig.2).

4. Binding studies on site-directed mutants of human H_1 receptor

In order to test these predictions, site-directed mutagenesis of the H_1 receptor was performed and the drug-binding properties of mutant receptors expressed in CHO cells were tested, as described before (7). Binding experiments, using both agonist (histamine) and antagonists, were performed on membranes derived from the mutated receptors. Fig. 2 shows the difference, indicated in ΔpK (ΔpKd for ^{3}H-mepyramine, ΔpKi for histamine and antagonists) between the mutant and native receptors. It allowed us to draw the following conclusions : a) the mutation of Asp_{107} leads to the complete loss of affinity for the radioligands, antagonists as well as histamine, underlining the crucial role of the ionic interaction with the ammonium ion; b) replacing Trp_{103} by Ala does not affect antagonists binding but leads to 20 fold decrease in histamine binding; c) neither amino acid of helix IV, i.e. Trp_{158} or Ser_{155} are important for binding, since a Δpk of 0.1 or 0.2 is close to the experimental error; d) Asn_{198} is extremely important for histamine but not for antagonists binding; e) interestingly, the mutation of Thr_{194} to Ala is unfavourable for histamine, but strengthens to some extent the binding of antagonists; f) the mutation of Phe_{199} leads again to a complete loss of binding of mepyramine. [Therefore no pKi could be measured for histamine]. The receptor can, however, still be activated but to a lesser extent, by histamine as shown by measuring a cellular function .

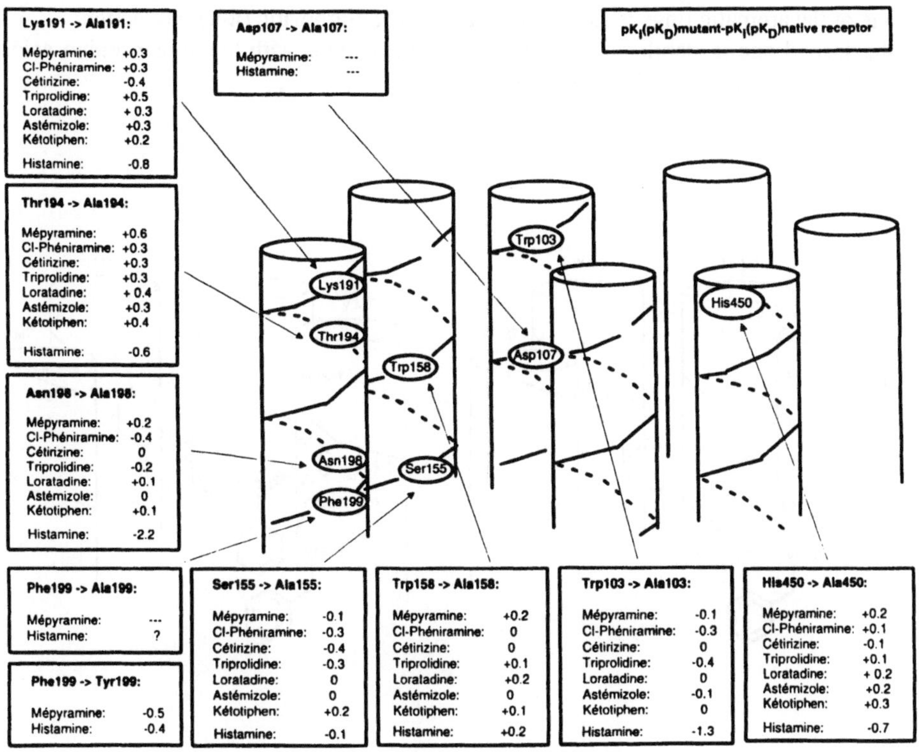

Fig.2 Affinities of histamine and various antagonists for mutant receptors

5. The agonist and antagonist binding sites on the H_1 receptor

The data presented above lead to the following picture for agonist (histamine) binding, according to a three-point attachment to the receptor.(Fig.3).

- a very strong ionic interaction with Asp_{107}, the ion pair being stabilized by Trp_{103}.
- Asn_{198} makes a hydrogen bond to the imidazole ring, probably via the NH_2 group to the N_π.
- Phe_{199} stabilizes the imidazole by an aryl-aryl interaction, probably in a perpendicular manner.
- In this arrangement, Thr_{194} should only lead to a weak stabilization owing to an H bond towards the α-helix.

The interaction with the antagonists shows significant differences with histamine. With a model of benzylhydrylpiperazine (Fig.4), the binding mode involves the following features :

- a strong ionic interaction with Asp_{107}. This time, no stabilization by Trp_{103} is required.
- an aryl-aryl interaction with Phe_{199} and one of the aromatic groups.
- in this arrangement, the second aryl group is located at the interface of helix IV and V, in close vicinity to Thr_{194} which probably explains some steric hindrance which disappears by mutation of the Thr to Ala.

This means that the two aromatic rings of the antagonists can clearly be differentiated :

- one of the aromatic rings is stabilized by an aromatic-aromatic (π-π) interaction.
- the second one is stabilized by hydrophobic interactions with the apolar amino acids which define this area of the receptor.

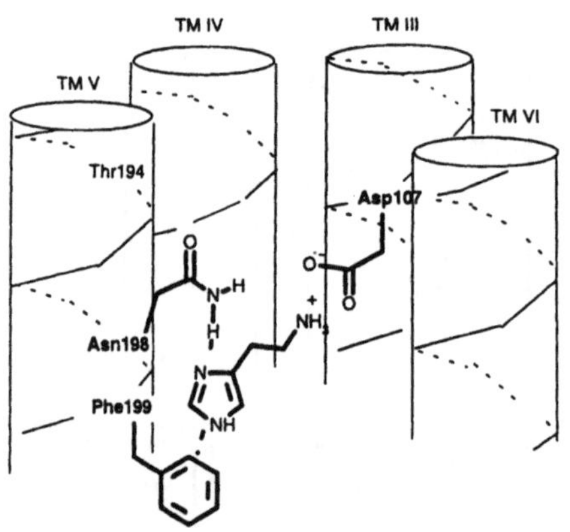

Fig.3 : Binding mode of histamine

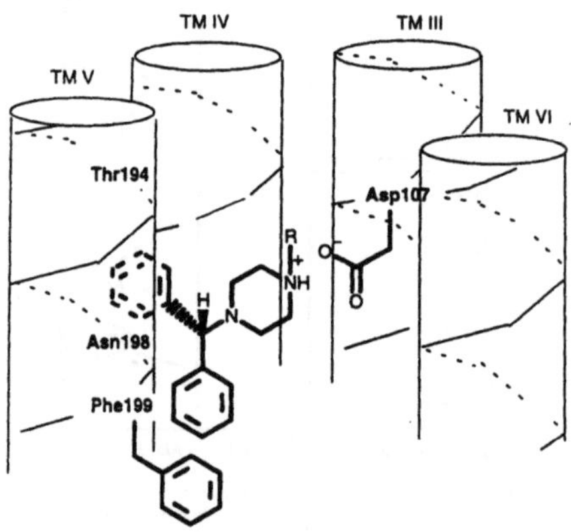

Fig.4 : Binding mode of antagonists

6. Design of new antihistaminic drugs

On the basis of these data, a model structure for ideal H_1 antagonists can be proposed. According to this model, the antagonist must fit in a way suitable for interaction with Asp_{107} through a basic nitrogen and via a spacer arm located in the vicinity of the hydrophobic pocket centered on Phe_{199}. The affinity is increased by a second interaction thanks to an aliphatic group, or substituted aromatic. This will dock in a stereospecific way in an hydrophobic pocket delineated by the contact zone between helix IV and V. In addition, we know that the antagonist sits in the upper part of the receptor, in close proximity to the surface. As a consequence of these constraints, we think that new, high affinity ligands could take advantage by H-bonding, of the presence of two relatively polar amino acids, i.e. Thr_{194} and Asn_{198}, fairly isolated in a hydrophobic environment. In addition, one should not forget the presence of Ser_{155}, deeply buried inside the receptor in close vicinity to Phe_{199}. Although none of the known antagonists interact with this serine, it might eventually be available for an additional hydrogen bond.

7. Conclusions

The work presented here illustrates a rational approach to drug design, based on two complementary and synergic experimental basis. It maximally uses all informations present in structure-activity relationships and derived from superpositions of antagonist structures. The pharmacophore, which emerges from these studies, can thus be confronted to a model of the H_1 receptor to draw hypothesis on potentially important amino acid residues. Challenging the hypothesis with experimental data on site-directed mutagenized receptor molecules then provides a model for drug-receptor interactions and opens the way to designing improved antagonists.

8. References

1. Cooper, D.G., Young, R.C., Durant, G.J., Ganellin, C.R. (1990) Histamine receptors, in C.Hansch, P.G. Sammes, J.B.Taylor, J.C.Emmett (eds.), *Comprehensive Medicinal Chemistry*, Pergamon Press, Oxford, pp. 323-421.
2. Differding,E., Gillard,M., Moguilevsky,N., Varsalona,F. Noyer,M., Daliers,J., Goldstein,S., Neuwels,M.A., Lassoie,M.A., Guillaume,J.P., Bascour,M., Bollen,A., Hénichart,J.P. (1996). Mutagénèse du récepteur histaminique H_1 humain et design de nouveaux agents antihistaminiques. *J. Pharmacol. Belg.* (in press).
3. Moguilevsky,N., Varsalona,F., Noyer,M., Gillard,M., Guillaume,J.P., Garcia,L., Szpirer,C., Szpirer,J., Bollen,A. (1994) Stable expression of human H_1 histamine receptor cDNA in Chinese Hamster Ovary cells, *Eur. J. Biochem.* **224**, 489-495.
4. Kite,J. and Doolittle,R.F. (1982) A simple method for displaying the hydropathic character of a protein. *J. Mol. Biol.* **157**, 105-132.
5. Baldwin,J.M. (1993) The probable arrangement of the helices in G-protein-coupled receptors. *EMBO J.* **12**, 1693-1703.
6. Henderson,R. Baldwin,J.M., Ceska,T.A., Zemlin,F., Beckmann,E., Downing,K.H. (1990) Model for the structure of bacteriorhodopsin based on high-resolution electron cryomicroscopy. *J. Mol. Biol.* **213**, 899-929.
7. Moguilevsky,N., Varsalona,F., Guillaume,J.P., Noyer,M., Gillard,M., Daliers,J., Hénichart,J.P., Bollen,A. (1995) Pharmacological and functional characterization of the wild-type and site-directed mutants of the human H_1 histamine receptor stably expressed in CHO cells. *J. Receptor Sign. Trans. Res.* **15**, 91-102.

BIOLOGIC RESPONSE OF ENDOTHELIAL CELLS : MODULATION OF ADHESION AND MULTIPLICATION BY CYTOKINES AND PROSPECTS FOR THE PRODUCTION OF A VACCINE AGAINST THE RICKETTSIA *Cowdria ruminantium.*

T. MARIQUE, D. BLANKAERT, A. RASCHELLA, V. HENDRICK, B. DECLERCK, I. TEIXERA-GUERRA, N. VACHIERY, P. TOTTE, C. ALLOIN, M. CHERLET, P. ZILIMWABAGABO, D. PARENT*, C. KIRKPATRICK*, J.P. VAN VOOREN*, AND J. WÉRENNE (+)
*Laboratory of Animal Cell Technology, Agronomy and Bioengineering Dept, Faculty of Sciences, and * Erasme Hospital, Faculty of Medicine Université Libre de Bruxelles, 50 Av. Roosevelt, Brussels, Belgium.*

1. Introduction

Endothelial cells play a crucial role in a number of pathological conditions and in their regulation. Angiogenesis mechanisms are particularly relevant to the understanding of these different situations (1-2). Cytokines play also an important role in the regulation of pathophysiological responses of endothelial cells and in their overall behaviour.
Two important pathological situations for which better knowledge of endothelial properties could permit the identification of therapeutical solutions attracted our interest.
1. Cowdriosis or Heartwater a deadly disease prevalent in cattle in the south of the Sahara and which is caused by the multiplication of an obligate intracellular bacteria, the rickketsia *Cowdria ruminantium*, for which the final cellular targets are endothelial cells from the brain capillary vessels (3). 2. Kaposi sarcoma (KS), an angioproliferative disease prevalent in AIDS patients and characterized by the dissemination of vascular lesions on the skin or mucosal tissues. (4-5).
Knowing that matrix metalloproteionases (MMPases) activities affect vascular permeability and cellular mobility of endothelial cells (6) , we adressed the two following related questions :
1. could we find conditions, while enabling in vitro maintenance of spindle like cells obtained from Kaposi sarcoma that we characterized here as of endothelial origin, in order to select potential treatment (including using cytokines) to reduce growth and invasiveness of the KS cells ? 2. could we find conditions permiting maintenance of endothelial cell cultures in spinner bioreactors for the extended period of time (more than two weeks) necessary to produce enough of *Cowdria ruminantium* variants to be tested as potential attenuated vaccine which was up to now unavailable ?

2. Material and methods.

Routinely , infection with the rickettsia *Cowdria ruminantium* (Senegal stock originally

+ Corresponding author.

obtained from Dr F. Jongejan from Utrecht University), was operated and revealed as previously described (6) in different cells of interest grown as described earlier : HUVEC cells (7- 8), KS cells (4), pSV1 cells / HUVEC transformed with SV40 T antigen, kindly given to us by V. Faffeur from Pasteur Institute - Lille (9), 1/2C10 cells, HUVEC-Human Osteosarcoma cell line provided to us by the group of Prof. Kattinger-Vienna, and BUEC cells first originating from Dr F. Jongejan - Utrecht (7).
-Cytodex 1, 2 and 3 were generally used used, but in some experiments Cytocell or Cultispher S microcarriers have been tested also. Cultures of HUVEC in tissue culture flasks of 80 cm2 were maintained for 48 h in fresh medium which is then harvested and kept as "conditioned medium" (CM) . Its activation resulted from acidification until pH2 with 10 M HCl for 1 h at 4°, after which neutralization is quickly operated with 10 M NaOH. - *TNFα* (Human recombinant TNFα) at 10^8 cytotoxic units / mg was received from Dr M. Bochaus (Hoffman LaRoche, Basel, Switzerland. - *PMA* and - *TGFβ* (Human natural TGF β) were obtained from Promega . During treatment with all those factors, medium was changed every day and collected after 24 and 48 h incubation without serum before testing MMPases by Zymography (10). Antibodies against TGFβ (Monoclonals against 300 pg TGFβ 1, 2 and 3 , or rabbit polyclonals used at a 1/200 dilution ,were from Promega). For immunofluorescence analysis, the cells (HUVEC, BUEC or Spindle like cells) were treated first with Factor VIII rabbit antibodies (1/150) and second by fluorescein conjugated immunoglobulins (DAKO).

3. Results.

3.1. EXPRESSION OF FACTOR VIII IN THE CELLS USED.

All the cells used in the present study - BUEC, HUVEC, KS (Fig 1 A, B, C) and 1/2 C 10 (Fig 1 D) and psV1 (data not shown) expressed Factor VIII when tested with rabbit antibodies detected by fluorescein conjugated immunoglobulins. The expression is rather intense and in fact less diffuse in BUEC and KS cells, than in HUVEC and in the derived 1 / 2 C 10 cells. This indicated that, at least at the early passage used here, KS cells were originating from endothelial related tissues.

3.2. EFFECT OF CONDITIONED MEDIUM, PMA AND TGFβ ON ENDOTHELIAL CELLS GROWTH, MOBILITY AND METALLOPROTEINASES EXPRESSION.

Conditions stabilizing endothelial cells growing on microcarriers could be highly valuable to develop their mass production, and looking for factors active in this respect, we have observed by Zymographic analysis in pSV1 cells (Fig. 2) that under treatment with normal conditioned medium , endothelial cells express 92 kD MMPases and that this expression is increased by PMA and highly significantly reduced in presence of activated conditioned medium. Similar effects were observed with the different endothelial cells studied (BUEC,HUVEC and KS cells - data not shown)..
Table 1 indicates that the growth of BUEC and KS cells is inhibited by conditioned medium, especially when activated. Anti TGFb reverse this inhibitory action, showing the important role of this cytokine on endothelial cells behaviour.

3.3. COLONIZATION OF MICROCARRIERS BY BUEC AND KS CELLS AND ITS INHIBITION BY ACTIVATED CONDITIONED MEDIUM.

Cytodex were more efficiently colonized than Cytocell and Cultispher, and BUEC cells spread better on Cytodex 3 than on Cytodex 2, after seeding them on a confluent monolayer culture (data not shown). Therefore, using Cytodex 3, we showed that in 48 hours, the microcarriers have been invaded by KS or BUEC cells (up to 30-35 % saturation when seeded on a confluent monolayer) , while conditioned medium, especially after activation, is readily inhinbiting the colonization (Table 2). Therefore mobility and spreading of endothelial cells are inhibited in parallel to the reduction of the expression of the 92 kD MMPase.

A B

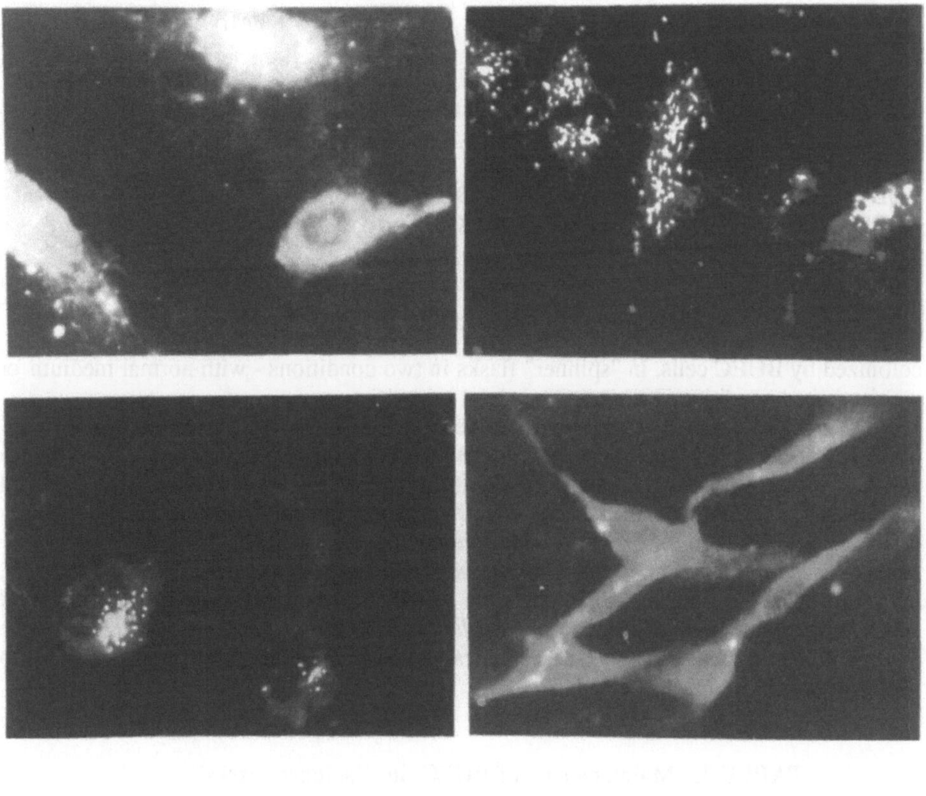

C D

Fig 1 Expression of Factor VIII in cells in culture
A : BUEC - B : HUVEC - C : 1/2 C10 - D : KS cells

In order to produce , in mass amount, the *Cowdria ruminantium* stock that we intend to test as an attenuated vaccine, we followed first the behavior of Cytodex 3 microcarriers

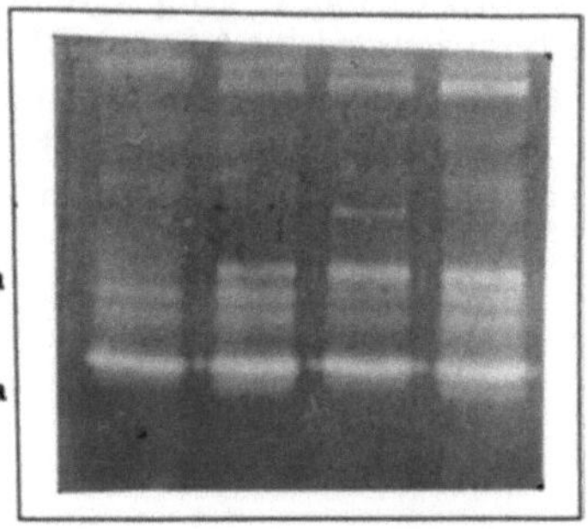

Fig. 2. Effect of PMA / Conditioned medium on MMPases expression in psV1 cells
1 : ACM - 2 : NM - 3 : ACM + PMA - 4 : NM + PMA

TABLE 1 Effect of conditioned medium and TGFβ on the growth of cells (%)

	NM	CM	ACM	ACM+Ab
BUEC:	100	77	53	94
KAPOSI :	100	59	41	92

TABLE 2 Effect of activated medium on the colonization of Cytodex 3 by cells (%).

	NM	CM	ACM
BUEC :	36	28	13
KAPOSI :	32	23	8

colonized by BUEC cells, in "spinner" flasks in two conditions - with normal medium or with activated conditioned medium.
We observed (Table 3) that cells seeded on confluent BUEC monolayers, when incubated in normal culture medium, detached readily from the microcarriers, and after a week, of incubation, all cells have died. On the contrary, when incubated with activated conditioned medium, most of the cells were maintained in healthy conditions for the long period of time required for *Cowdria ruminantium* development.
Appropriate conditions for an efficient vaccine production can therefore be now foreseeen. Indeed different endothelial cells could be tested and may be used in order to find appropraite attenuation conditions for *Cowdria ruminantium*, and TGFb which is as we showed protecting the cells against mechanical damage could be tested for the scaling up of the process in bioreactor.

TABLE 3. Maintenance of BUEC in ''spinner flasks".

Day (cells / ml .10^{-3})	0	1	2	3	4	5	6
NM	106	80	24	10	6	4	2
ACM	80	76	62	72	74	80	82

4. Discussion.

Kaposi Sarcoma (KS) cells that we isolated from AIDS patients biopsies, express Factor VIII, like HUVEC, and the derived 1/2C10 cell line as well as BUEC, indicating their endothelial origin.pSV1 cells (SV40 T antigen transformed HUVEC), like other endothelial cells, express 92 kD MMPase at high level especially when stimulated by PMA and also as observed before by TNF, or by both agents (4). TGFβ and activated conditioned medium reduced the expression of 92 kD MMPase, appearing therefore as a potential cell mobility inhibitor of interest for KS treatment. TGFβ appears also for the same reason as a valuable maintenance factor for Cowdriosis vaccine production. Therefore, the scale up of endothelial cell production process, recently proposed by S. Duvar et al (11), could be adapted for infection with *Cowdria ruminantium*, as we suggested recentky (14).

Abreviations : ACM / Activated Conditioned Medium / CM : Conditioned Medium / NM / Normal Medium, BUEC / Bovine Umbilical Endothelial Cells, HUVEC / Human Umbilical Vein Endothelial Cells, KS / Kaposi sarcoma, MMPases /Metalloproteinases, PMA / Phorbol Myristate Acetate, TGF / Tumor Growth Factor , TNF / Tumor Necrosis Factor.

Acknowledgements

This work was supported in part by the Belgian National Funds for Scientific Research (FRSM N° SM 3-4535-94 F).

5. References

1. Fajardo, L.F. (1989) The complexity of endothelial cells. Am. J. Clin. Pathol. 92 : 241-250
2. Folkman, J. and Shing, Y. (1992) Angiogenesis. J. Biol. Chem. 267 : 10.931-10.934
3. Totté, P., Jongejan, F., de Gee, A.L.W. and Wérenne, J. (1994) Production of interferon alpha in *Cowdria ruminantium* infected cattle and its effect on infected endothelial cell culture, Infection and Immun. 62 : 2.600-2604.
4. Blankaert, D., Parent, D., Farber, C.M., Van Vooren, J.P., Liesnard, C. and Wérenne, J. Metalloproteinases and TIMPs production by AIDS Kaposi sarcoma derived cells (1994) , in Animal Cell Technology , Eds . Butterworth-Heinemann, Oxford, 638-642
5. Simonart, T. , Noel, J.C., Liesnard, C. Parent, D., Heenen, H. Brancart, B., Van Vooren, J.P., Farber, C.M., Blankaert, D. and Wérenne, J. (1996) Kaposi's sarcoma and Herpes 8 : a word of caution, Dermatology, 193 : 272.
6. Coulombe, B., Severin, J.E., Marique, T., Pontons, A., Wérenne, J. and Skup, D. (1989) The tissue inhibitor of Metalloproteinase (TIMP) : Gene expression in mammalian cells, in Animal Cell Technology , Eds Butterworth, 187-191.
7. Totté, P., Blankaert, D., Marique, T., Kirkpatrick, C., Van Vooren, J.P. and Wérenne, J. (1993) Bovine and human endothelial cell growth on collagen microcarriers and their infection with the rickettsiae *Cowdria ruminantium., Revue Elev. Méd. vét. trop.* 46: 153-156.
8. Gimbrone, Jr M.A. , Cotran, R.S. and Folkman, J. (1974) Human vascular endothelial cells in culture. Growth and DNA synthesis. J. Cell. Biol. 60 : 673-682
9. Lassale, P., Lagrou, C., Wietzerbin, J., Stehelin, D., Tonnel, A.B., and Capron, A. (1992) Human endothelial cells transfected by SV40 T antigen : characterization and potential use as a source of normal endothelial factors. Eur. J. Immunol. 22 : 425-431
10. Marique, T., Blankaert, D., and Wérenne, J. (1991) Production of collagenase by anchorage dependent cells in culture, in Animal cell Technology , Eds. Butterworth, 543-546.
11. Duvar, S., Mühting, J., Mohr, H. and Lehman, J. (1996) Scale up cultivation of primary human umbilical vein endothelial cells on microcarriers from spinner vessels to bioreactor fermentation, Cytotechnology, 21: 61-72.
12. Totté, P., Blankaert, D., Marique, T., Vachiery, N., Teixeira-Guerra., I., Kirckpatrick, C., Van Vooren, J.P., and Wérenne, J. (1994) Endothelial cells and their infection with *Cowdria ruminantium* : prospects for vaccine production , in Animal Cell Technology , Eds. Butterworth , 643-645.
13. Totté, P., Vachiery, N., Martinez, D., Trap, I., Ballingall, K.T., Mac Hugh, N.D., Bensaid, A. and Wérenne, J. (1996) Recombinant bovine interferon γ inhibits the growth of *Cowdria ruminantium, but fails to induce MHCII following infection of endothelial cells* , Vet. Immunol. and Immunopathol. 53 : 61-71.
14. Alloin, C., Marique, T., Paul, T., Lowagie, S., Zilimwabagabo, P. Skup, D. and Wérenne, J. ((1996) Metabolic engineering and control of mammalian cells for increased adhesion properties : Improvements for a rickettsia vaccine production, Cytotechnology, 21 abst, 3.

EFFECTS OF CHEMICALS ON THE IMMUNE RESPONSE

Akio AMETANI, Hiroko WATANABE,
Satoshi HACHIMURA, Kwang-Young LEE, and
Shuichi KAMINOGAWA
Dept. Applied Biological Chemistry, The University of Tokyo
1-1-1 Yayoi, Bunkyo-ku, Tokyo 113, Japan

ABSTRACT. We have tested the effects of various environmental chemicals on proliferation of murine antigen-primed lymphocytes and their secretion of cytokines. Lymph node (LN) cells from mice immunized with bovine α_{s1}-casein were incubated with this antigen in the presence of β-benzene hexachloride (β-BHC), *p,p'*-dichlorodiphenyldichloroethylene (*p,p'*-DDE), *p*-nitrophenyl 2,4,6-trichlorophenyl ether (CNP), D-α-aminobenzylpenicillin (Amp), and *o*-acetylcitric acid tributyl ester (ATBC), and tested for proliferation and secretion of interleukin (IL-) 2, IL-4, IL-10 and interferon γ. The results of the proliferation assays indicated that more than 20 ppm of CNP and DDE strongly inhibited the proliferation of LN T cells. All of the chemicals tested affected cytokine secretion by LN cells in response to antigen. The effects of these chemicals on cytokine secretion varied depending upon the cytokine examined and were independent of the inhibitory effect on cell proliferation. These results suggest that costimulation of T lymphocytes with certain environmental chemicals can modulate the antigen-specific cytokine responses of these cells, even at concentrations which are considered to have no toxicity and no inhibitory effect on the proliferative response of T cells.

1. Introduction

Cellular responses induced by a single stimulus, even if indispensable, are sensitively influenced by other simultaneous stimuli. In the acquired immune system, specific T and B lymphocytes are activated through a mechanism triggered by molecular interactions of antigens with receptors on these cells. This activation may be further enhanced or inhibited, depending upon costimulation by signals other than an antigen signal. Immune disorders such as allergy and autoimmunity may be related to a combination of antigens and other coexisting stimuli. We have addressed the question of whether various environmental molecules can affect the antigen-specific immune response *in vitro* and in vivo. In this report we show effects of chemicals on responses of T cells *in vitro*. Two types of effects on these responses can be considered. First, T cells and B cells may respond directly to chemicals or chemically modified molecules. Allergy to penicillin and metals are cases of the latter and specific T cells have been isolated from patients allergic to these [1, 2]. Second, chemicals may modulate the antigen-specific response of the lymphocytes, although this type of modulation has not been well characterized except for toxicity. In this study we focused on the in vitro modulation of T cell proliferative and cytokine secretion responses induced by chemicals. Both inhibition and enhancement of T cell proliferation may relate to immune disorders such as defects in immunity and hyperimmunity, respectively. A change in cytokine secretion patterns may also affect the *in vivo* immune response. Since cytokines secreted from T cells each have a significant biological function, altered regulation of cytokine secretion may induce an adverse immune response.

2. Materials and Methods

We evaluated the effects of the following chemicals: β-benzene hexachloride (β-BHC), *p,p'*-dichlorodiphenyldichloroethylene (*p,p'*-DDE), *p*-nitrophenyl 2,4,6-trichlorophenyl ether (CNP), D-α-aminobenzylpenicillin (Amp), and *o*-acetylcitric acid tributyl ester (ATBC). C3H/He mice were immunized with bovine milk α_{s1}-casein and complete Freund adjuvant. Ten days later we collected lymph node (LN) cells which contained T cells specific for α_{s1}-casein. The cells were incubated with this antigen in the presence or absence of the chemicals. T cell proliferation was determined by measuring [^{3}H]-thymidine incorporation and by MTT assays. We also measured the levels of interleukin (IL-) 2, IL-4, IL-10 and interferon (IFN-) γ in supernatants of the cell cultures by sandwich ELISA. The antibodies used were JES6-1A12 and JES6-5H4 for IL-2, BVD4-1D11 and BVD6-24G2 for IL-4, JES5-2A5 and SXC-1 for IL-10, and R4-6A2 and XMG1.2 for IFN-γ.

3. Results

The response of LN cells determined by our methods can be attributed to the response of T cells. The lowest concentrations of each chemical that we tested in assays with LN cells were those corresponding to the highest concentrations found in some foods in Japan [3]. BHC and DDE are organic chloride pesticides. CNP is a diphenyl ether pesticide. Amp is an antibiotic. ATBC is a plasticizer used in the production of plastic food containers.

TABLE 1. Summary of the effects of chemicals on LN T cell responses

Chemicals	Proliferation		Cytokine secretion			
	DNA synthesis	Numbers of cells	IL-2	IL-4	IL-10	IFN-γ
Amp	–	–	+	±	+	–
BHC	–	–	↓	–	↑+	↓+
CNP	↓	↓	+	↓	+	+
DDE	↓+	↓	+	–	–	–
ATBC	↓+	+	↓	+	↓+	+

The symbols used indicate: –, no effect; +, inhibition and enhancement at different concentrations; ↓, dose-dependent inhibition; ↑, dose-dependent enhancement; ±, weak effect; arrow with +, inhibition or enhancement at some concentration, and no effect at other concentrations.

A summary of the results is shown in Table 1. The proliferative response of the LN cells was determined by measuring thymidine incorporation and by MTT assays which indicate the amount of DNA synthesis and the number of viable cells, respectively. In the absence of the chemicals, the LN cells proliferated well in response to stimulation with α_{s1}-casein. Also, in the absence of the chemicals IL-2, IL-4 and IFN-γ were secreted by the LN cells in response to this antigen. IL-10 secretion was not simply dependent upon antigenic stimulation of the LN cells. Both thymidine incorporation and MTT assays indicated that when CNP was added to the cultures, cell proliferation was inhibited to an extent dependent upon the concentration of CNP present. CNP inhibited IL-4 secretion in a manner similar to its inhibition of cell proliferation. Levels of secretion of IL-2, IL-10 and IFN-γ were affected, but were not simply dependent upon the concentration of CNP present. In the case of DDE as well, LN cell proliferation was inhibited. However, DDE affected IL-2 secretion, but not IL-4, IL-10 or IFN-γ secretion. Another organic chloride pesticide, BHC, did not affect the proliferative response of the T cells, or IL-4 secretion, but it slightly inhibited IL-2 and IFN-γ secretion. IL-10 secretion was increased, but was not simply dependent upon the concentration of BHC present. Amp did not affect the proliferative response of the T cells, or the levels of IL-4 and IFN-γ secretion, while it inhibited IL-2 and IL-10 secretion. The plasticizer, ATBC, appeared to inhibit the proliferative response of the T cells, but the effect was not simply dependent upon the concentration of ATBC present. Levels of IL-2 and IL-10 secretion were decreased in the presence of ATBC, while the patterns of IL-4 and IFN-γ secretion differed from those of IL-2 and IL-10. Considering that we cultured the cells in the plastic culture plates, it is possible that ATBC may have plasticized the plates, thereby influencing the cellular response. Thus, although effects of ATBC on the cellular response were evident in our assays, these effects may be anomalous.

4. Discussion

Among the chemicals tested CNP and DDE inhibited T cell proliferation, and the effect was simply dose-dependent in each instance. However, BHC and Amp did not affect cell proliferation at any of the concentrations tested. The effects of these chemicals on cytokine secretion were heterogeneous. Even when inhibition of the T cell proliferative response was clearly dependent upon the concentration of the chemicals, the effect on cytokine secretion varied depending upon the cytokine examined. That is, in some instances a dose-dependent inhibition or enhancement was observed, while in other instances inhibition and enhancement having a minimum or a maximum at a certain dose was evident.

Inhibition of cell proliferation by some kinds of chemicals should be regarded to be related to cellular damage. Such damage can either inhibit or enhance cytokine secretion. Inhibition of cytokine secretion results from a decrease in cytokine production due to a defect in transcription, translation, translocation and/or modifications relating to protein biosynthesis. On the other hand, damage of plasma membranes can enhance permeation of intracellular cytokines. Therefore, the in vitro effects of chemicals on cytokine secretion are considered to be more complicated than those relating to cell proliferation. Our results suggest that some chemicals may modulate the in vivo immune response by affecting both the proliferative and cytokine secretion responses of T cells.

Although BHC and Amp did not affect LN T cell proliferation, cytokine secretion was srongly affected. Since the *in vivo* immunological response is defined as a total of the activation states of lymphocytes and the balance in production of various cytokines, we suppose that these chemicals have the potentials to modify the immune response.

5. Acknowledgments

This study was supported by a Grant-in-Aid from the Ministry of Education, Culture and Science of Japan.

6. References

1. Brander, C., Mauri-Hellweg, D., Bettens, F., Rolli, H., Goldman, M., and Pichler, W.J.: Heterogeneous T cell responses to β-lactam-modified self-structures are observed in penicillin-allergic individuals, *J. Immunol.* **155** (1995), 2670-2678.
2. Moulon, C., Vollmer, J., and Weltzien, H.-U.: Characterization of processing requirements and metal cross-reactivities in T cell clones from patients with allergic contact dermatitis to nickel, *Eur. J. Immunol.* **25** (1995), 3308-3315.
3. Uemura, S., Kawamura, H., Tsuji, M., Tomita, S., and Maeda, S.: Zanryu Noyaku Databook, Sanseido, Tokyo, 1992 (in Japanese).

MOLECULAR MECHANISMS OF *SELAGINELLA TAMARISCINA* FOR ANTINEOPLASTIC ACTIVITY IN HUMAN LEUKEMIA CELL LINE U937

IN SEON LEE[1], SOO UN KIM[2], YUN SIK LEE[3], SUNG HUI PARK[4] AND IN JA RHEE[4]

[1]Dep. of Food Science and Technology, Keimyung University, Taegu 704-701 Korea, [2]Dep. of Agricultural Chemistry, Seoul National University, Seoul 441-744, Korea, [3]National Institute of Bioscience and Human Technology, Tsukuba, Japan, [4]College of Pharmacy, Taegu Hyosung Catholic University, Kyungsan 713-702, Korea

Summary

In order to study the antitumoral effect of *Selaginella tamariscina* extracts, the cytotoxicities to human histiocytic lymphoma(U937) and lymphocyte were measured by MTT method. The water extract of *Selaginella tamariscina* was partitioned into chloroform ($CHCl_3$), ethylacetate (EtOAc), n-butanol (BuOH) and water (H_2O), successively. $CHCl_3$, EtOAc and BuOH fractions of *Selaginella tamariscina* had the cytotoxicity to the U937 cells but no effect on the lymphocyte under the same conditions. The expression of p53 tumor suppressor gene was evaluated by nothern blotting method. The increased expression of p53 was only induced by water fraction but $CHCl_3$, EtOAc and BuOH fractions. To confirm the correlation between role of p53 and cell cycle phase for sensitivity of *Selaginella tamariscina*, the human colon carcinoma cell line HT-29 and the human ovarian cancer cell line A2780 were treated for 24h with 50 μg/ml of each fraction. In 2780 cells, the water fraction significantly showed the G1 arrest compared with the other fractions. These results suggest that the antitumoral effect of water fraction is related to p53 and G1 progression.

Introduction

Cancer is a disease characterized by uncontrolled growth of malignant cells. At the current rate of cancer incidences, human will eventually have much more cancers than ever before. Therefore, urgent need for more effective anticancer drug is evident. Some of the medicinal plants have a long history of providing antitumor agents with novel structures and unique mechanisms, many of those natural antitumor agents have also served as new leads for further development of active analogs of clinical importance. Several vinca alkaloids, podophyllotoxins, and taxol have currently been used in the treatment of various cancers[1]. *Selaginella tamariscina* is a traditional medicinal plant in the Orient. Lee et al. have reported that amentoflavon isolated from *Selaginella tamariscina* shows effective cytotoxicity against p388 and MKN45[2]. In order to study the antitumoral effect of *Selaginella tamariscina* extracts, the cytotoxicities to human histiocytic lymphoma U937 and lymphocyte were measured by MTT method.

A 53,000 dalton protein called p53, was noted to mutate in human cancer in 1987. Further studies have shown that p53 is a tumor suppressor gene which may be one of the most frequently altered genes in cancer of all types. Accumulated evidence suggests that p53 acts by binding DNA and activating transcription. Expression of p53 after transfection with a constitutive promoter, blocks cellular proliferation and leads to an apoptotic cell death[3]. In order to study the antitumoral mechanisms of *Selaginella tamariscina* extracts, the expression of p53 tumor suppressor gene was evaluated by northern blotting.

The tumor suppressor p53 is a transcription factor which has been identified as a participant in the cellular DNA damage response resulting in ether "G1 arrest" or "apoptosis". The mechanism by which p53 induction results in growth suppression is not firmly established, but a potential explanation was provided by the demonstration that p53 transcriptionally activates the production of a MW 21,000 dalton protein (WAF1) which was simultaneously discovered as a potent inhibitor of cyclin-dependent kinases (CIP1).

In the present studies, we sought to determine whether the water extract of *Selaginella tamariscina* induced p53 overexpression in U937 cell line and significantly expressed the G1 arrest in the human ovarian cell line A2780. These observations suggest a rationale for the design of new anticancer drug.

2. Materials and Methods

2.1. The preparation of the fractions from *Selaginella tamarascina* [4]

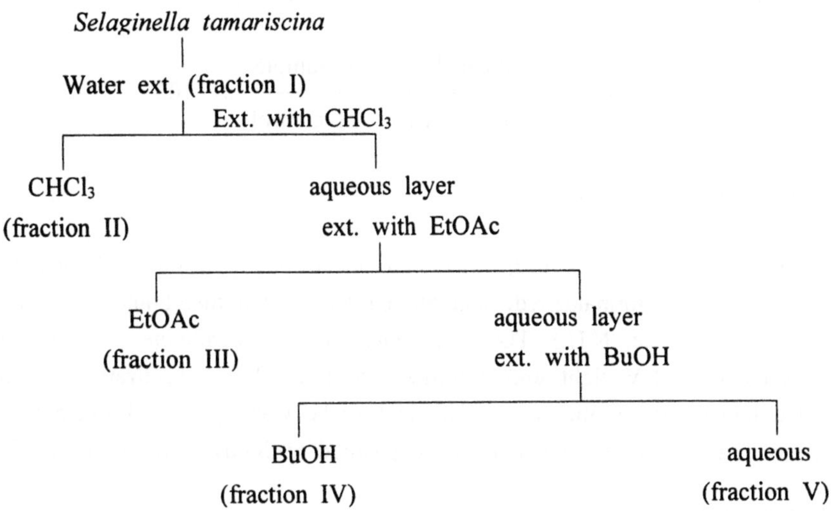

2.2. Cell lines and cell culture condition

The used cell lines were human histiocytic leukemia cell line U937, human ovarian cancer cell line A2780, human colon cell line HT-29, human lymphocyte(primary culture)

All cell lines were cultivated in RPMI 1640 medium (GIBCO BRL, USA) supplemented with 10% fetal bovin serum(FBS,Gibco BRL, USA), 100 units/ml of penicillin and 100 μg/ml of streptomycin and were maintained at 37℃ in an humidified atmosphere of 5% CO_2, 95% air incubator.

2.3. Isolation of lymphocyte[8]

Heparinized peripheral blood drawn from a healthy donor was diluted 1:3 in phosphate buffed saline(PBS). The suspension was placed on Histopaque 1077 in a 50ml conical tube and then centrifuged at 400 xg for 30min at room temperature. Lymphocyte were collected at the interface and then cultured in RPMI1640 medium in a humidified atmosphere of 95% air and 5% CO_2 at 37 ℃

2.4. MTT assay

Human U937 cells was suspended to a cell density of $1x10^5$ cells per ml in RPMI 1640 medium. Kwonback preparation was added to 24$\mu\ell$ of the cell suspension to a final concentration and then cultured in a volum 150$\mu\ell$/well

96-well microtiter plate for 44 hours. 50$\mu\ell$/well of MTT solution were added to each well and after 4 hours plates were centrifuged at 2,000 rpm for 10 min. The absorbance was read at 550nm (Elisa Reader). We calculated the survival percentage(S.P.) of each value vs the control [5-8].

$$S.P.=\frac{\text{Optical density of sample}}{\text{Optiical density of control}}\times 100$$

2.5. Northern blot analysis

Extracted RNA samples (10μg/ml) were electrophoresed on 1.2% agarose gel containing 0.66M formaldehyde and blotted onto nylon membrane (Scheicher and Schuell Inc., Keene, N.H.). The RNA bound to the membrane was cross linked by exposure to UV light and hybridized with a SDS hybridization buffer. ^{32}P labeled human p53 probe was added, and probe was hybridized overnight at 65 ℃, membrane were washed and then exposured to X-ray film for 5 days(9).

2.6. Flow cytometry

To analyze DNA content, cells were harvested, washed in ice-cold PBS, and fixed in 70% ethanol. The fixed cells were treated with RNase A for 30 min at 37℃, and cellular DNA was stained with 50 mg/ml propidium iodide. Cell cycle distributions were determined by using a Becton Dickinson fluorescence-activataed cell analyzer. Data were interpreted using the SFIT model program provided by the manufacturer. Results represent the mean of triplicate determination in which a minimum of 50,000 cells were assayed for each determination.

3. Results and discussion

U937 was tested with various concentration of *Selaginella tamariscina* fractions. As shown in Table 1, cytotoxicity effect begins at 10μg/ml $CHCl_3$ fraction(survival percents 79.7%), BuOH fraction(survival percents 89.1%) and 100μg/ml EtOAc fraction(survival percents 66.3%), and it's effect was dose dependent. While all of *Selaginella tamariscina* fractions did not show any effect on lymphocyte(Table 2).

Table 1. Cytotoxic effect of *Selaginella tamariscina* on human leukemia U937 cells by MTT assay

S.tamariscina	Conc. μl/ml	O.D.±S.D.	S.p.(%)
Water extract (fraction Ⅰ)	1	0.623±0.023	99.7
	10	0.623±0.029	99.8
	100	0.570±0.020	91.3[b]
$CHCl_3$ (fraction Ⅱ)	1	0.507±0.011	79.7[b]
	10	0.361±0.023	53.7[b]
	100	0.349±0.093	51.6[b]
EtOAc (fraction Ⅲ)	1	0.622±0.041	100.3
	10	0.569±0.037	95.1[a]
	100	0.432±0.037	66.3[b]
BuOH (fraction Ⅳ)	1	0.628±0.047	100.5
	10	0.556±0.072	89.1
	100	0.522±0.044	83.6[b]
H_2O (fraction Ⅴ)	1	0.620±0.010	99.1
	10	0.591±0.024	94.6
	100	0.579±0.028	92.6[a]
Control		0.624±0.028	100

[a]$p<0.05$, [b]$p<0.01$: significant to the control

Table 2. Cytotoxic effect of *Selaginella tamariscina* on human lymphocyte by MTT assay

S.tamariscina	Conc. μl/ml	O.D.±S.D.	S.p.(%)
$CHCL_3$ (fraction Ⅱ)	1	0.631±0.035	97.1
	10	0.636±0.013	97.9
	100	0.681±0.015	102.2
EtOAc (fraction Ⅲ)	1	0.623±0.045	95.7
	10	0.833±0.021	90.2
	100	0.637±0.032	98.1
BuOH (fraction Ⅳ)	1	0.647±0.020	99.5
	10	0.618±0.025	95.1
	100	0.591±0.028	90.9

The increased expression of p53 was induced by water fraction of *Selaginella tamariscina* but no expression by the other fractions (data not shown). The water fraction of *Selaginella tamariscina* showed also G1 arrest in the human ovarian cancer A2780 cells (Fig. 2) but did not in the human colon cancer HT-29 cells (Fig.1).

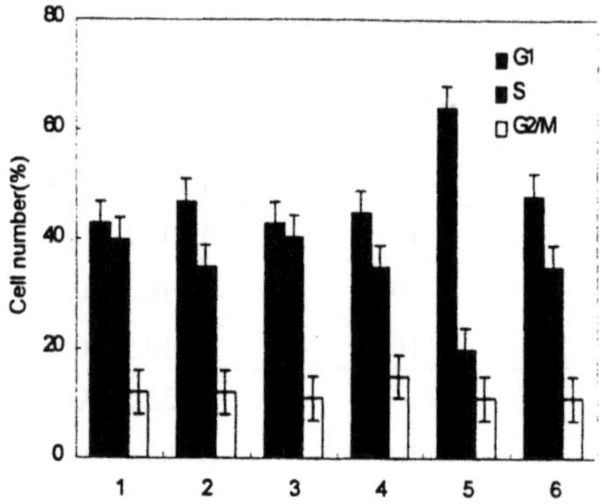

1. None 2. H_2O 3. $CHCl_3$ 4. EtOAc 5. BuOH 6. H_2O

Figure 1. Effects of *Selaginella tamariscina* extracts on cell cycle distribution in HT-29 cells.

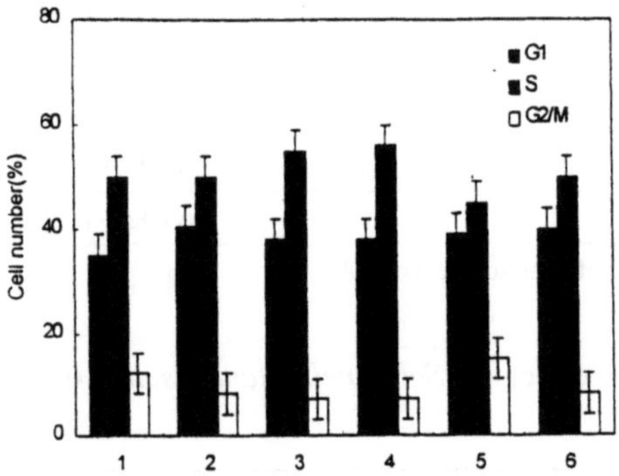

1. None 2. H_2O 3. $CHCl_3$ 4. EtOAc 5. BuOH 6. H_2O *$p<0.05$ vs None

Figure 2. Effects of *Selaginella tamariscina* extracts on cell cycle distribution in A2780 cells.

The herb of *Selaginella tamariscina* is one of the most important herbal medicines, have being used to treat tumor in the oriental medicines. Some papers described isolation of flavonoids as active principles for the antitumor activity [2]. The water extract of *Selaginella tamariscina* was partitioned into chloroform($CHCl_3$), ethylacetate(EtOAc), n-butanol(BuOH), and water(H_2O) successively. And the cytotoxicities to human histiocytic lymphoma cell line U937 and lymphocyte were measured by MTT method. $CHCl_3$, EtOAc, and BuOH fractions of *Selaginella tamariscina* showed the effective cytotoxicity but they had no effect on the cytotoxicity of lymphocyte[4]. In order to elucidate mechanisms of the antitumor activity, the effect on p53 expression was studied. The expression of p53 tumor suppressor gene was evaluated by northern blotting method. The increased expression of p53 was induced by *Selaginella tamariscina* fraction V(water fraction) but no expression of p53 was induced by $CHCl_3$, EtOAc, and BuOH fractions of *Selaginella tamariscina.* The northern blot analysis of the antitumor fractions showed that the antitumoral effect of the water extract was related to p53[10].

4. Conclusion

Results of the present study show that the increased expression of p53 induced by water extract of *Selaginella tamariscina* in U937 cells. We have also demonstrated that the water extract of *Selaginella tamariscina* occurs G1 arrest in the human ovarian cancer A2780 cells. Our results indicate that the antitumoral effect of water extract of *Selaginella tamariscin* was related to p53 and G1 progression in cancer cell lines.

References

1. Chang, C.J., Ashendel, C.L., Chan, T., Geahlen, R.L., and McLaughlin, J.L. (1990) Discovery of national antitumor drugs :new strategies, *J. Nat.. Prod.* **53**, 23-25.
2. Lee, I.R, Song, J.Y., and Lee, Y.S. (1992) Cytotoxicity of Folkloric Medicine in Murine and Human Cancer Cells, *Kor. J. Pharmacogn* **23**, 132-134.
3. Miller, C. and Koenffler, H.P. (1993) p53 mutations in human cancer Leukemia 7/suppl. 2. 18-20.
4. Rhee, I.J. and Park, S.H.(1994) Effect of selaginella tamariscina on U937 cytotoxicity, *J. Korean Soc. Food Nutr.* **23**, 799-804.
5. Carmichael, J., DeGraff, A.F., Gazder, J.D., Minna, J.D., and Mitchell J.B. (1987) Evaluatio n of a tetrazolium-based semiautomatic colorimetric assay ; Assessment of radiosensitivity, *Cancer Res.* **47**, 943-945.
6. Carmichael, J., DeGraff, A.F., Gazder, J.D., Minna, J.D., and Mitchell J.B. (1987) Evaluatio n of a tetrazolium-based semiautomatic colorimetric assay ; Assessment of radiosensitivity, *Cancer Res.* **47**, 935-937.
7. Park, J.G., Kramer, B.S., Steinberg, S.M., Carmichael,J., Collins J.M., Minna, J.D., and Gaxder, A.F. (1987) Chemosensitivity testing of human coloretal carcinoma cell lines using a tetrazolium-based colorimetric assay, *Cancer Res.*, **47**, 5875-5878.
8. Mosman, T. (1983) Rapid colorimetric assay for cellular growth and survival : application to proliferation and cytotoxic assay, *J. Immunol Methodes.*, **65**, 55-57.
9. Maluish A. E. and Strong D. M. : Lymphocyte proliferation, In Rose, F. H. and Fahey J. L.(eds.) (1986), Mannual of clinical laboratory immunology (3rd.ed.), American Society for Microbiology, Washington,D.C.,pp 274
10. Kataoka, S., Naito, M., Tomida, A., and Tsuruo, T.(1994) Resistance to antitumor agent induced apoptosis in a mutant of human myeloid leukemia U937 cells, *Exp. Cell Res.*, **215**, 199-204.

Differential Effect of Antigen Analogs with T cell Inhibitory Activity on *in vivo* Antibody Response

Masako Toda, Mamoru Totsuka, Tohru Yoshioka,
Akio Ametani, and Shuichi Kaminogawa
Department of Applied Biological Chemistry, The University of Tokyo, 1-1-1 Yayoi, Bunkyo-ku, Tokyo 113, JAPAN

Abstract The efficacy of T-cell receptor antagonist peptides in inhibition of antibody responses *in vivo* was evaluated. Substitution analogs of peptide 119-133 (p119-133) of bovine β-lactoglobulin, R124Q (^{124}Arg→Gln) and D129S (^{129}Asp→Ser), have been shown to display TCR antagonist activity for three of four p119-133-specific T-cell clones. Both D129S and R124Q, when co-administered with p119-133 to mice, showed an inhibitory effect *in vivo* on T-cell priming and/or activation by p119-133, as evidenced by the decreased proliferation of lymph node T cells in response to p119-133. Production of p119-133-specific antibody was inhibited by D129S when mice were injected with p119-133 plus D129S at the first immunization, while co-immunization with R124Q instead led to increase in production of p119-133-specific antibodies. Anti-p119-133 antibodies showed cross-reactivity to R124Q but not to D129S, suggesting that p119-133-specific B cells are efficiently activated by R124Q and R124Q-specific T cells. The present study suggests the necessity of selecting TCR antagonist peptides which escape recognition by pathogenic antibodies against allergens or autoantigens, for the purpose of inhibiting an undesirable antibody response.

1. Introduction

The first step of T-cell activation is the recognition of peptide-MHC complexes by the antigen specific T-cell receptor (TCR). Recently, antigenic peptide analogs with a single substitution involving TCR contact residues have been shown to act as antigen-specific inhibitors of T-cell activation [1]. Peptides with this activity, called TCR antagonists, are expected to be useful as antigen-specific means to down-regulate undesirable immune responses. Several reports have shown inhibitory effects of TCR antagonist peptides on the onset of T-cell mediated experimental autoimmune diseases, such as experimental allergic encephalomyelitis, following the administration of an analog peptide together with pathogenic peptides *in vivo* [2-4]. In antibody-mediated immune diseases, such as type I allergy, inhibition of pathogenic antibody production as well as the T-cell response is crucial to the remitting of symptoms. However, the efficacy of TCR antagonists in the treatment of antibody-mediated immune diseases, has not yet been determined. Aiming to establish a suitable method for application of TCR antagonist peptides to treatment of antibody-mediated diseases, we investigated the difference in effectiveness of two distinct TCR antagonists to inhibit antibody production*in vivo*.

β-Lactoglobulin (β-LG) is known to be one of the major allergenic factors involved in cow's milk allergy. Peptide p119-133 corresponding to the region 119-133 of β-LG includes a dominant determinant recognized by T cells in C57BL/6 mice. In this study, we compared the effects of two analogs of p119-133 *in vivo* on (1) priming of T cells to p119-133 and (2) anti-p119-133 antibody production, by co-administering each analog and p119-133 to C57BL/6 (H-2^b) mice.

2. Materials and Methods

2.1 Mice

Six-week-old female C57BL/6N mice were purchased from CLEA Japan (Tokyo, Japan).

2.2 Peptides

A peptide corresponding to residues 119-133 of bovine β-LG (p119-133; CQCLVRTPEVDDEAL) and its analog peptides R124Q (CQCLVQTPEVDDEAL) and D129S (CQCLVRTPEVSDEAL) were synthesized using a peptide synthesizer (model 430A; Perkin-Elmer, Norfolk, CT, U.S.A.) and purified by reversed-phase HPLC. A peptide corresponding to residues 136-155 of bovine α_{s1}-casein, P10 (IGVNQELAYFYPELFRQFYQ), was purchased from Chiron Mimotopes (Clayton, Australia).

2.3 Lymph-node cell proliferation assay

Mice were immunized subcutaneously in the hind footpads with antigens in the form of an emulsion with CFA-H37Ra (Difco Laboratories, Detroit, MI, U.S.A.). Seven days later, lymph-node cells were removed and cultured in the presence of various concentrations of the peptides for three days. Each well was then pulsed with 1 μCi of [^{3}H] thymidine. Cells were harvested 20 hr later, and the uptake of [^{3}H]thymidine was assessed by liquid scintillation counting.

2.4 Antibody production assay

Mice were immunized intraperitoneally with p119-133 alone or with p119-133 plus an analog peptide emulsified in CFA-H37Ra. Two weeks after the first immunization, the mice were boosted with p119-133 alone emulsified in IFA (Difco Laboratories, Detroit, MI, U.S.A.). On the seventh day following the second immunization, the mice were bled and the sera were prepared. Anti-p119-133 antibody levels in the sera were measured by enzyme-linked immunosorbent assay (ELISA).

2.5 ELISA

Microtiter plates (Maxi Sorp F96; Nunc, Roskilde, Denmark), which had been coated with 10 μg/ml of p119-133 by overnight incubation at 4 ℃, were washed with PBS containing 0.05% Tween 20 (PBS-TW) and incubated with the blocking reagent (1.0% BSA in 0.1 N citrate buffer, pH 6.4) at RT for 2 hr. After removing the solution, each serum diluted 100-1000 times in PBS was added to the plate wells, then the coated wells were rinsed and treated with alkaline phosphatase-labeled goat anti-mouse Ig (M+G+A) (Cappel; Organon Teknika, Durham, NC, U.S.A.) in PBS-TW for 2 hr. After adding enzyme substrate (*p*-nitrophenyl-phosphate in a diethanolamine-HCl buffer at pH 9.8), the absorbance at

405 nm was measured. The titer of anti-p119-133 antibody was calculated by using an anti-β-LG monoclonal antibody specific for p119-133, LG13.1, as standard.

2.6 Cross-reactivity analysis of anti-p119-133 antibodies
The ability of anti-p119-133 antibodies to bind R124Q or D129S was examined by an inhibition ELISA. A 96-well microtiterplate was coated with p119-133 by incubation with the peptide solution at 4 ℃ overnight. Each coated well was then incubated with the blocking reagent at RT for 2 hr. Another plate was treated only with the blocking reagent in the same manner. The sera from the mice immunized with p119-133 alone were diluted 200 times with PBS-TW and allowed to react to various concentrations of p119-133 in the liquid phase in the latter plate at RT for 2 hr. These reaction mixtures were added to the former plate after the coating and blocking treatment, and incubated at 4 ℃ overnight. Binding of free antibodies in the mixtures to p119-133 on the solid phase was detected by the same procedures as those described for the determination of anti-p119-133 antibody titer.

3. Results and Discussion

3.1 R124Q and D129S inhibited T-cell priming to p119-133 *in vivo*
In previous studies, analogs of p119-133, R124Q and D129S, have been shown to function as TCR antagonists for distinct three of four p119-133-specific T-cell clones. Moreover, both of the analog peptides inhibited the proliferative response to p119-133 *in vitro* of polyclonal T cells derived from the immunized with p119-133 (data not shown).

First, we examined the effect of co-administration of each TCR antagonist with p119-133 on priming of T cells to p119-133 *in vivo*. Lymph node cells from the mice immunized with p119-133 and those from the mice co-immunized with p119-133 plus P10 (at a 1:1 molecular ratio) proliferated vigorously in response to *in vitro* stimulation with p119-133. The magnitude of the response was almost the same for both groups. On the contrary, lymph-node cells from the mice co-immunized with p119-133 plus R124Q or D129S (at a 1:1 molecular ratio) showed less proliferation in response to p119-133 compared with

Fig.1 Inhibitory effect of R124Q or D129S on the priming of lymph node cells to p119-133 *in vivo*. C57BL/6 mice were immunized sc with 10 μg of p119-133 alone (closed circle), 10 μg of p119-133 plus 10 μg of R124Q (open square), 10 μg of p119-133 plus 10 μg of D129S (closed square), or 10 μg of p119-133 plus 10 μg of P10 (open triangle). After 7 days, the lymph node cells were prepared and cultured in the presence of various concentrations of p119-133. The data are shown as the ratio of average cpm of each point to that of background cpm (Stimulation index).

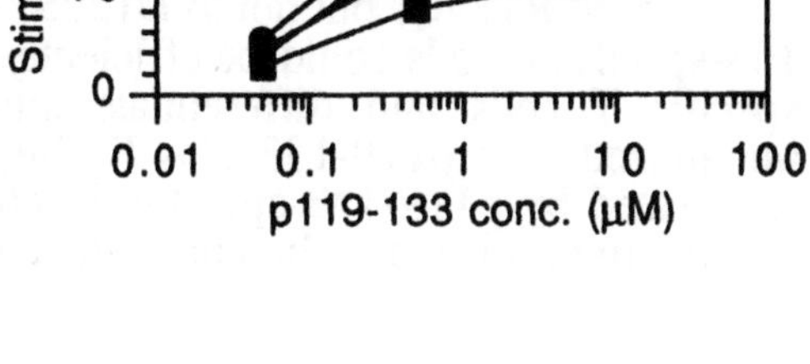

those from the mice immunized with p119-133 alone (Fig 1). These results indicate that priming and subsequent propagation of T cells specific for p119-133 was inhibited in the presence of R124Q or D129S. The I-A^b binding peptide control, P10 did not show the inhibition, indicating that the inhibitory effects of R124Q and D129S are not due to competition between p119-133 and each analog peptide for MHC binding.

3.2 D129S but not R124Q inhibited the production of anti-p119-133 antibodies *in vivo*

Next, the effect of R124Q and D129S on anti-p119-133 antibody production *in vivo* was investigated. The levels of anti-p119-133 antibodies in the mice co-immunized with p119-133 plus D129S (at a 1:10 molecular ratio) at the first immunization were greatly decreased compared with that from the mice immunized with p119-133 alone. The extent of inhibition of anti-p119-133 antibody production was 74.1%. However, in the mice immunized with p119-133 plus R124Q or P10 (at a 1:10 molecular ratio), the levels of anti- p119-133 antibodies were not decreased (Fig 2).

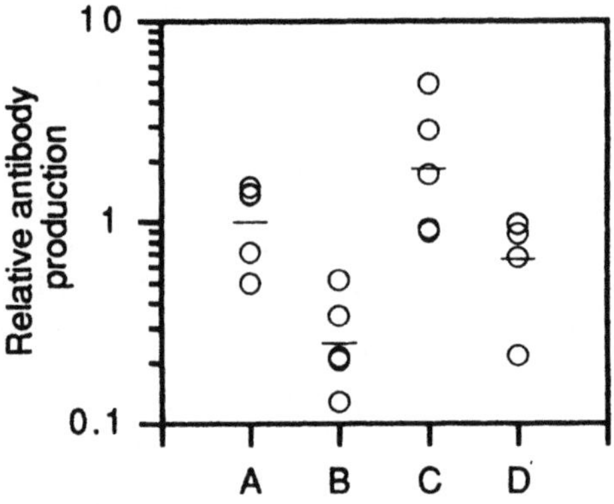

Fig.2 Different effect of R124Q and D129S on the production of anti-p119-133 antibody *in vivo*.
C57BL/6 mice were immunized ip with 10 μg of p119-133 alone (A), 10 μg of p119-133 plus 100 μg of D129S (B), 10 μg of p119-133 plus 100 μg of R124Q (C), or 10 μg of p119-133 plus 100 μg of P10 (D). Two weeks later, each group of mice was boosted ip with 10 μg of p119-133 in IFA. The anti-p119-133 antibody titer one week after the boosting was measured by ELISA. The data are shown as relative amount of the specific antibodies for an individual mouse (circles) and the average in each group (bars).

3.3 Anti-p119-133 antibody has cross-reactivity to R124Q but not to D129S.

To investigate the mechanism responsible for the differential effect of the two analogs on specific antibody production, we next examined the cross-reactivity of anti-p119-133 antibodies to R124Q or D129S by an inhibition ELISA. As shown in Fig. 3, R124Q but not D129S in the liquid phase inhibited the binding of anti-p119-133 antibodies to p119-133 on the solid phase. Anti-p119-133 antibodies showed cross-reactivity to R124Q, but not to D129S, suggesting that most of the p119-133-specific B cells could be efficiently activated by R124Q and R124Q-specific T cells and differentiate into memory B cells in the mice immunized with p119-133 plus R124Q. These memory B cells may be reactivated by p119-133-specific T cells to produce antibodies after the second immunization. In contrast, D129S-specific T cells elicited by immunization with p119-133 plus D129S do not help to generate p119-133-specific memory B cells. Therefore, the difference in the number of p119-133-specific memory B cells elicited in response to the first immunization is likely to be responsible for the different inhibitory effects of D129S and R124Q. Our results suggest the necessity of selecting TCR antagonist peptides which lack reactivity to pathogenic antibodies against allergens or autoantigens, for the purpose of inhibiting an undesirable antibody response.

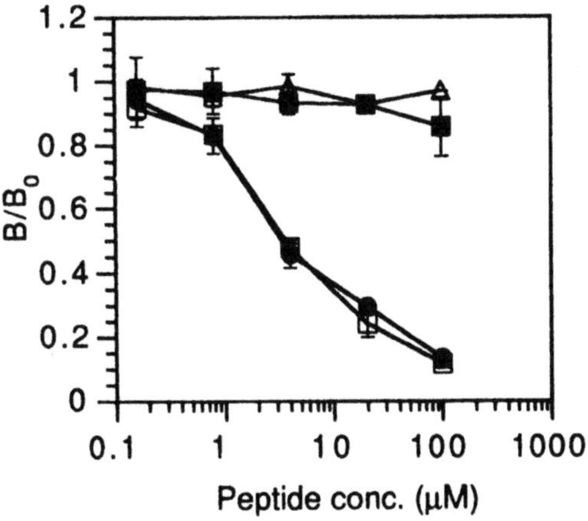

Fig.3 Inhibition of binding of anti-p119-133 antibodies to R124Q or D129S.
ELISA plates were coated with p119-133. p119-133 (closed circle), R124Q (open square), D129S (closed square) or P10 (open triangle) was used as the inhibitor. B and B_0 show the absorbance in the presence and absence of the inhibitory peptide, respectively. Each point represents the mean value from three independent experiments.

4. References

[1] De Magistris, M.T., Alexander, J., Coggeshall, M., Altman, A., Gaeta, F.C.A., Grey, H.M. and Sette, A. (1992) Cell, **68**, 625-634.

[2] Karin, N., Mitchell, D.J., Brocke, S., Ling, N. and Steinman, L. (1994) J. Exp. Med., **180**, 2227-2237.

[3] Franco, A., Southwood, S., Arrhenius, T., Kuchroo, V.K., Grey, H.M., Sette, A. and Ishioka, G.Y. (1994) Eur. J. Immunol., **24**, 940-946.

[4] Kuchroo, V.K., Greer, J.M., Kaul, D., Ishioka, G., Franco, A., Sette, A., Sobel, R.A. and Lees, M.B. (1994) J. Immunol., **153**, 3326-3336.

INCREASE OF INTRACELLULAR Ca^{2+} LEVEL INDUCES OSTEOCLAST- LIKE CELL FORMATION IN COCULTURE OF MOUSE BONE MARROW CELLS AND OSTEOBLASTIC CELLS

Osteoclast -like cell formation by ionophores

MASAMICHI TAKAMI, JE-TAE WOO AND KAZUO NAGAI
Department of Bioengineering, Tokyo Institute of Technology, 4259 Nagatuta-cho, Midori-ku, Yokohama 226, Japan

Ionomycin and A23187 which are known as calcium ionophores, induced osteoclastic differentiation in the coculture of mouse bone marrow cells and primary osteoblasic cells. Osteoclast-like cells induced by these agents appeared as multinucleated giant cells with tartrate resistant acid phosphatase activity, and formed resorption lacunae on dentine slices. Calcium ionophores are reported to increase intracellular Ca^{2+} level. Therefore, these results suggests that osteoclastic differentiation induced by calcium ionophore is a result of the increase of intracellular Ca^{2+} level, and that high Ca^{2+} level may be a part of signaling pathways that induce differentiation of osteoclast progenitor to osteoclast.

1. Introduction

Osteoclasts are multinucleated giant cells that play a critical role in bone resorption [6]. It is known that they are derived from mononuclear precursors in hematopoietic tissues, and that dihydroxy vitamin D [$1,25(OH)_2D_3$], parathyroid hormone (PTH), PTH related peptides (PTHrP), and prostagrandin E_2 stimulate osteoclastic differentiation in vitro and in vivo [3,4]. However, the details of osteoclast differentiation mechanisms are still not known. Takahashi et al. established an in vitro system of mouse bone marrow culture, in which osteoclast-like cells (OCLs) are formed efficiently within 6 days of culture [1]. In the present study, we used this culture system and demonstlated that calcium ionophores induce osteoclastic differentiation.

2. Materials and methods

2.1 CHEMICALS

Ionomycin and A23187 were purchased from Sigma Chemical Company (St. Louis, MO). Sodium tartrate was purchased from Wako Pure Chemical Industries Ltd. (Osaka, Japan)

2.2. ANIMAL

Five- to eight week old male mice, ddY strain, were obtained from Sankyo Labo Service Co. (Tokyo, Japan).

2.3. COCULTURE SYSTEM

The mice were killed by cervical dislocation under light ether anesthesia, and femola and tibiae were removed. The bone ends were cut off with scissors and marrow cavity was flushed with 1 ml α-minimal essential medium (α-MEM) by injecting at one end of the bone using a 25G needle. The bone marrow was twice washed with a-MEM. Primary osteoblastic cells were isolated from new-born ddy mouse calvaria. A total of 20 to 30 calvaria was collected and subjected to routine five sequential digestions, using 0.1% collagenase (Wako Pure Chemical Industries Ltd., Osaka, Japan) and 0.2% dispase (Godo Shusei, Tokyo, Japan). Osteoblastic cells isolated in fractions 2 to 5 were combined. Bone marrow cells and osteoblastic cells were cocultured in α-MEM containing 10 % fetal calf serum (CSL Limited, Victoria, Australia) at 5×10^5 cells/ml and 2×10^4 cells/well respectively in 48 well plates (CORNING Inc., Corning, NY) (0.4 ml/well). All cultures were maintained at 37 °C in a humidified atmosphere of 5 % CO_2 in air [1, 2].

2.4. DETERMINATION AND CHARACTERIZATION OF OCLS

After being cultured for indicated times, the adherent cells were fixed and stained for tartrate resistant acid phosphatase (TRAP) in the presence of 50 mM sodium tartrate as described [4]. TRAP-positive multinucleated cells which had more than 5 nuclear were counted as OCLs. To determine the bone resorbing activity, OCLs were induced in collagen gel-coated plate and treated with 0.1 % collagenase to collect OCLs and osteoblastic cells. Suspension of OCLs and osteoblastic cells was kept on dentine slices in 96 well plates containing 0.2 ml α-MEM supplemented with 10 % FCS. After incubation for 24 hours, cells were removed from dentine slices, and the resorbed pits were stained with Mayer's hematoxylin solution.

3. Results

At the beginning of coculture, OCLs were not detected in the culture. When cells were treated with ionomycin (1 μM) or A23187 (0.4 μM) from 3rd to 6th day of culture, OCL formation was induced on 6th day of culture (Fig. 1A, B). In the control cultures, no OCLs were detected (Fig. 1C). In addition, the number of OCLs was significantly reduced when ionomycin or A23187 was added before 3rd day of culture or after 4th day of culture (data not shown).

Table 1 shows effect of ionomycin or A23187 on OCL formation. These agents induced formation of OCLs in a dose-dependent manner. Maximal TRAP-positive OCL formation was observed at 1 μM ionomycin or 0.4 μM A23187. The number of TRAP-positive MNCs formed in cultures treated with 1 μM ionomycin or 0.4 μM A23187 were 390 ± 35 and 287 ± 25 per well, respectively. However, no OCLs were formed at higher concentrations, because of the toxicity.

In order to examine the bone resorption abilities of OCLs, suspensions of OCLs were kept on dentine slices for 2 days, and bone resorption pits were detected by staining with hematoxylin solution. Fig. 2 shows resorption pits formed on dentine slices on which ionomycin-induced OCLs were cultured. OCLs induced by A23187 formed similar resorption lacunae on dentine slices (data not shown). For these results, it is clear that calcium ionophores are able to induce OCL formation.

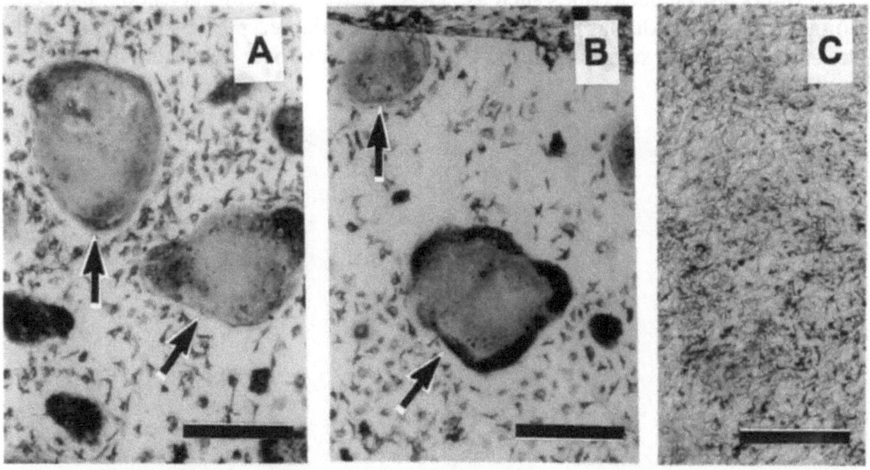

Fig. 1 OCL formation induced by ionomycin or A23187 (x 1000). Mouse bone marrow cells and osteoblastic cells were cocultured for 3 days. After exchanging the medium, cells were treated with 1 μM ionomycin (A), 0.4 μM A23187 (B) or MeOH (C) for additional 3 days. Cell were fixed and stained for TRAP. Arrows indicate OCLs. All bars indicate 250 μm.

TABLE 1. Dose response for ionomycin or A23187 on OCL formation.

Treatment	Conc. (μM)	Number of MNCs / well
Ionomycin	1.0	390 ± 35
	0.5	144 ± 22
	0.25	57 ± 6
A23187	0.4	287 ± 25
	0.2	201 ± 15
	0.1	63 ± 8
Vehicle	—	0 ± 0

Mouse bone marrow cells and osteoblastic cells were cocultured for 3 days. After exchanging the medium, cells were treated with indicated concentrations of ionomycin, A23187 or vehicle (MeOH) for additional 3 days. Cells were fixed and stained for TRAP. Data are expressed as means ± SD of 4 cultures.

4. Discussion

In this study, we found that, calcium ionophores are the new agents to induce OCL formation in mouse coculture system. The OCLs induced by these agents presented several osteoclast characteristics such as, TRAP activity, multi-nucleation and bone resorption activity. Although expression of calcitonin receptor is remained to be detected, these results suggest that the OCLs induced by these agents are the same as the osteoclasts.

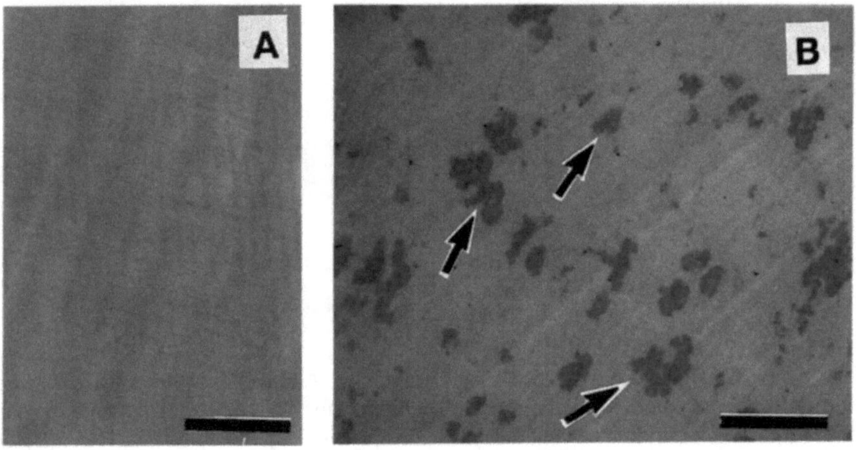

Fig. 2 Dentine slices on which OCLs were cultured (x 500). Control cells (A) or OCLs induced by ionomycin (B) were cultured on dentine slices. After culturing for 24 hours, slices were removed from the cells and stained with hematoxylin solution. Arrows indicate resorption lacunae. All bars indicate 250 μm.

It has been reported that $1,25(OH)_2D_3$, PTH, PTHrP, IL-1, IL-6, and PGE_2 induce OCL formation, and it has been thought that the target cells of these hormones and cytokines may be osteoblastic cells. Therefore, it is believed that there are three independent signal pathways to induce osteoclast formation such as $1,25(OH)_2D_3$ receptor mediated pathway, gp130 mediated pathway and cAMP mediated pathway [3, 4]. In our results, calcium ionophores induced OCL formation instead of these cytokines or hormones. It is known that calcium ionophores cause the increase of intracellular Ca^{2+} level. Thus, the increase of intracellular Ca^{2+} level might be a new signal pathway to induce osteoclast differentiation from its precursor cell. Kaji *et al.* have reported that high extracellular calcium stimulated OCL formation in mouse bone cell cultures [5]. This report also support our results. However, it is not clear whether the target cells of calcium ionophores are osteoclast precursor cells or osteoblastic cells. Further studies are required to clear this question.

5. References

1. Takahashi, N., Yamana, H., Yoshiki, S., Roodman G. D., Mundy, G. R., Jones S. J., Boyde, A., and Suda, T.: Osteoclast-like cell formation and its regulation by osteotropic hormones in mouse bone marrow cultures. *Endocrinology* **122**

(1988), 1373-1382.
2. Takahashi, N., Akatsu, T., Udagawa, N., Sasaki, T., Yamaguchi, A., Moseley, J. M., Martin, T. J., and Suda, T.: Osteoblastic cells are involved in osteoclast formation, *Endocrinology* **123** (1988), 2600-2602.
3. Suda, T., Takahashi, N., and Martin, T. J.: Modulation of osteoclast differentiation, *Endocrine Rev.* **13** (1992), 66-80.
4. Suda, T., Takahashi, N., and Martin, T. J.: Modulation of osteoclast differentiation: Update 1995, *Endocrine Rev.* **4** (1995), 266-270.
5. Kaji, H., Sugimoto, T., Kanatani, M., and Chihara, K.: High extracellular calcium stimulates osteoclast-like cell formation and bone resorbing activity in the presence of osteoblastic cells. *J. Bone Miner. Res.* **11** (1996), 912-920.
6. Väänäen, H. K., Salo, J., and Lehenkari, P.: Mechanism of osteoclast mediated bone resorption. *J. Bone Miner. Metab.* **14** (1996), 187-192.

NEW ANGIOGENESIS INHIBITORS OF MICROBIAL ORIGIN

T. SHIBATA, T. OTSUKA, H. TERANO AND M. OKUHARA
Exploratory Research Laboratories, Fujisawa Pharmaceutical Co., Ltd.
5-2-3 Tokodai, Tsukuba, Ibaraki 300-26, Japan

Angiogenesis, the growth of new blood vessels, occurs in various physiological conditions such as embryonic development, ovulation and wound healing. Angiogenesis is, however, strictly controlled under normal conditions in the adult. On the other hand, uncontrolled angiogenesis is often associated with diseases such as diabetic retinopathy, rheumatoid arthritis and solid tumor. Particularly, angiogenesis is necessary for the continued growth of solid tumor [1]. Folkman introduced the new concept that an antiangiogenesis strategy might represent a novel therapeutic approach for the treatment of solid tumor [2]. This has led to a search for angiogenesis inhibitors which might be used therapeutically.

It is now well known that angiogenesis is the process of new blood vessel formation by endothelial cells, including a series of sequential steps [3] as follows: (i) degradation of basement membrane by proteolytic enzymes; (ii) endothelial cell migration; (iii) endothelial cell proliferation; (iv) tube formation; (v) biosynthesis of basement membrane. Accordingly, compounds which can inhibit one of these steps might suppress angiogenesis.

To search for new angiogenesis inhibitors, we screened inhibitors of endothelial cell proliferation from fermentation products and found WF16775A2 and FR111142. Furthermore, we chemically modified FR111142 and discovered a derivative with more potent antiangiogenic activity, FR118487. Here we will briefly overview the biological activities of these compounds.

1. WF16775A2

WF16775A2 (Fig. 1) was isolated from the cultured broth of a fungus, *Chaetasbolisia erysiphoides* No. 16775 [4]. The inhibitory effects on the proliferation of various cells *in vitro* were examined. The concentration required for 50% inhibition (IC_{50} value) for human umbilical vein endothelial cells (HUVEC), EL-4 mouse lymphoma cells and Meth A mouse fibrosarcoma cells was 0.02, 0.84 and 0.64 μg/ml, respectively.

Chick chorioallantoic membrane (CAM) assay was used to evaluate the antiangiogenic effect *in vivo* as described [5]. WF16775A2 at doses of 0.1 - 2.5 μg/egg inhibited capillary

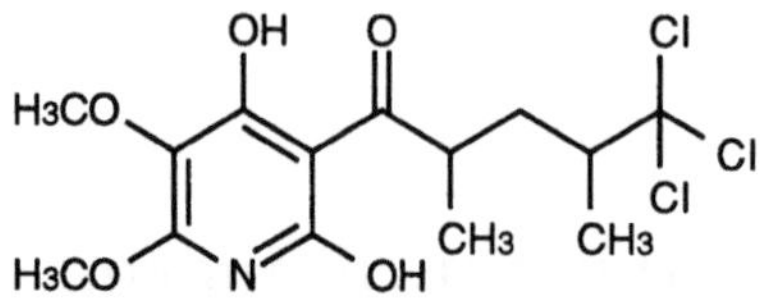

Figure 1. Structures of WF16775A2.

growth in the CAM and induced avascular zones in a dose-dependent manner, although it was toxic to the embryo at 5 μg/egg (data not shown). Based on this result, the antitumor effect against mouse solid tumor Meth A was examined. Intraperitoneal administration of WF16775A2 at doses of 1 - 3 mg/kg showed moderate antitumor activity as follows: 32% inhibition at 1 mg/kg and 27% inhibition at 3 mg/kg on day 14 (data not shown). It was difficult to solubilize in physiological saline. In addition, the effective range was rather narrow. Considering these results, an improvement of solubility and a chemical modification which exhibits antiangiogenic activities more effectively *in vivo* might be required.

2. FR111142

FR111142 (Fig. 2) was isolated from the cultured broth of a fungus, *Scolecobasidium arenarium* F-2015 [6]. The effects of FR111142 on the proliferation of various *in vitro* cultured cells were examined. FR111142 specifically inhibited the growth of endothelial cells at low concentrations (Table 1), although its effect was cytostatic.

Antiangiogenic activity *in vivo* was evaluated in CAM assay. FR111142 displayed potent angiogenesis inhibitory activity and apparently caused no damage to the growth of the embryo (Table 2). When administered systemically, FR111142 at doses of 3.3 - 30 mg/kg remarkably suppressed the growth of mouse solid tumor Meth A, but was not effective against mouse leukemia P388 in ascitic form whose growth does not require angiogenesis (data not shown). In addition, as shown in Table 1, FR111142 inhibited the

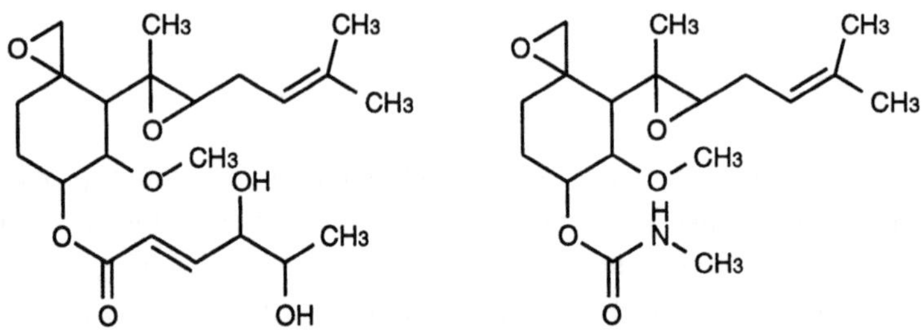

I II

Figure 2. Structures of FR111142 (*I*) and FR118487 (*II*).

Table 1. Effects of FR111142 and FR118487 on proliferation of various cells *in vitro*.

Cell lines	IC_{50} value (M)	
	FR111142	FR118487
HUVEC	1.2×10^{-9}	2.9×10^{-10}
Calf pulmonary artery endothelial cell	2.4×10^{-9}	1.5×10^{-9}
P388 mouse leukemia cell	1.0×10^{-4}	8.8×10^{-5}
Meth A mouse fibrosarcama cell	6.0×10^{-5}	1.8×10^{-4}
Colon 38 mouse adenocarcinoma cell	1.0×10^{-4}	1.8×10^{-4}
BHK-21 baby hamster kidney cell	2.4×10^{-5}	4.4×10^{-5}

Table 2. Inhibitory effects of FR111142 and FR118487 on angiogenesis in CAM assay. Ethylene-vinyl acetate copolymer (EVA) pellets containing FR118487 were placed on the CAM of 5-day embryo. After 2 days, antiangiogenic response was evaluated by measuring an avascular zone in the CAM around the pellet.

Dose (mg/egg)	No. of CAMs (avascular/total)	
	FR111142	FR118487
0	0 / 20	0 / 30
0.5	—	9 / 16
1	2 / 15	10 / 15
5	7 / 16	11 / 18
10	13 / 18	13 / 18
50	14 / 18	18 / 20
100	13 / 16	26 / 28
500	16 / 16	20 / 20

growth of such tumor cells much less than endothelial cells *in vitro*. These results suggest that the antitumor effect of FR111142 may be due to its antiangiogenic action, probably by the inhibition of endothelial cell proliferation.

3. FR118487

FR118487 (Fig. 2) was a semisynthetic derivative of FR111142 [7]. FR118487 had about 5 - 10 times more potent antiangiogenic activity than FR111142 in the endothelial cell proliferation assay *in vitro* (Table 1) and CAM assay *in vivo* (Table 2). Further, the inhibitory effect on angiogenesis in the rabbit cornea induced by an angiogenic factor was evaluated as described [8]. Local administration of FR118487 suppressed capillary growth in the cornea in a dose dependent manner (Fig. 3). A similar inhibitory effect was observed by systemic administration (data not shown).

In vivo antitumor activity was determined in the experimental tumor system in mice. FR118487 was quite effective against mouse solid tumors, Meth A fibrosarcoma and colon 38 adenocarcinoma, at doses of 0.3 - 5 mg/kg (Fig. 4). Further, when administered subcutaneously to mice, it produced slight weight loss at high dose (about 10% body weight of control mice). Moreover, recent studies revealed that FR118487 inhibited the vascular endothelial growth factor (VEGF)-mediated *in vitro* growth of endothelial cells and markedly inhibited metastasis in mouse ovarian carcinoma OV2944 derived tumor line as well as primary tumor growth through mechanism(s) including the suppression of VEGF function *in vivo* [9]. These results suggest that FR118487 may be a beneficial drug for the treatment of solid tumors.

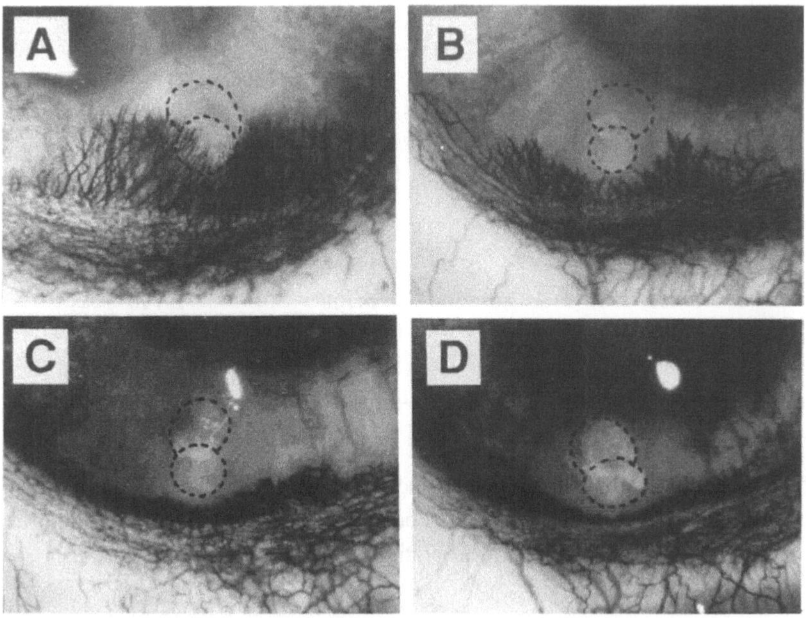

Figure 3. Inhibition of angiogenesis by FR118487 in a rabbit cornea. An EVA pellet containing 300 μg of endothelial cell growth supplement (ECGS) and 30 μg of heparin was implanted into the rabbit corneal micropocket. An empty EVA pellet (A) or an EVA pellet containing 20 μg (B), 100 μg (C) or 500 μg (D) of FR118487 was also implanted close to the ECGS pellet. After 10 days, inhibition of angiogenesis was examined by stereomicroscopy.

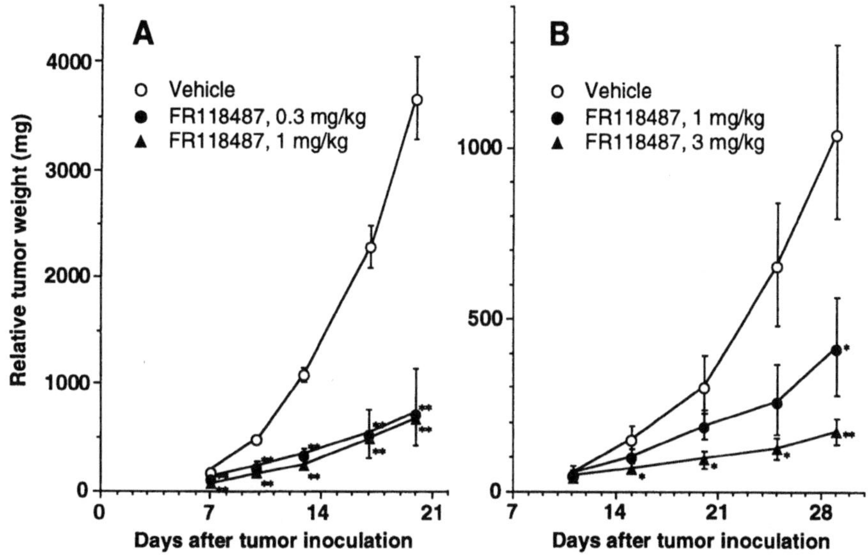

Figure 4. Antitumor effect of FR118487 on Meth A fibrosarcoma (A) and colon 38 adenocarcinoma (B) in mice. Meth A was implanted intradermally into Balb/c mice. Colon 38 was implanted subcutaneously into BDF_1 mice. Twenty-four hours after tumor implantation, doses of FR118487 were administered subcutaneously to mice on days 1-4, 7-11 and 14-18 in the test on Meth A, or on days 8, 11, 13, 15, 20, 25 and 27 in the test on colon 38. FR118487 was suspended in sesame oil containing aluminium stearate (3%, w/w) and lactic acid-glycolic acid (50:50) copolymer. Data represent mean ± S.E. (n=8). * $P < 0.05$ and ** $P < 0.01$ vs vehicle.

4. References

1. Gimbrone, M.A. Jr., Leapman, S., Cotran, R.S., and Folkman, J. (1972) Tumor dormancy *in vivo* by prevention of neovascularization, *J. Exp. Med.* **136**, 261-276.
2. Folkman, J. (1972) Anti-angiogenesis: new concept for therapy of solid tumors, *Ann. Surg.* **175**, 409-416.
3. Klagsbrun, M. and Folkman, J. (1990) *Angiogenesis*, in M.B. Sporn and A.B. Roberts (eds.), *Handbook of Experimental Pharmacology* Vol. 95/II, Springer-Verlag, Berlin, pp. 549-586
4. Otsuka, T., Takase, S., Terano, H., and Okuhara, M. (1992) New angiogenesis inhibitors, WF-16775A_1 and A_2, *J. Antibiot.* **45**, 1970-1973
5. Tanaka, N.G., Sakamoto, N., Tohgo, A., Nishiyama, Y., and Ogawa, H. (1986) Inhibitory effects of anti-angiogenic agents on neovascularization and growth of the chorioallantoic membrane (CAM) - The possibility of a new CAM assay for angiogenesis inhibition, *Exp. Pathol.* **30**, 143-150
6. Otsuka, T., Shibata, T., Tsurumi, Y., Takase, S., Okuhara, M., Terano, H., Kohsaka, M., and Imanaka, H. (1992) A new angiogenesis inhibitor, FR-111142, *J. Antibiot.* **45**, 348-354
7. Otsuka, T., Ohkawa, T., Shibata, T., Oku, T., Okuhara, M., Terano, H., Kohsaka, M., and Imanaka, H. (1991) A new potent angiogenesis inhibitor, FR-118487, *J. Microbiol. Biotech.* **1**, 163-168
8. Gimbrone, M.A. Jr., Cotran, R.S., and Folkman, J. (1974) An experimental model using rabbit cornea, *J. Natl. Cancer Inst.* **52**, 413-427
9. Mu, J., Abe, Y., Tsutsui, T., Yamamoto, N., Tai, X.-G., Niwa, O., Tsujimura, T., Sato, B., Terano, H., Fujiwara, H., and Hamaoka, T. (1996) Inhibition of growth and metastasis of ovarian carcinoma by administrating a drug capable of interfering with vascular endothelial growth factor activity, *Jpn. J. Cancer Res.* **87**, 963-971

HYDROCORTISONE INDUCES A FORMATION OF BASOPHILIC GRANULES IN THE HUMAN LEUKEMIA CELL LINE KU812

Takashi HARA, Sanetaka SHIRAHATA, and Hirofumi TACHIBANA*
Graduate School of Genetic Resources Technology, Kyushu University,
**Department of Food Science and Technology, Kyushu University,*
6-10-1 Hakozaki, Higashi-ku, Fukuoka 812, Japan

ABSTRACT The human leukemia cell line KU812 exhibits a potential to differentiate into basophilic cells and has been described as a basophilic precursor cell line (1-6). KU812 cells produce inflammatory mediators such as histamine and arachidonic acid metabolites. In addition, the cell line expresses high affinity IgE receptor (FcεRI) which is central to an allergic responses, but formation of cytoplasmic granules has not been accomplished. When basophilic differentiation was induced in basophilic lineage cells, the synthesis of inflammatory mediators are increased and granulated cells are appeared in culture.

Previously, we found that hydrocortisone can enhance the expression of FcεRI on KU812 cells. In this study, the effect of hydrocortisone on the formation of basophilic granules in KU812 cells was examined. When the cells were cultured with 1μM hydrocortisone for 3 weeks, granulated cells that were stained metachromatically with Wright-Giemsa dye were observed. In addition, the cellular histamine content of hydrocortisone-treated KU812 cells markedly increased after 3 weeks. These results suggest that hydrocortisone induces basophilic differentiation of KU812 cells.

INTRODUCTION

Basophils and mast cells play a central role in allergic reactions and inflammation. Both types of cells express FcεRI, high affinity IgE receptor, on the cell surface and have cytoplasmic granules containing many potent inflammatory mediators such as histamine proteases, chemotactic factors, metabolites of arachidonic acid and cytokines. When an allergen associates with FcεRI-bound IgE, it crosslinks FcεRI receptor molecules. The aggregation of FcεRI results in degranulation and release of many potent inflammatory mediators. These secreted mediator molecules attract and activate inflammatory.

Human leukemia cell line KU812 is characterized by a constitutive potential to differentiate into basophilic cells . Because KU812 cells can express FcεRI on the cell surface and produces histamine, this cell line is useful to study allergic reactions and basophilic differentiation (3,7). Previously, we reported that hydrocortisone can enhance the expression of FcεRI on KU812 cells. The effect of hydrocortisone as a potent agent which induce basophilic differentiation of KU812 cells was expected. In the present study, the influence of hydrocortisone on a formation of basophilic granules in KU812 was examined.

MATERIALS AND METHODS

Cells

KU812 cells were obtained from Japan Cancer Research Resources Bank (JCRB). The cells have been maintained in RPMI-1640 (NISSUI) supplemented with 10% fetal bovine serum (BIO WHITTAKER), 100 U/ml penicillin G (Meiji), 100 mg/ml streptomycin (Meiji) and 10 mM HEPES (WAKO) buffer. KU812 cells were kept in a humidified atmosphere of 5% CO_2 at 37℃.

Stimulation of KU812 cells with hydrocortisone

Hydrocortisone was dissolved in ethanol at 200 μM and these solution were diluted 200-fold into medium so that the final ethanol concentration was 0.5%. KU812 cells were cultured in the presence or absence of 1μM hydrocortisone or ethanol alone (0.5%). Cells were inoculated at 2.5×10^5 cells/ml and maintained for 21 days with medium changes every 4 days. On 21 days, cultures were examined for the appearance of granulated cells (detected by Wright-Giemsa staining; basophilic granules were stained dark reddish-purple), and total histamine values.

Histamine measurement

After cultivation with 1μM hydrocortisone for 21 days, total histamine content was assessed by sonicated cell lysates. Cells were washed twice and sonicated in phosphate buffer sarine. Histamine content in the cell lysates was measured with fluorometric analysis. HPLC equipped with C_{18} column was used for separation of histamine in the cell lysates. Histamine was coupled with o-phtalaldehyde at 40℃ and gives rise to fluorometric complex which can be measured on fluorometer.

RESULTS

Figure-1 shows the Wright-Giemsa staining of KU812 cells treated with or without 1μM hydrocortisone. At the start of each experiment, none of the cells were stained metachromatically with Wright-Giemsa dye. When the KU812 cells were cultured for up to 14 days, granulated cells were not detected. After 21 days cultivation with 1μM hydrocortisone, KU812 cells gave rise to granulated cells that expressed basophilic granules stained dark reddish-porpure by Wright-Giemsa staining (less than 0.1%).

Table-1 shows the cellular content of histamine in KU812 cells after 21 days culture. Under hydrocortisone-free culture condition, a constant level of about 20 to 30 ng histamine / 10^6 cells was maintained in the KU812 cells. The cellular content of histamine in KU812 cells substantially increased (by 10- to 20-fold) after exposure to 1μM hydrocortisone whereas ethanol supplemented culture condition shows any effect on synthesis of histamine in KU812 cells. These data suggests that hydrocortisone is a potent agent to induce a formation of basophilic granulation in KU812 cells.

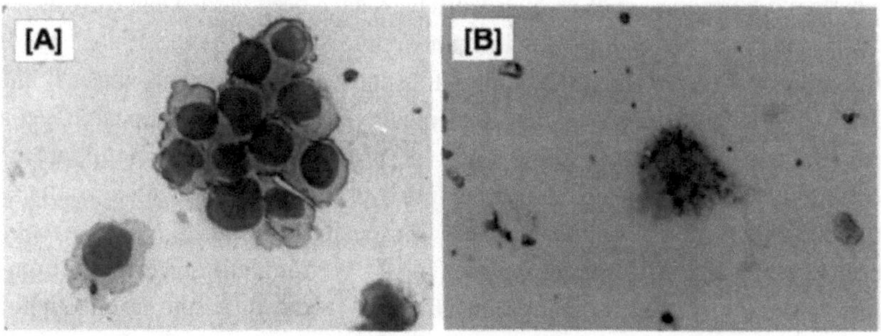

Figure 1. Wright-Giemsa staining of KU812 cells (X400)

[A] Cells cultured in control medium.
[B] Cells cultured with 1 μM hydrocortisone for 21 days.
<Cytoplasmic granules were stained dark reddish-purpure.>

Table 1. Histamine content in cell lysates of KU812 cells after 21 days cultivation with hydrocortisone

	histamine content [ng/10^6 cells]
control	**24.7±4.1**
hydrocortisone [1 μM]	**376.0±64.7**

The values were measured with fluorometric analysis using column HPLC. The method is based on the coupling of histamine with o-phthalaldehyde. Results represent the mean±SD of triplicate determinations.

DISCUSSION

Mature basophil expresses FcεRI and possesses cytoplasmic granules which contain inflammatory mediators such as histamine. Previously, we found that hydrocortisone enhances the expression of FcεRI on KU812 cells. In this study, we describe hydrocortisone induces a) the appearance of granulated cells and b) the increase in intracellular histamine content in KU812 cells. These results suggest that hydrocortisone induces basophilic differentiation of KU812 cells. It has been reported that recombinant human interleukin-3 (IL-3) had similar effects on KU812 cells (3). IL-3 is known to promote the differentiation and growth of basophils from cultures of human bone marrow cells (8-11). Therefore, it is possible that KU812 cells can differentiate into basophilic cells through more than two signaling pathways.

Glucocorticoid hormones, such as hydrocortisone, have been used extensively as potent anti-inflammatory agents and general immunosuppressants. Hydrocortisone binds to the glucocorticoid receptor and this complex translocates to the nucleus where it bind to directly to specific target DNA regulatory elements (glucocorticoid-responsive element) and activates transcription. In some types of blood cells such as lymphocyte and monocyte, glucocorticoids induce the transcription of the IκB gene, which results in an increased rate of IκB protein synthesis. IκB is a specific inhibitor of NFκB, a regulator of immune system and inflammation genes (14). By stimulating IκB production, the glucocorticoids may prevent inflammation (15,16). Meanwhile, the results shown in this study suggest that the genes responsible for expression of FcεRI as well as granulation and production of histamine may be also controlled under grucocorticoid-responsive element in the human basophil precursor cell line. Besides the immunosuppressive and anti-inflammatory effects, it is possible that hydrocortisone can play an important role in allergic reactions through differentiation of basophil precursor cells and modulation of basophil function.

REFERENCES

1 Kishi, K. 1985. Leuk. Res. 9 : 381.

2 T. Fukuda, K. Kishi, Y. Ohnishi, and A. Shibata. 1987 Blood 70 : 612.

3 P. Valent, J. Besemer, K. Kishi, R. Kaltenbrunner, B. Kuhn, D. Maurer, K. Lechner, and P. Bettelheim. 1990 J. Immunol. 145 : 1885.

4 J-F. Grauchat, S. Henchoz, G. Mazzei, J-P. Aubry, T. Brunner, H. Blasey, P. Life, D. Talabot, L. Flores-Romo, J. Thompson, K. Kishi, J. Butterfield, C. Dahinden, and J-Y. Bonnefoy. 1993 Nature 365 : 340.

5 Almlof, I., K. Nilsson, V. Johansson, E. Akerblom, H. Slotte, S. Ahlstedt, and P. Matsson. 1988 Scand. J. Immunol. 28 : 293.

6 Matsson, P., I. Almlof, K.Nilsson, and S. Ahlstedt. 1989 Int. Arch. Allegy Appl.

Immunol. 88 : 122.

7 N. Kawata, M. Tanaka, T. Suwaki, Y. Tanimoto, R. Soda, K. Takahashi, I. Kimura. 1996 Int. Arch. Allergy Immunol. 109 : 27.

8 P. Valent, G. Schmidt, J. Besemer, P. Mayer, G. Zenken, E. Liehl, W. Hinterberger, K. Lechner, D. Maurer, and P. Bettelheim. 1989 Blood 73 : 1763.

9 Kirshenbaum, A. S., J. P. Goff, S. A. Dreskin, A-M. Irani, L. B. Schwartz, and D. D. Metcalfe. 1989. J. Immunol. 142 : 2424.

10 Kirshenbaum, A. S., J. P. Goff, and D. D. Metcalfe. 1989. Immunology 68 : 120.

11 Kirshenbaum, A. S., S. D. Dreskin, and D. D. Metcalfe. 1989. J. Immunol. Methods 123 : 55.

12 Jonat, C., Rahmsdorf, H., Park, K., Cato, A., Gebel, S., Ponta, H., and Herrlich, P. 1990 Cell 62 : 1189.

13 Yang-Yen, H., Chambard, J., Sun, Y., Smeal, T., Schmidt, T., Drouin, J., and Karin, M. 1990 Cell 62 : 1205.

14 Baeuerle, P. A. and Baltimore, D. 1988. Science 242 : 540.

15 R. I. Scheinman, P. C. Cogswell, A. K. Lofquist, and A. S. Baldwin Jr. 1995 Science 270 : 283.

16 N. Auphan, J. A. DiDonato, C. Rosette, A. Helmberg, and M. Karin. 1995 Science 270 : 286.

CYTOTOXICITY TESTING OF TOMATINE

DAIO KOZUKA[1)], SHIRO YAMASHOJI[2)], YUKIKO CHIBA[3)] and KENJI ISSHIKI[4)]
[1) 4)]National Food Research Institute, Tsukuba, Ibaraki 305, Japan
[2)]Kobe Gakuin Women's Junior College, Kobe, Hyogo 653, Japan
[3)]The Japanese Canned Food Inspection Association, Yokohama, Kanagawa 231, Japan

Abstract

Glycoalkaloids , e.g. solanine and tomatine are found in potato, tomato and other plants. They are toxic to animals. It is difficult to analyze them in food. We developed a bioassay of tomatine. A combination method of HepG2 cell with a chemiluminescent was more easy to operate and rapid than others. Recently, it has been reported that tomato contains dehydrotomatine, which has C-5-C-6 double bond in the aglycon. We purified crude tomatine and obtained pure tomatine and dehydrotomatine. We measured the cytotoxicity of these compounds with chemiluminescent assay. As the animal cells for bioassay, HepG2, NIH3T3 and U937 cells were used. Cell viability in the presence of pure tomatine was just a little lower than that in the presence of pure dehydrotomatine.

1. Introduction

The analysis of glycoalkaloids in foods is a difficult problem to solve. Tomatine is a steroidal glycoalkaloid which has activity toxic to humans and other animals. The analyses were difficult because of its complex chemical structure and high molecular weight. As tomatine does not have any chromophoric groups, it has been determined photometrically after reaction with chromogenic reagent[1]. The separation and quantification of tomatine in tomato by HPLC has been investigated after hydrolyzing or making tomatine derivatives with chemical reagents[2]. In the case of GC or GC/MS, tomatine was decomposed by acid, and the products ware analyzed[3]. However, these methods for measuring tomatine was complicated, time-consuming and unsatisfactory as they were interfered with by other tomato components. We paid attention to the cytotoxicity of tomatine. For the detection of the cytotoxicity, we examined several methods; *e.g.* Alamr Blue reduction assay[4], MTT reduction assay[5], WST -1 reduction assay[6] and chemiluminescence assay[7]. As a result, we found chemiluminescence method using HepG2 cell was suitable for detecting the cytotoxicity of tomatine and was more rapid and easier than the chemical detection method of tomatine.

In recent studies, it has been reported that two peaks of approximately equal area were observed when analyzing tomatine standard with HPLC method. From LC/MS analysis, it was shown that the first peak was dehydrotomatine, which had C-5-C-6 double bond in the aglycon[8]. A comparative study of tomatine and dehydrotomatine toxicity hasn't been done. In this study, with chemiluminescent assay, we examined the comparative study of tomatine and dehydrotomatine toxicity on animal cells.

2. Materials and Methods

2.1. Sample Materials

For this study, we needed pure tomatine and dehydrotomatine . We purified commercial tomatine (Sigma Chemical Co.) to obtain pure tomatine and dehydrotomatine. One thousand μg/ml standard solution of each pure tomatine and dehydrotomatine was obtained, through each standard dissolved in ethanol. Test solutions (0.02 to 20 μg/ml) obtained by diluting with MEM medium (phenol red free) were used for cytotoxicity assay.

2.2. Cell Lines and Media

HepG2 and NIH3T3 cells were cultured in DME medium containing 10 % FCS and 0.4 % L - glutamine. U937 cells were cultured in RPMI1640 medium containing 10 % FCS. These cells were incubated in a CO_2 incubator at 37 °C in humidified atmosphere containing 5 % CO_2 in air.

2.3. Preparation of Cell Suspensions

Prepare cell suspensions from confluent cultures in 100 mm culture dishes by washing cells on dish with PBS, incubating under normal culture conditions with PBS containing 0.05 % trypsin for 5 min until cells begin to detach from substratum, drawing off trypsin solution, and flushing cells from the substratum with fresh medium. Count number of cells in suspension on hemocytometer and dilute suspension of cells to 5 x 10^4 cells /ml.

2.4. Detection of Cytotoxicity of Tomatine

Chemiluminescent cytotoxicity assay was carried out as shown in (Fig. 2, [7]).

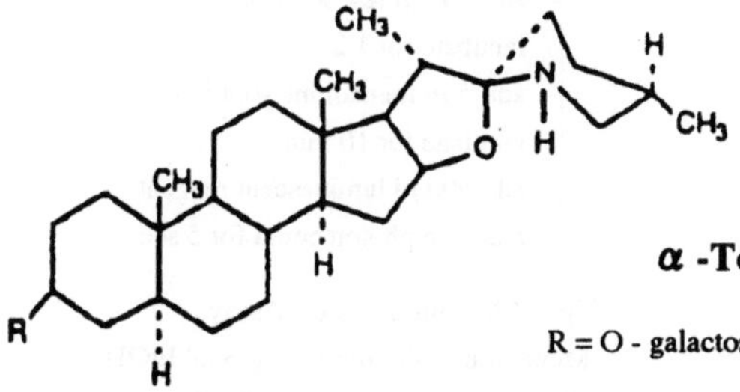

α -Tomatine

R = O - galactose - glucose - xylose
|
glucose

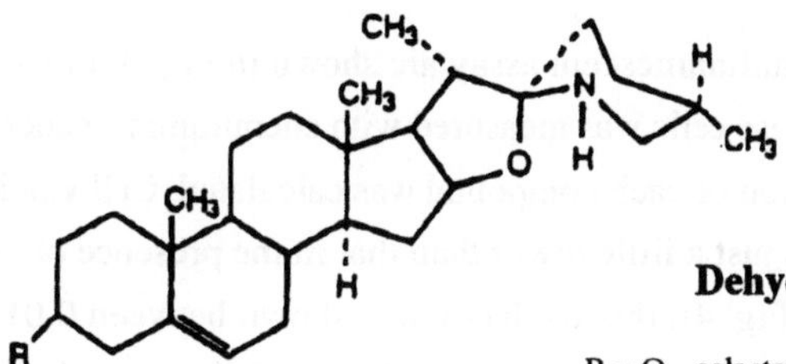

Dehydrotomatine

R = O - galactose - glucose - xylose
|
glucose

Fig. 1 Structures of tomato glycoalkaloids.

Cells

- preincubate
- treat with trypsin
- centrifuge
- suspend in MEM (phenol red free)
  cell density: 4 x 10^5 cells/ml
- put 250 μl into each well
- add 250 μl test solution
- incubate for 1 h
- add 5 μl menadione solution
- incubate for 10 min
- add 500 μl luminescent reagent
- measure photon count for 5 sec

Fig.2. Chemiluminescent Assay.
Menadione solution: 17 mg/5 ml EtOH.
Luminescent reagent: TDPO 50 mg,
pyrene 5 mg/50 ml acetonitrile.

3. Results and Discussion

The results of chemiluminescent assay are shown in Fig. 3, Fig. 4 and Fig. 5. Luminescent intensity of these cells was measured with chemiluminescence. Cell viability (%) on various concentration of each compound was calculated. Cell viability in the presence of pure tomatine was just a little lower than that in the presence of pure dehydrotomatine. On NIH3T3 cell (Fig. 4), this tendency was shown between 0.01 μg/ml and 10 μg/ml, but, on HepG2 cell (Fig. 3) and U937 cell (Fig. 5), between 0.5 μg/ml and 2.5 μg/ml. U937 cell was less sensitive to pure dehydrotomatine than pure tomatine between 0.5 μg/ml and 2.5 μg/ml.

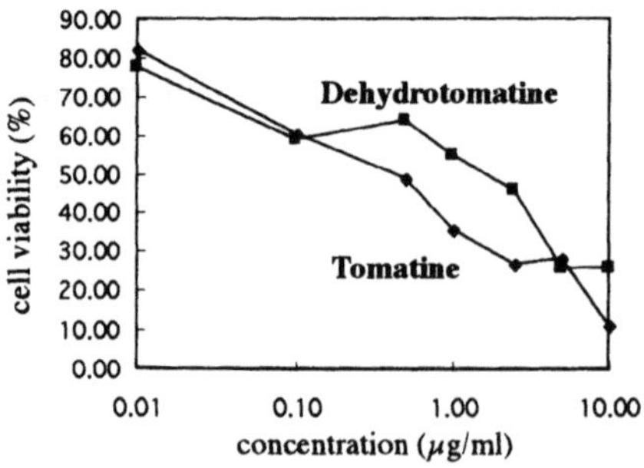

Fig. 3 Cytotoxic effects of tomato glycoalkaloids in HepG2 cell

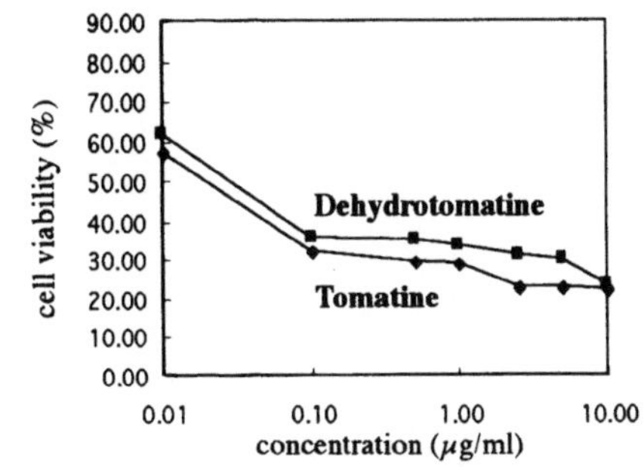

Fig. 4 Cytotoxic effects of tomato glycoalkaloids in NIH3T3 cell

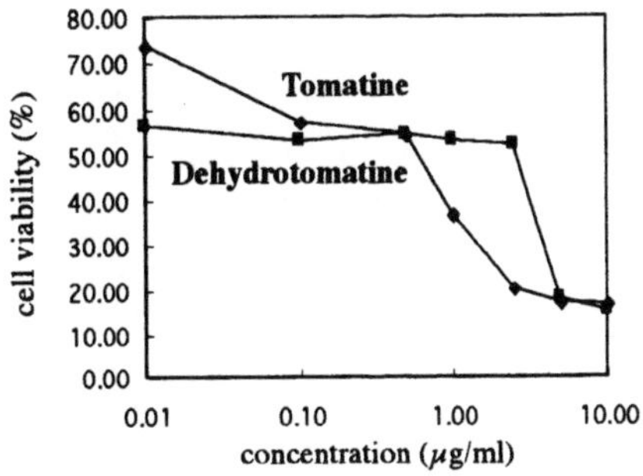

Fig. 5 Cytotoxic effects of tomato glycoalkaloids in U937 cell

5.References

1. Roddick, J. G. and Butcher, D. N. (1972) Isolation of tomatine from cultured excised roots and callus tissues of tomato, *Phytochemistry*, **11**, 2019 - 2424.
2. Takagi, K., Toyoda, M., Shimizu, M., Satoh, T. and Saito, Y. (1994) Determination of tomatine in foods by liquid chromatography after derivatization, *J. Chromatogr., A.*, **659**, 127 - 131.
3. Lawson, D. R., Erb, W. A. and Miller, A. R. (1992) Analysis of Solanum Alkaloids using internal standardization and capillary gas chromatography, *J. Agric. Food Chem.*, **40**, 2186 - 2191.
4. Page, B., Page, M., and Noel, C. (1993) A new fluorometric assay for cytotoxic measuremnets in vito, *International Journal of Oncology*, **3**, 474 - 476.
5. Ally, M. C., Scudiero, D. A., Monks, A., Hursey, L., Czerwinski, M. J., Fine, D. L., Abbot, B. J., Mayo, J. G., Sheomaker, R. H, and Boyd, M. R. (1988) Feasbility of drug screening with panels of human tumor cell lines using microculture tetrazolium assay, *Cancer Research*, **48**, 589 - 601.
6. Ishiyama, M., Shiga, M., Sasamoto, K., Mizoguchi, M. and Pin-gang, H. (1991) A new sulfonated tetrazolium salt that produces a high water-soluble formazan dye, *Chem. Pharm. Bull.*, **41**, 1118 - 1122.
7. Yamashoji, S., Nishimoto, F., Usuda, M., Kubota, H. and Isshiki, K. (1992) Application of the chemiluminescent assay to cytotoxicity test: Detection of minadione-catalyzes H_2O_2 production by viable cells, *Anal.Biochem.*, **207**, 255 - 260.
8. Rodney, J. B., Lewis, B. P., Lance, R. P. and Steve, V (1994) High - Performance Liquid Chromatographic Determination of the Tomato Glycoalkaloid, Tomatine, in Green and Red Tomatoes, *J. Agric. Food Chem.*, **42**, 2824 - 2829

MANNOSYLERYTHRITOL LIPID INDUCED DIFFERENTIATION OF HUMAN MYELOGENOUS LEUKEMIA CELL LINE K-562

HIROKO ISODA, HIROSHI SHINMOTO,* DAI KITAMOTO,** MASATOSHI MATSUMURA AND TADAATSU NAKAHARA
Institute of Applied Biochemistry, University of Tsukuba, 1-1-1 Tennodai, Tsukuba, Ibaraki, 305, Japan
**National Food Research Institute, MAFF, 2-1-2 Kannondai, Tsukuba, Ibaraki, 305, Japan*
***National Chemical Laboratory for Industry, AIST, MITI, 1-1 Higashi, Tsukuba, Ibaraki, 305, Japan*

1. Introduction

Mannosylerythritol lipid, which is a microbial extracellular glycolipid, was produced by *Candida antarctica* T-34 from soybean oil and was produced in concentrations of about 40g per liter of culture broth (Kitamoto et al., 1993). MEL were identified as 4-*O*-(di-*O*-acetyl-di-*O*-alkanoyl-β-D-mannopyranosyl)-erythritol (Fig.1).

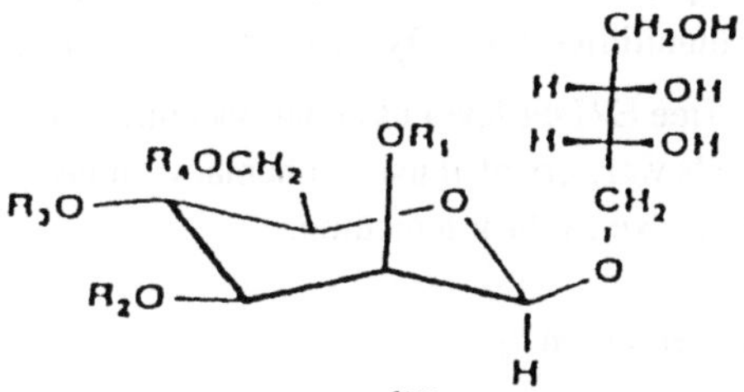

(Kitamoto et al., 1990)

On the other hand, ganglioside and glycosphingolipids (GSL), the ubiquitous membrane components are reported to modulate the cell growth, adhesion, and transmembrane signaling, because of dramatic changes in GSLs composition and metabolism observed during oncogenesis, differentiation, and oncogenic transformation (Hakomori et al., 1992). In addition, specific types of ganglioside are reported to inhibit receptor-associated tyrosine kinase activity, and therefore, gangliosides in the lipid bilayer surrounding the receptors are considered to have an important role in modulation of cell proliferation through direct or indirect interaction with receptor-associated tyrosine kinase (Nojiri et al., 1991). Although MEL contains no components of cell membrane, such structure is similar to ganglioside and both of them are amphiphilic compounds. In order to investigatewhether MEL have an ability to induce the differentiation of cells, human myelogenous leukemia cell line K562 was used in this study. The K562 cell line was established by Lozzio and Lozzio (1975) from a patient with chronic myeloid leukemia in blast cell transformation, and this line is a useful model system which are multipotent with respect to the erythrocytic, granulocytic, monocytic and megakaryocytic differentiation.

In this study, we show that MEL induced granulocytic differentiation of K562 cells and increased the differentiation associated characteristics such as NBT reducing ability, Fc receptor expression and phagocytic activities of the K562 cells, and that the change of glycolipids on plasma membranes and protein phosphorylation regulation in the cells.

2. Materials and Methods

2.1. Preparation of MEL

MEL was prepared by the method of Kitamoto et al.(1990).

2.2. Cells and Cell growth

The human myelogenous leukemia cell line K562 was obtained from Riken Cell Bank (Tsukuba, Ibaraki, Japan). The cells were maintained in RPMI -10% FBS containing medium at 37C in humidified 5% CO_2/95% air. For MEL treatment, the cells were cultured in a serum-free ERDF(Kyokuto Pharmaceutical Kogyo Co. Tokyo, Japan) -ITES medium. The cells were counted using a hemacytometer, and the viability was estimated by the trypan blue dye exclusion method.

2.3. Determination of cell differentiation

For cytochemical assessment of cell differentiation, specific and nonspecific leukocyte esterase activities were determined cytochemically by the esterase double-staining method for α-naphtyl acetate esterase (NAE) and naphthol AS-D-chloroacetate esterase (NCAE) using reagents for determination of esterase (SIGMA CHEMICAL Co. Ltd. St. Louis, USA). Specific and nonspecific esterase activitieswere examined on days 3. The NBT-reducing ability was determined by the modified method of Takeda et al. (1988). The expression of Fc receptors on the cell surface was assayed by EA rosette formation. The phagocytic activity of cells was measured by counting the percentage of the cells that phagocytosed yeast particles.

2.4. Extraction and Determination of Glycosphingolipid (GSLs) Composition

GSLs from the cell were extracted and determined by the modified method of Tsuruoka et al. (1993).

2.5. Western Analysis of Tyrosine Phosphorylation

Tyrosine phosphorylation was analysed by western blotting by the method of Rollet et al. (1994).

3. Results

3.1. Effect of MEL on Insulin-Dependent Cell Growth in Human Myelogenous Leukemia Cell Line K562

The effects of MEL on the proliferation of K562 were examined. After 2-days of culture, MEL inhibited the proliferation of K562 cells (Fig.2).

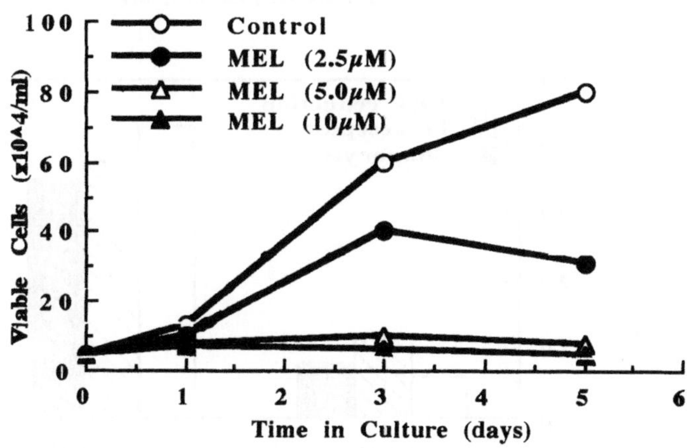

Fig.2. Effect of MEL on insulin-dependent cell growth.

On the second day of cultivation, morphological changes of K562 such as granules and adherence were observed in the culture treated with MEL.

3.2. Determination of Cell Differentiation

K562 cells treated with MEL highly expressed NCAE activities and which correspond to indicate specificity for the granulocytic lineage (Fig. 3). As shown in Fig. 4, MEL induced the occurrence of NBT reducing ability, expression of Fc receptors, and phagocytic activities in HL60 cells significantly .

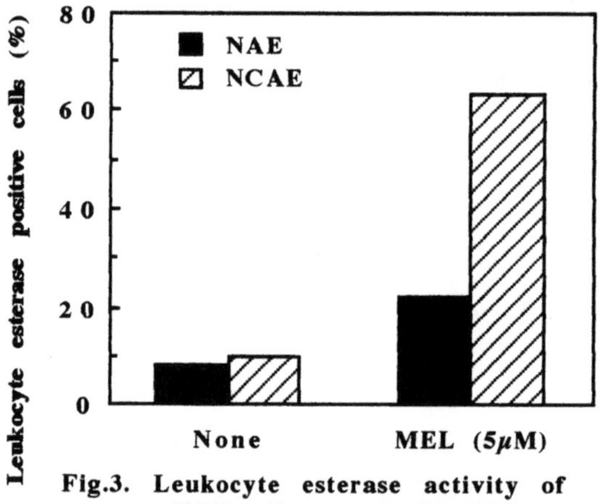

Fig.3. Leukocyte esterase activity of K562 cells treated with MEL.

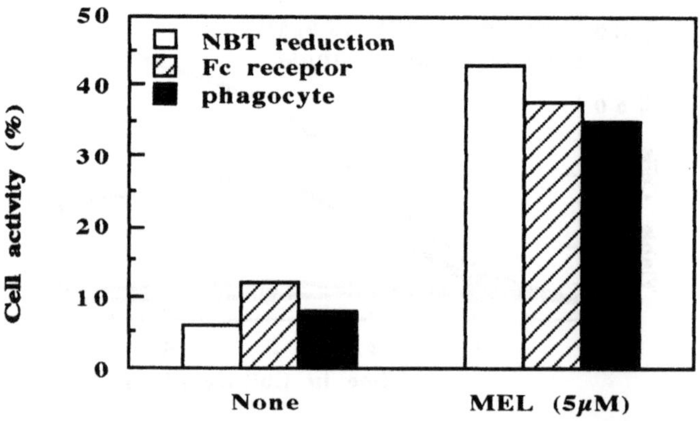

Fig.4. Differentiation-associated characteristic properties of K562 cells treated with MEL.

3.3. GSLs Composition of MEL Treated K562 cells

The major GSLs of K562 cells were identified as GM3 ganglioside and LacCer by its TLC mobility. LacCer was increased in MEL treated cells compare to non-treated K562 cells. The relative GM3/LacCer ratio of non-treated, MEL treated cells were 1.75, 0.57 on day 5 respectively.

3.4. Tyrosine phosphorylation of MEL treated K562 cells

The insulin-dependent tyrosine phosphorylation of 95-kDa and 135kDa proteins were observed in both non-treated and LacCer treated K562 cells. On the other hand, such tyrosine phosphorylations were clearly inhibited in MEL treated cells.

4. Discussion

In this study, we examined whether the microbial extracellular glycolipid, MEL has an ability to induce human leukemia cell line K562 to cell differentiation. MEL was found to inhibit the insulin-dependent growth of cells, and induced its morphological change. The result of leukocyte esterase activity showed that K562 cells treated with MEL expressed naphthol AS-D-chloroacetate esterase activity and this esterase activity indicated the specificity for the granulocytic lineage. MEL markedly increased the differentiation-associated characteristics in granulocytes respectively. The composition of cell surface GSLs changed such that the ratio of LacCer/GM3 increased in K562 cells treated with MEL compared to non-treated cells. Furthermore, the insulin-dependent phosphorylation of tyrosine in K562 cells were clearly inhibited by MEL, but not by LacCer. These results suggest that the differentiation-inducing activity of MEL is not due to a simple detergent-like effect but due to a specific action on the plasma membrane, and that MEL directly inhibited the protein kinase activities in K562 cells, then inhibited insulin-dependent cell growth and induced differentiation stage.

5. References

D. Kitamoto et al., *Biosci. Biotech. Biochem.*, *54, 37-40* (1990).
S. Hakomori et al., *Advances in Lipid Research*, 1992, *Vol. 25, pp. 147-162.*
H. Nojiri et al., *J. of Biological Chemistry, 266, 4531-4537* (1991).
T. Tsuruoka et al., *J. Biol. Chem. 268, 2211-2216* (1993).
E. Rollet et al., *J. of Immunology 153, 353-363* (1994).

COFFEE'S INHIBITORY ACTION ON THE INVASION OF HEPATOMA CELLS CO-CULTURED WITH MESOTHELIAL CELLS

T. FURUSE, H. SHIOMI, Y. MIURA AND K. YAGASAKI
Department of Applied Biological Science, Tokyo Noko University
Fuchu, Tokyo 183, Japan.

ABSTRACT. The invasion of cancer cells is an important and characteristic step of cancerous metastasis. In the present study, the action of coffee extract and its components on the invasion of cancer cells was studied by means of co-culture system of AH109A, a rat ascites hepatoma cell line, with mesentery-derived mesothelial cells. Coffee extract was found to dose-dependently suppress the invasion of AH109A cells. Rats were given oral intubation of coffee extract, blood was taken at appropriate time intervals and serum was prepared. Coffee-loaded rat serum also suppressed the invasion, suggesting that effective component(s) in coffee could be absorbed from the guts. Caffeic acid, a known component in coffee, was capable of suppressing the invasion of AH109A cells. Caffeic acid-loaded rat serum also showed a suppressive effect on the invasion, indicating that caffeic acid is one of effective components in coffee and absorbable across the guts.

1. Introduction

Cancer cells have two biological properties, that is, endless proliferation and metastasis. The invasion of cancer cells is an important and characteristic step of metastasis [1]. The inhibition of these biological events by some means must prolong the life span of hosts with cancers. We have recently reported that green tea extract suppresses the invasion of hepatoma cells in culture [2]. In the present study, the action of another popular beverage, coffee, on the invasion of hepatoma cells was studied in a cell culture system.

2. Materials and Methods

Cell culture and invasion assay. As model cancer cells, an ascites hepatoma cell line of

AH109A was used. The hepatoma cells were provided by SRL, Tokyo, Japan. The coffee's effect on the invasion of AH109A cells was measured by the co-culture system of Akedo *et al.* [3] with slight modifications as described previously [2]. Briefly, primary cultured mesothelial cells (M cells) from messentery of rats were seeded at a density of 1.5 x 10^5 cells/60 mmϕ culture dish with 2 mm grids and cultured in 3 ml of DM-160 containing 10% calf serum (10% CS/DM-160) for 7-10 days to attain a confluent state. AH109A cells (2.4 x 10^5) were then seeded on the monolayers of M cells and cultured without or with test substances for 48 hr. Invaded AH109A cells and colonies underneath M cell layers were counted with a phase-contrast microscope. Usually at least 10 areas were counted and the invasive activity of AH109A was indicated by the number of invaded cells and colonies/cm^2.

Preparation of coffee extract and coffee extract-loaded rat serum. Boiled water (150 ml) was added to 10 g of coffee powder and the filtrate was concentated to 15 ml by evaporation at reduced pressure. After being sterilized by membrane filtration, the coffee extract was added to experimental media at concentrations of 0-1%. To obtain coffee-loaded serum, rats fasted overnight were given oral intubation of concentrated coffee extract (1 ml/100 g body weight), blood was withdrawn 0, 0.5, 1, 2 and 3 hr after the intubation, serum was prepared, and the effect of coffee extract-loaded rat serum on the invasion was tested *in vitro* after sterilization.

Statistical analysis. Data were expressed as means ± standard errors. Statistical analysis was done using a one way analysis of variance, followed by Tukey's Q test.

3. Results and Discussion

The effect of coffee extract on the invasion of AH109A cells is illustrated in Figure 1. Coffee extract dose-dependently suppressed the invasion of hepatoma cells underneath M cell layers. It is important to know whether or not effective component(s) would be absorbed from the gastrointestinal tract. Figure 2 shows the effect of coffee-loaded rat serum on the invasion of AH109A cells. Coffee-loaded serum commenced to suppress the invasion 30 min after oral intubation of coffee extract and significant suppression lasted for up to 3 hr after the intubation, suggesting that effective component(s) in coffee could be absorbed across the guts. To identify the effective component(s), the effect of several well-known components in coffee on the invasion was investigated by adding to culture media of the invasion assay system. So far, caffeic acid has been found to suppress the invasion of AH109A underneath M cell layers and caffeic acid-loaded rat serum has also been found to suppress the AH109A invasion (data not shown), indicating that caffeic acid is a chemical entity in coffee suppressing the hepatoma invasion and absorbable across the gastrointestinal tract.

Caffeic acid has been reported to prevent experimental tumorigenesis [4]. In the

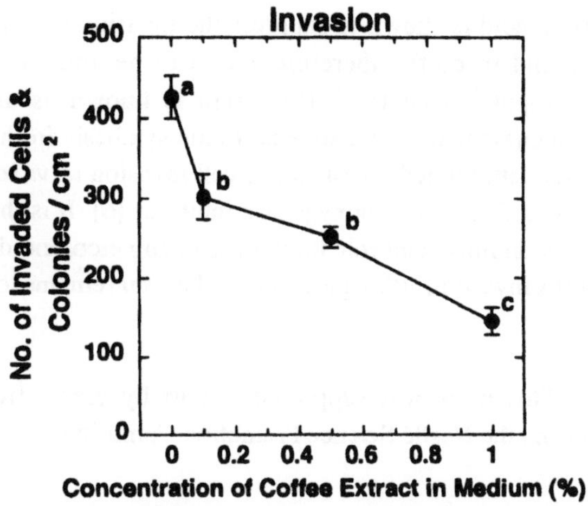

Fig. 1 Effect of Coffee Extract on Invasion of AH109A.

Each point and bar represents the mean and standard error for 10 areas. Values not sharing a common letter are significantly different at $p < 0.05$ by Tukey's Q test.

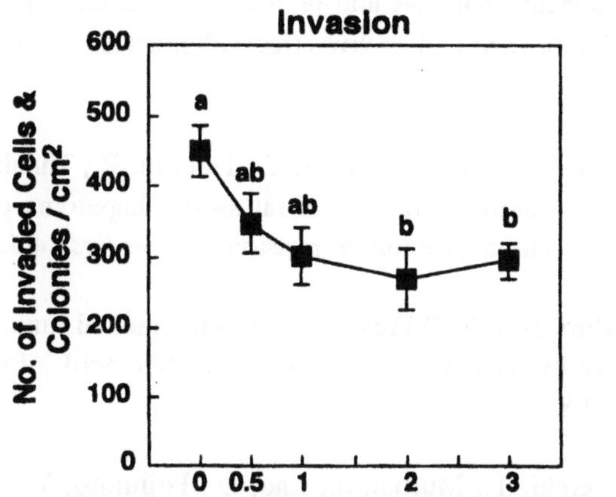

Fig. 2 Effect of Coffee Extract-loaded Rat Serum on Invasion of AH109A.

Each point and bar represents the mean and standard error for 10 areas. Values not sharing a common letter are significantly different at $p < 0.05$ by Tukey's Q test.

present study, caffeic acid is shown to suppress the invasion of hepatoma cells. This well-known compound in coffee therefore seems to be anti-cancerous from both preventive and theraputic aspects. Caffeic acid is known as an inhibitor of 5-lipoxygenase [5], an enzyme involved in leukotriene synthesis from arachidonic acid. Recently, linoleic acid-stimulated breast cancer cell invasion *in vitro* is reported to be blocked by esculetin, a 5- and 12-lipoxygenase inhibitor [6]. It is therefore likely that caffeic acid may play an important role through affecting eicosanoid production in the inhibition of AH109A invasion, although exact mechanism remains to be elucidated.

Acknowledgement This work was supported in part by grants from the All Japan Coffee Association and the Nestlé Science Promotion Committee.

4. References

1. Liotta, L., Rao, C.N., and Barsky, S.H. (1983) Tumor invasion and the extracellular matrix, *Lab. Invest.*, **49**, 636-649.

2. Miura, Y., Shiomi, H., Sakai, F., and Yagasaki, K. (1997) Assay systems for screening food components that have anti-proliferative and anti-invasive activity to rat ascites hepatoma cells: *In vitro* and *ex vivo* effects of green tea extract, *Cytotechnology*, in press.

3. Akedo, H., Shinkai, K., Mukai, M., Mori, Y., Tateishi, R., Tanaka, K., Yamamoto, R., and Morishita, T. (1986) Interaction of rat ascites hepatoma cells with cultured mesothelial cell layers: a model for tumor invasion, *Cancer Res.*, **46**, 2418-2422.

4. Tanaka, T. and Mori, H. (1995) Prevention of experimental tumorigenesis in colon, liver, and oral cavity by dietary chlorogenic and caffeic acids, *Proc. 16th Int. Sci. Colloq. Coffee*, **1**, 79-87.

5. Koshihara, Y., Neichi, T., Murota, S., Lao, Q., Fujimoto, Y., and Takahashi, T. (1983) Selective inhibition of 5-lipoxygenase by natural compounds isolated from Chinese plants, *Artemisia rubripes* Nakai, FEBS Lett., **158**, 41-44.

6. Liu, X.-H., Connolly, J.M., and Rose, D.P. (1996) Eicosanoids as mediators of linoleic acid-stimulated invasion and type IV collagenase production by a metastatic human breast cancer cell line, *Clin. Exp. Metastasis*, **14**, 145-152.

INHIBITORY EFFECTS OF TEA CATECHINS ON INVASION OF HUMAN FIBROSARCOMA HT1080 CELLS TO THE MONOLAYER OF HUMAN UMBILICAL VEIN ENDOTHELIAL CELLS

Mari Maeda-YAMAMOTO, Kazuhiro OSADA, Yuichi YAMAGUCHI and Kenkou TSUJI

National Research Institute of Vegetables, Ornamental Plants and Tea, Ministry of Agriculture, Forestry and Fisheries, 2769 Kanaya, Shizuoka 428, JAPAN

Abstract. It was reported that tea (*Camellia sinensis*) had various physiological modulative activities, and the major effectors were catechins (Kada, 1985) . Catechins are a group of polyphenolic compounds. Green tea contains many kinds of catechins, such as (-)-epigallocatechin gallate (EGCg), (-)-epigallocatechin (EGC), (-)-epicatechin gallate (ECg) and (-)-epicatechin (EC). Interactions between metastatic tumor cells and vein endothelium are important in the metastatic process. Recently, the screening protocol for inhibitory effectors of cancer metastasis was developed using human cell lines. So we investigated the effects of tea catechins on the invasion of human fibrosarcoma HT1080 cells to the monolayer of human umbilical endothelial cells (HUVECs). As a result, ECg and EGCg suppressed the invasion of HT1080 cells to the monolayer of HUVECs, but EC and EGC had no effects. By comparing the structure of these catechins, it was supposed that gallate residue in ECg and EGCg had a relation to the inhibitory effects. Futhermore, gallate, methyl gallate and lauryl gallate did not suppress the cancer invasion. From these results, it is suggested that the inhibitory effector of tea catechin on cancer invasion requires the structure that gallate is combined with catechin.

1.Introduction

In the process of cancer metastasis, it is important that a cancer cell adheres to and invades the endothelial cells in a blood vessel (Nicolson, 1988)(Fig.1). Therefore, it will become an important key to inhibit adhesion or invasion to prevent cancer metastasis. The inhibitory effect of tea on adhesion (Isemura, 1993) or invasion (Sazuka, 1995) of mouse carcinoma cells to blood vessels endothelium has been reported. It was reported that in recent studies tea had various physiological modulative activities, such as a carcinogenesis inhibitory effect (Wang, 1992), free-radical scavenging activity, antioxidative activity (Okuda ,1983), anti-hypertensive action (Yokozawa, 1994), anti-hypercholesterolemic action (Chisaka, 1988), anti-dental caries action (Sakanaka,1989), anti-viral action and intestinal flora amelioration action. So, we screened the invasive inhibitory factor of tea ingredients using human cell lines, and found the effects of tea catechins.

2.Materials and Method

2.1 MATERIAL AND CHEMICALS

(-)-Epicatechin (EC) , (-)-epigallocatechin (EGC), (-)-epicatechin gallate (ECg) and epigallocatechin gallate (EGCg) were kindly provided by Mitsui Norin Co. Ltd. (Fig.2). For use, these catechins were dissolved in PBS (phosphate buffered saline) and filtered through a 0.22 μ m nitrocellulose filter for sterilization.

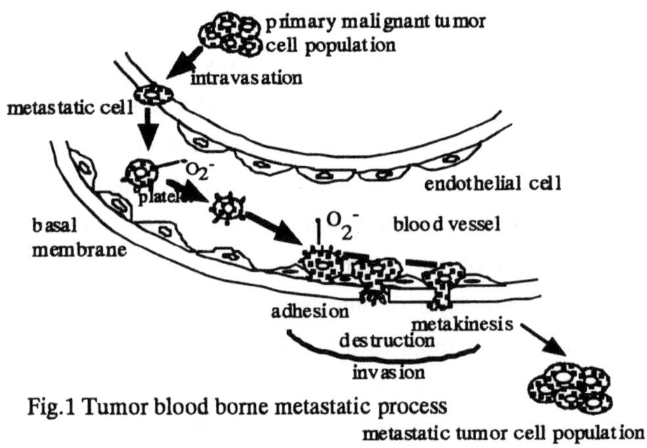

Fig.1 Tumor blood borne metastatic process

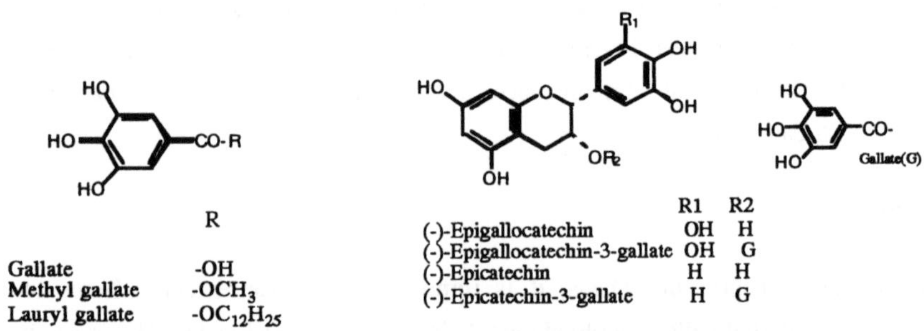

Fig.2 Chemical structure of tea catechins and polyphenols

2.2 CELLS AND CELL CULTURE

Human fibrosarcoma cell line HT1080 cells were routinely grown in ERDF medium supplemented with 5% fetal bovine serum (FBS). Human umbilical vein endothelial cell line HUVEC-C was grown in MCDB104 medium supplemented with 5% FBS and

HUVEC growth factor (HUVEC kit; MORINAGA institute of biological science) at 37 ℃ under humidified 5% CO_2/95% air. At the test for invasion of tumor cells to HUVEC, these cell lines were cultured with MCDB104 medium supplemented with 5% FBS and HUVEC-C growth factor. HT1080 and HUVEC-C were purchased from the Institute for Fermentation (Osaka,Japan).

2.3 MEASUREMENT OF INVASION OF TUMOR CELLS TO HUVEC (Fig.3)

Two hundred μL of HUVEC-C (1×10^6 cells/mL) was inoculated in each well of cell culture insert (8μm pore, Falcon 3182) pre-coated with 0.5% gelatin-PBS solution. Cells were pre-cultured for 24hrs, and then treated with LPS (1 μg/well) for 2hrs. For fluorescent labeling, HT1080 cells were treated with PKH2 (Zynaxis) previously, and added to the HUVEC-C monolayer after washing with PBS. HUVEC-C were co-cultured with tumor cells for 5hrs with catechin samples. After incubation, each lower phase of culture insert was treated with trypsin. Trypsinized invasive tumor cells were centrifuged and the pellet was suspended in 500μL of 1% Triton-X-100 solution. After sonication for 5min., fluorescence intensity of the cellular extract was measured with a fluorophotometer (cellular extract was excited with 490nm, and measured fluorescence intensity of 504nm). Cell viability was measured by WST-1 method (Dojin,Japan).

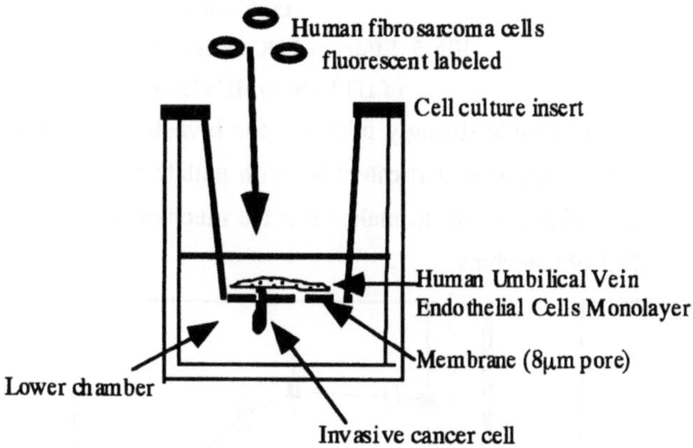

Fig.3 Cancer cell invasion assay

3.Results and Discussion

3.1 INHIBITORY EFFECTS OF TEA CATECHINS ON INVASION OF HT1080 CELLS TO HUVEC-C

To screen the preventive effectors on cancer metastasis, the inhibitory effects of tea catechins on invasion of cancer cells to HUVEC-C were examined. Figures 4 and 5 show the effects of catechins without gallate on HT1080 invasion to HUVEC-C. EC and EGC had no inhibitory effects.

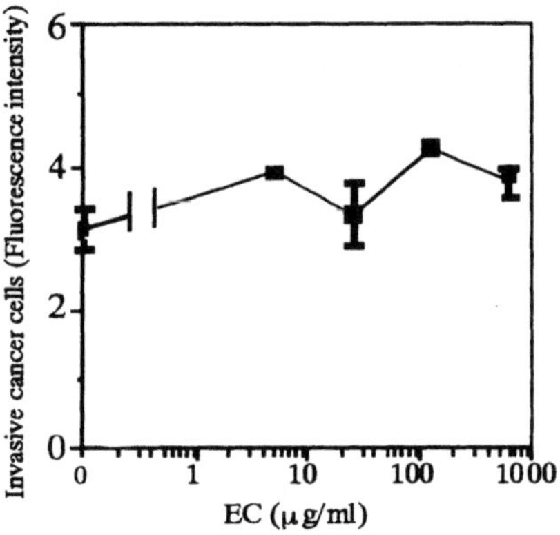

Fig.4 Effect of EC on invasion
of HT1080 to HUVEC-C

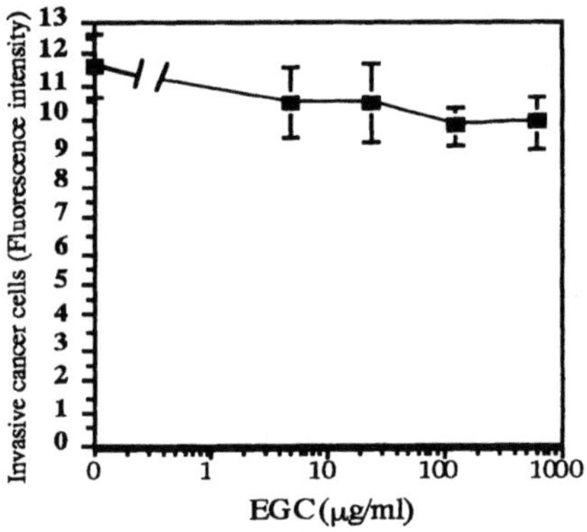

Fig.5 Effect of EGC on invasion
of HT1080 to HUVEC-C

As shown in Figures 6 and 7, ECg and EGCg strongly inhibited the invasion of human fibrosarcoma cells to HUVEC-C. We suggested that catechins with gallate are effective for prevention of cancer metastasis. Now, we try to make clear the mechanism of the cancer cell invasive inhibitory effect of catechins.

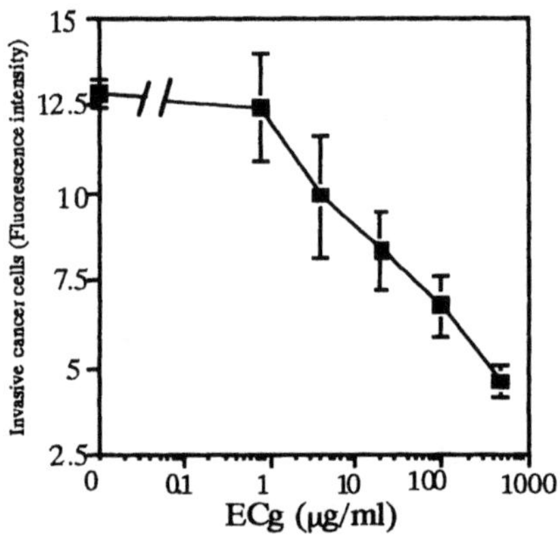

Fig.6 Inhibitory effect of ECg on invasion
of HT1080 to HUVEC-C

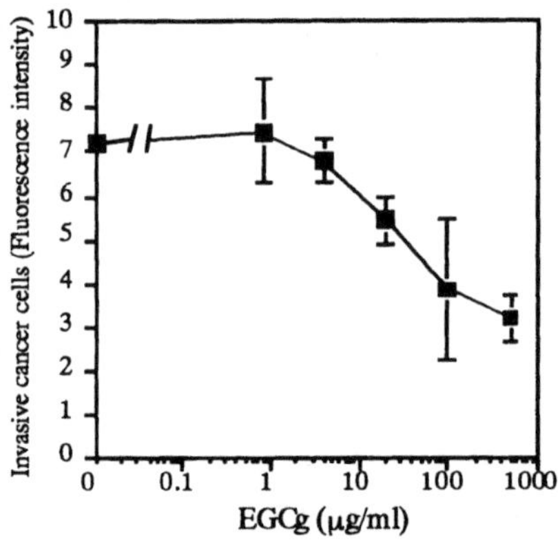

Fig.7 Inhibitory effect of EGCg on
invasion of HT1080 to HUVEC-C

Fig.8 shows the effect of ECg which has inhibitory activity, on the viability of HT1080 cells. ECg inhibited the growth of cells above 25μg/ml addition. From this result, the cancer invasive inhibitory effect of ECg was not caused by cytotoxicity.

3.2 EFFECTS OF GALLATES ON INVASION OF HT1080 CELLS TO HUVEC-C

We examined 3 kinds of gallates on invasion assay, but the gallates had no effect, as shown in Fig.10. From these results, it is suggested that the cancer invasive inhibitory effector of tea catechin requires the chemical structure that gallate is combined with catechin.

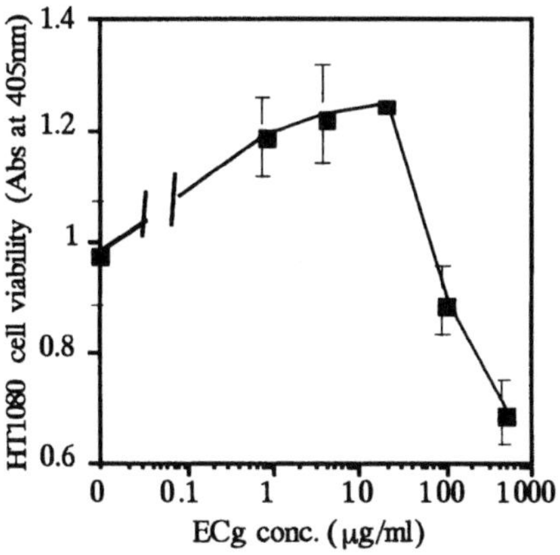

Fig.9 Effect of ECg on cell viability

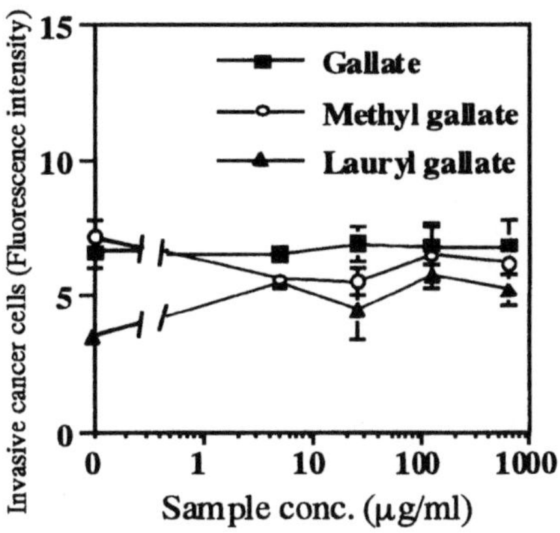

Fig.10 Effects of gallates on cancer invasion

4.References

Chisaka,T., Matsuda,H., Kubomura,Y., Mochizuki,M., Yamahara,J. Fujimura,H. (1988) The effects of crude drugs on experimental hypercholesteremia: mode of action of (-)-epigallocatechin gallate in tea leaves. *Chem. Pharm. Bull.*, **36**,227-233.

Isemura,M., Suzuki,Y., Satoh,K., Narumi,K., Motomiya,M. (1993) Effects of catechins on the mouse lung carcinoma cell adhesion to the endothelial cells, *Cell Biol.I nternational*. **17**, 559-564.

Kada,T., Kaneko,K., Matsuzaki,S., Hara,Y. (1985) Detection and chemical identification of natural bio-antimutagens: A case of the green tea factor,*Mutation Res.*, **150**,127-132.

Nicolson,G.L. (1988) Organ specificity of tumor metastasis: role of preferential adhesion, invasion and growth of malignant cells at specific sites, *Cancer and Metastasis Rev*. **7**, 143-188.

Okuda,T.,Kimura,Y., Yoshida,T., Hatano, T., Okuda,H., Arichi,S. (1983) Studies on the activities of tannins and related compounds from medicinal plants and drugs. I. Inhibitory effects of lipid peroxidation in mitochondria and microsomes of liver. *Chem. Pharm. Bull.*, **32**,1625-1631.

Sakanaka,S.,Sato,T., Kim,M., Taniguchi,M.,Yamamoto,T. (1989) Antibacterial subatances in japanese green tea extract against *Streptococcus mutans*, a cariogenic bacterium.*Agric. Biol. Chem.*, **53**,2307-2311.

Sazuka,M., Murakami,S., Isemura,M., Satoh,K., Nukiwa,T. (1995) Inhibitory effects of green tea infusion on in vitro invasion and in vivo metastasis of mouse lung carcinoma cells, *Cancer Lett.*, **98**,27-31.

Wang, Z-Y., Hong,J-Y., Reuhl, K.R., Conney,A.H., Yang,C.S. (1992) Inhibition of N-nitrosodiethylamine- and 4-(methylnitrosoamine)-1-(3-pyridil)-1-butanone-induced tumorigenesis in A/J mice by green tea. *Cancer Res.*,**52**,1943-1947.

Yokozawa,T., Okura,H., Sakanaka,S., Ishigaki,S., Kim,M. (1994) Depressor effect of tannin in green tea on rats with renal hypertension. *Biosci. Biotech. Biochem.*, **58**, 855-858.

NORZOANTHAMINE SUPPRESSES PTH-STIMULATED IL-6 PRODUCTION IN VITRO AND BONE RESORPTION IN VIVO

TOMOKO TSUJI*, KOHJI YAMAGUCHI, MIKA YADA, KAORU YAMADA AND DAISUKE UEMURA,
Sagami Chemical Research Center
4-4-1, Nishi-Ohnuma Sagamihara Kanagawa 229, Japan

1. INTRODUCTION

Interleukin 6 (IL-6) is known as a mediator of bone resorption caused by estrogen deficiency[1]. IL-6 is produced by osteoblasts and involved in bone metabolism through osteoclast formation. Ovariectomy causes an increase of IL-6 level in bone marrow and loss of bone weight in normal mice, however ovariectomy in IL-6 defficient mice was reported not to cause loss of bone weight [2,3]. Therefore, inhibition of IL-6 production or IL-6 activity is expected to suppress bone weight loss under the estrogen defficient condition. In vitro experimental system, IL-6 production is enhanced by PTH (parathyroid hormon) and suppressed by estrogen[4]. To find a novel osteoporotic drug, we searched for compounds with a suppressive effect on PTH-stimulated IL-6 production in MC3T3-E1, pre-osteoblastic cell line.
In the course of our screening, norzoanthamine[5] (Fig.1), alkaloid from a colonial zoanthid, was found to have a potent inhibitory activity for PTH-stimulated IL-6 production in MC3T3-E1. Norzoanthamine proved to suppress loss of bone weight, strength, ductility and elasticity in ovariectimised mice.

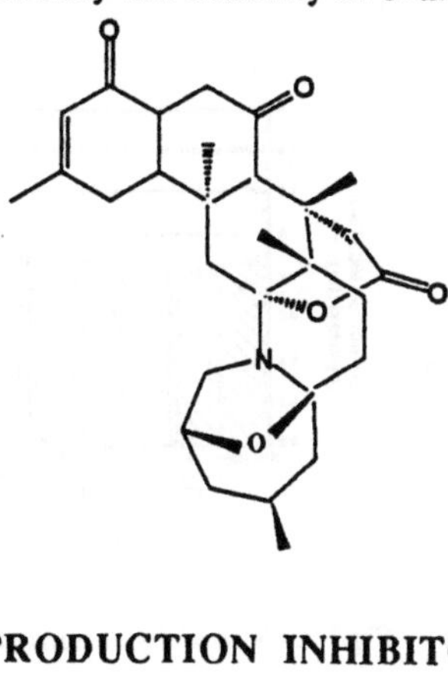

Fig. 1. Chemical structure of norzoanthamine.

2. IL-6 PRODUCTION INHIBITORY ACTIVITY IN VITRO

The assay method of IL-6 production inhibitory activity is shown in Fig.2 MC3T3E1, preosteoblastic cells produce IL-6 in response to PTH(20ng/ml) stimulation. However this IL-6 production was dose-dependently inhibited by norzoanthamine as shown in

Fig.3-1. When the amount of IL-6 produced by PTH stimulation was 100(%), 50(%) inhibition was given by 4.5 μg/ml of norzoanthamine. Norzoanthamine did not show any cytotoxicities for MC3T3E1 (Fig.3-2) nor for MH-6O* (data not shown) at a concentration less than 100 μg/ml.

*MH-60 cell is hybridoma of which growth is IL-6 dependent and is used for IL-6 measurement[6].

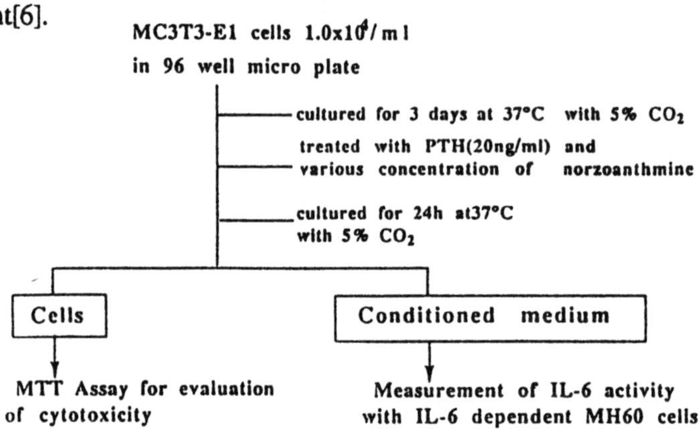

Fig. 2. Assay method of IL-6 production inhibitory activity.

MC3T3E1 cells were cultured to be confluent and stimulated by PTH (20 ng/ml) under the condition with or without norzoanthamine hydrochloride. After 24 hours' incubation, IL-6 content in the conditioned medium was measured by MH60 proliferation test (IL-6 bioassay). The cytotoxicity of norzoanthamine hydrochloride for MC3T3E1 was tested by MTT assay.

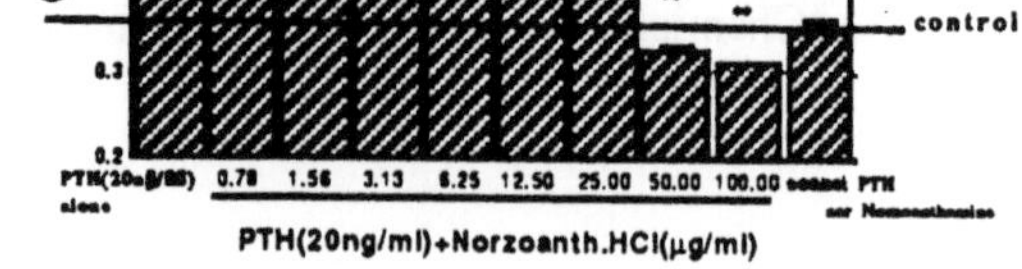

Fig.3-1

Fig.3-1. IL-6 production inhibitory activity of norzoanthamine.

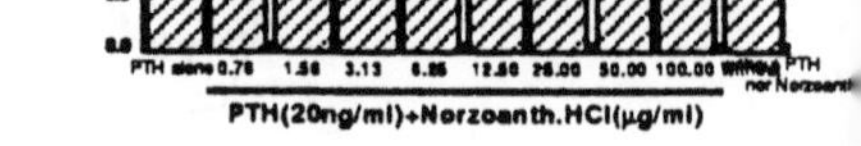

Fig.3-2. Cytotoxicity of norzoanthamine for MC3T3

3. ANTI-BONE RESORPTIVE ACTIVITY OF NORZOANTHAMINE IN VIVO.

3.1. MEASUREMENT OF BONE WEIGHT AND BIOMECHANICAL INDEXES IN OSTEOPOROTIC MODEL MICE.

4 week-old female ddY mice were ovariectomized to be experimental osteoporotic models and the same aged female ddY mice were sham operated to be the control. The ovariectomized (OVX) mice were divided into 4 groups (5 mice / group) and 0.016, 0.08 and 0.4 mg/kg/day of norzoanthamine hydrochloride was intraperitoneally administered to three OVX groups, respectively. One OVX group and a sham operated group were injected vehicle only. Norzoanthamine hydrochloride was resolved into 0.2 ml distilled water for 1 shot and 5 shots/week (from monday to friday; 1 shot per day) was continued for 4 weeks. After 4weeks' administration, the mice were sacrificed to take the femors. One femor of each mouse was cleaned up to remove the muscle and dried at 60 °C overnight to be weighed. Another femor was used for the measurement of biomechanical index by three-point bending test (Fig.4) without cleaning or drying. Body weight and uterus weight were also measured as the indexes of toxicity and sex hormonal action, respectively.

3.2. EFFECT OF NORZOANTHAMINE HYDROCHLORIDE ON BONE WEIGHT LOSS IN OVX MICE.

Bone weight was expressed as [dry weight/ body weight (mg/g)] to correct the differences caused by body size. As shown in Fig.5, the OVX group significantly lost bone weight and administration of norzoanthamine hydrochloride clearly prevented bone weight loss. The most effective dose was 0.08 mg/kg/day. Administration of 17-β-estradiol was confirmed to be effective in the prevention of bone weight loss in OVX mice (data not shown). Body weight and uterus weight were not affected by the administration of norzoanthamine (data not shown).

3.3. EFFECT OF NORZOANTHAMINE HYDROCHLORIDE ON BIOMECHANICAL INDEXES.

The failure load (N) [the ultimate power loaded when the bone was broken], the yield energy (mJ) [the total energy loaded until the bone was broken] and the stiffness (N/mm) [the load (N) necessary for 1 mm of bending] were measured as biomechanical indexes. The failure load, the yield energy and the stiffness are the indexes of bone strength, bone ductility and bone elasticity, respectively. As shown in Fig.6-1, 6-2 and 6-3, administration of norzoanthamine hydrochloride significantly prevented the decrease of bone strength, ductility and elasticity caused by OVX.

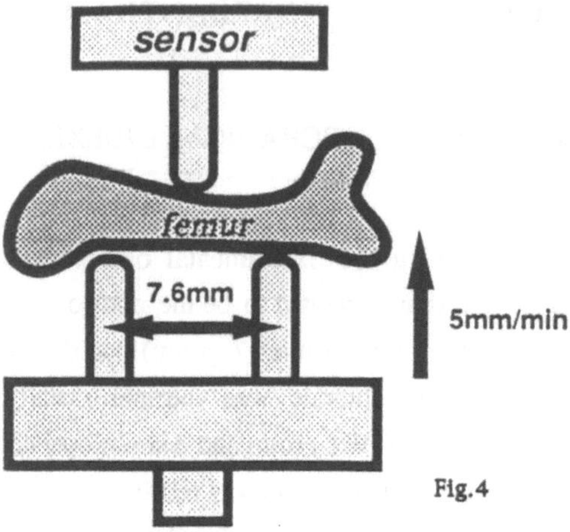

Fig.4

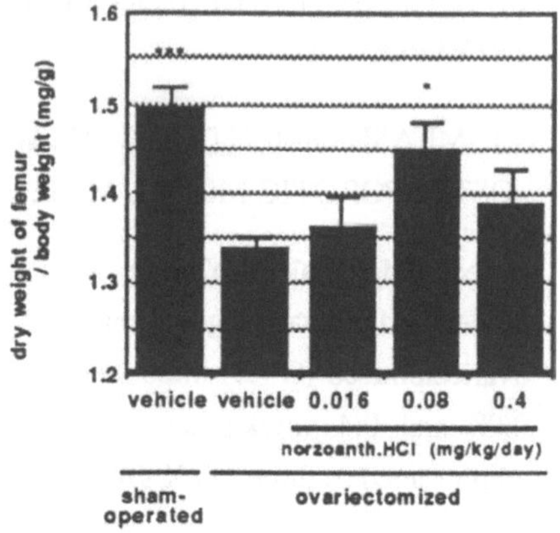

Fig.5

Fig.4. Three-point bending test. Bone strength tester, model TK-252C (MUROMACHI KIKAI Co. Ltd) was used for the measurement of bone biomechanical parameters.

Fig.5. Effect of norzoanthamine hydrochloride on the femur weight of ovariectomized mice. n=5, *;$p<0.05$, ***;$p<0.005$ vs. OVX-vehicle.

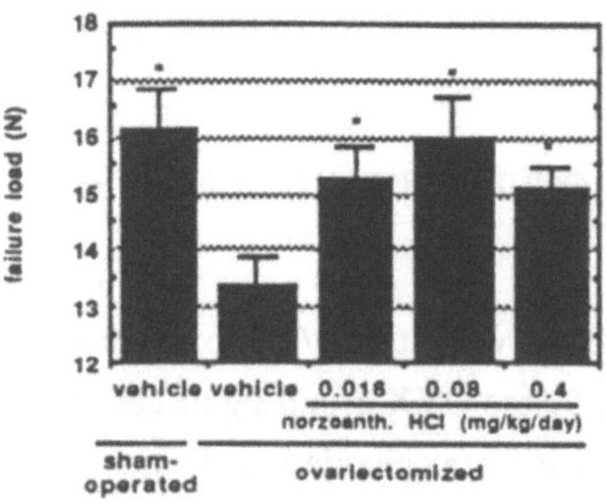

Fig. 6-1

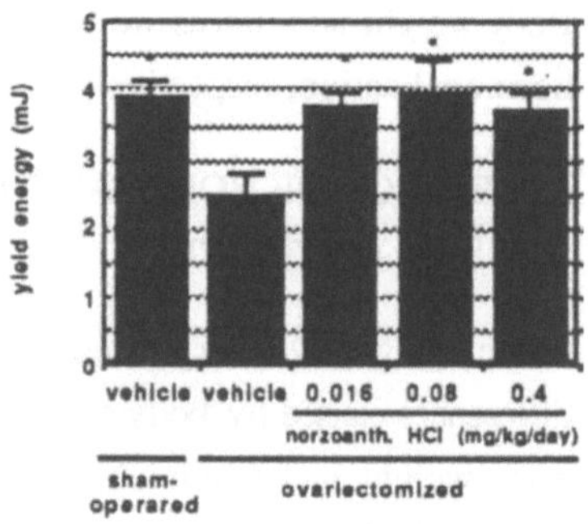

Fig. 6-2

Fig. 6 Effect of norzoanthamine hydrochloride on the bone biomechanical parameters in ovariectomized mice. Four-weeks-old female ddY mice were ovariectomized and intraperitonealy treated with 0.016, 0.08, 0.4 mg/kg of norzoanthamine hydrochloride and vehicle only for 4weeks. Sham-operated mice were treated with vehicle only as control. Bone biomechanical parameters were measured with the three-point bending machine shown in Fig. 4. The femurs were frozen and stored at -20°C until being measured. Fig.6-1; Effect of norzoanthamine hydrochloride on the failure load (N). The failure load (N) is a measure of the strength of the bone. n=5, *; $p<0.05$ vs. OVX-vehicle. Fig.6-2; Effect of norzoanthamine hydrochloride on the yield energy (mJ) in ovariectomized mice. The yield energy (mJ) is an indicator of the ductility of the bone. n=5, *; $p<0.05$ vs OVX-vehicle. Fig.6-3; Effect of norzoanthamine hydrochloride on the stiffness (N/mm). The stiffness (N/mm) is an indicator of the elasticity of the bone. n=5,

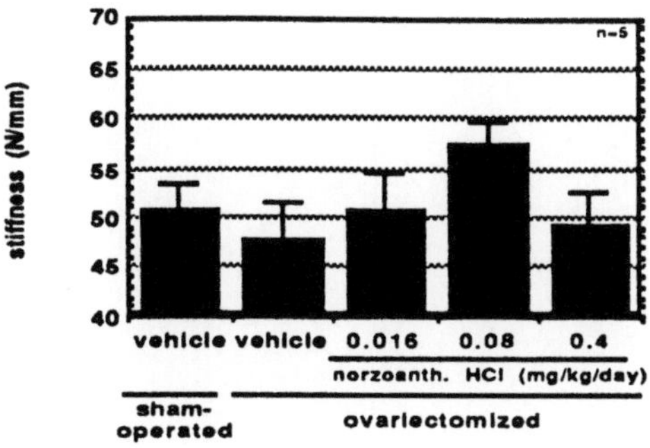

Fig.6-3

4. CONCLUSION

Norzoathamine was found to be an inhibitor of PTH-stimulated IL-6 production in *in vitro* assay system and showed a preventive effect on the bone weight loss and decline of biomechanical integrity of bone in OVX mice. This potent anti-bone resorptive activity *in vivo* can not be explained only by the data involving IL-6 production, and some other mechanism remains unknown. Norzoanthamine hydrochloride, showing a clear anti-resorptive activity *in vivo* without affecting the body weight or the uterine weight, can be a good candidate for new osteoporotic medicines.

5. REFERENCES

1) Ishimi, Y. Miyaura, C. Jin, C.H. Akatsu, T. Abe, E. Nakamura, Y. Yamaguchi, A. Yoshiki, S. Matsuda, T. Hirano, T. Kishimoto, T. and Suda, T. (1990) IL-6 is produced by osteoblasts and induces bone resorption, J. Immunology, 145, 3297-3303.

2) Masuzawa, T. Miyaura, C. Onoe, Y. Kusano, K. Ohta, H. Nozawa, S. and Suda, T. (1994) Estrogen deficiency stimulates B lymphopoiesis in mouse bone marrow, J. Clin. Invest., 94, 1090-1097.

3) Poli, V. Balena, R. Fattori, E. Markatos, A. Yamamoto, M. Tanaka, H. Ciliberto, G. Rodan, G. A. and Costantini, F. (1994) Interleukin-6 deficient mice are protected from bone loss caused by estrogen depletion, EMBO J. 13, 1189-1196,

4) Girasole, G. Jilka, R.L. Passeri, G. Boswell, S. Boder, G. Williams, D.C. and Manologas, S.C. (1992) 17b-estradiol inhibits interleukin-6 production by bone marrow-derived stroma cells and osteoclast in vitro: A potent mechanism for the antiosteoporotic effect of estrogens, J. Clin. Invest, 89, 883-891.

5) Fukuzawa, S. Hayashi, Y. Uemura, D. Nagatsu, A. Yamada, K. and Ijuin, K. (1995) The isolation and structures of five new alkaloids, norzoanthamine, oxyzoanthamine, norzoanthaminone, cyclozoanthamine and epinorzoanthamine, Heterocyclic Comm. 1, 207.

6) Matsuda, T. Hirano, T. and Kishimoto, T. (1988) Establishment of an interleikin 6 (IL-6)/B cell stimulatory factor 2-dependent cell line and preparation of anti-IL 6 monoclonal antibodies, Eur. J. Immunol. 18, 951-956.

PREPARATION AND BIOLOGICAL ACTIVITIES OF HIGH MOLECULAR WEIGHT SUBSTANCES FROM DRIED SAFFLOWER PETALS

Immunomodulators purified from Safflower extracts

TETSUYA WAKABAYASHI, TAKAO KATAOKA, JE-TAE WOO, and KAZUO NAGAI*

Department of Bioengineering, Tokyo Institute of Technology, 4259 Nagatsuta-cho, Midori-ku, Yokohama 226, Japan

In the course of screening for immunomodulators, we found a significant blastogenic activity specific for splenic B cells in the extracts of safflower (*Carthamus tinctorius* L.). Active fractions termed SF1 and SF2 were prepared from dried petals of safflower through boiling water extraction, ethanol precipitation and Sepharose CL-2B column chromatography. The elution profiles of the gel filtration indicated that the molecular weights of SF1 and SF2 were more than 100 kD. Major components of SF1 and SF2 seem to be polysaccharides, and structural analysis of alditol acetate derivatives by GC-MS revealed some differences between SF1 and SF2 in the sugar component. Biological activities of SF1 and SF2 on B cells and macrophages were examined in comparison with lipopolysaccharides (LPS). SF1 and SF2 induced both the proliferation and the IgM production of B cells to the equivalent level as those induced by LPS. In macrophages, SF1 and SF2 effectively stimulated the production of NO. However, SF1 stimulated the production of IL-1, IL-6, and TNF as strong as LPS, while SF2 induced them only weakly or not at all. Thus, these results suggest that SF1 and SF2 activate B cells and macrophages in different mechanisms from LPS.

*Corresponding author

Abbreviations: FCS, fetal calf serum; LPS, lipopolysaccharides; PEC, peritoneal exudate cells; TdR, thymidine; TG, thioglycollate

1. Introduction

To obtain immunomodulators, we have established an assay system using mouse spleen cells, and screened the natural products such as the microbial metabolites and plant exracts [1]. Recently, we have found a strong blastogenic activity in the safflower extracts. In this paper, we have described the preparation and biological activities of the high molecular weight substances from the safflower extracts. Safflower is currently used for food materials such as cooking oil and a source for natural red dyes. Moreover, safflower petals are also used as tradiotional drugs for women's diseases, poor circulation and menopausal disorders.

2. Materials and Methods

2.1. Mice

C57BL/6 and C3H/HeN mice (female, 6 wk-old) were purchased from Charles River Co. Ltd. (Yokohama, Japan). C3H/HeJ mice (female, 6 wk-old) were purchased from Sankyo Labo Service (Tokyo, Japan). Six to eight wk-old mice were used for experiments.

2.2. Culture medium

RPMI 1640 medium (Life Technologies Inc., Grand Island, NY) supplemented with 10% (v/v) fetal calf serum (FCS; Bioserum, Victoria, Australia), 50 mM 2-mercaptoethanol, 50 μg/ml kanamycin, and 8 μg/ml tylosin tartrate was used for experiments.

2.3. Preparation of SF1 and SF2 from dried safflower petals

Preparation steps are shown in Fig. 1. Dried petals of safflower (*Carthamus tinctorius* L.) (Kampo Science fundation, Tokyo, Japan) were extracted with boiling water for 15 min, and compressed with the gauze. The resultant extracts were centrifuged at 10,000 × *g* for 20 min, and the supernatants were precipitated with ethanol at a final concentration of 70%. The precipitates were collected by centrifugation (10,000 × *g*, 20 min), and washed three times with 70% ethanol. After being freeze-dried, the resultants were dissolved in water, and the aqueous solution was adjusted to pH 3 by HCl. The insoluble materials were removed by centrifugation (10,000 × *g*, 20 min), and the recovered supernatants were again adjusted to pH 7 by NaOH. Active components were collected by 70% ethanol precipitation and freeze-dried. The dried ethanol precipitates were dissolved in 10 mM Tris-HCl (pH 7.4) and applied to Sepharose CL-2B column (2.5 × 90 cm) equilibrated with 10 mM Tris-HCl (pH 7.4). The samples were eluted with the same buffer at a flow rate of 1.63 ml/min, and 5 ml-fractions were collected. SF1 and SF2 were collected by 70% ethanol precipitation and freeze-dried.

2.4. Assay for blastogenic activity

Spleen cells from C57BL/6 mice were treated with 0.83% ammonium chloride, 10 mM Tris-HCl (pH 7.5) for 2 min to remove red blood cells. Spleen cells (1×10^6 cells/ml, 100 μl) were cultured with test samples for 48 h in flat-bottom microtiter plates. For the last 4 h of incubation, the cells were pulse-labeled with 9.25 kBq/well of [^{3}H] thymidine (TdR, ICN Biomedicals Inc., Costa Mesa, CA). The cells were harvested on glass-fiber filters (Whatman GF/A) and the radioactivity was counted in a scintillation counter.

2.5. Measurement of IgM

Spleen cells from C57BL/6 mice were prepared as described above. Spleen cells (1×10^6 cells/ml, 100 μl) were cultured with test samples for 72 h in flat-bottom microtiter plates, and the supernatants were collected. The amount of IgM in the supernatants was determined by the ELISA method [2]. Round-bottom microtiter plates were coated with anti-mouse IgM (H&L) A107US (American Qualex, California, USA). Culture supernatants or standard mouse IgM were diluted with 0.1M $NaHCO_3$ containing 0.02% NaN_3 and incubated in the anti-IgM-coated plates. Bound IgM was detected with horseradish peroxidase-conjugated anti-mouse IgM (H&L) A107PS (American Qualex) in 0.1M $NaNO_3$ containing 1% BSA.

2.6. Isolation of macrophages

C3H/HeN mice were injected i.p. with 2.5 ml of 3.8% thioglycollate (TG) medium. After 4 days, peritoneal exudate cells (PEC) were collected. PEC (1×10^6 cells/ml, 1 ml) were cultured in the presence of test samples in 24-well plates for 24 h. The culture supernatants were harvested, and measured for the following biological activities.

2.7. Measurement of IL-1

Thymocytes from C3H/HeJ mice (5×10^6 cells/ml, 100 ml) were cultured with serially diluted PEC culture supernatants or standard mouse rIL-1 in the presence of 2 μg/ml of PHA for 48 h in flat-bottom microtiter plates, and pulse-labeled with 9.25 kBq/well of [^{3}H] TdR for the last 4 h.

2.8. Measurement of IL-6

IL-6-dependent mouse hybridoma clone, MH60.BSF2 [3], was cultured with optimally diluted PEC culture supernatants or standard mouse rIL-6 for 48 h in a final volume of 200 μl with 2.5×10^4 cells/ml in flat-bottom microtiter plates, and the cell viability was measured by MTT method.

2.9. Measurement of TNF-α

Mouse fibrosarcoma L929 (7.5×10^5 cells/ml, 100 μl) was cultured with serially diluted PEC culture supernatants in the presence of 2 μg/ml of actinomycin D in flat-bottom microtiter plates for 18 h. Adherent cells were stained with 0.2% crystal violet, and absorbance at 595 nm was measured. One unit of TNF was defined as the concentration that induces 50% cell lysis.

2.10. Measurement of NO

The amount of NO was determined by measuring nitrite, as described previously [4]. One hundred μl of culture supernatants or dilution thereof were mixed with an equal volume of 10% Griess-Romijn Nitrite Reagent (Wako Pure Chemical Industries Ltd., Osaka, Japan), and absorbance at 540 nm was measured. Total nitrite were determined by using the standard curve of $NaNO_2$ (0.1 to 100 μM).

3. Results

3.1. Effects of SF1 and SF2 on splenic B cells

We examined the effects of SF1 and SF2 on splenic B cells in comparison with lipopolysaccharides (LPS). Freshly isolated spleen cells were incubated in the presence of test samples for 48 h, and the cells were pulse-labeled with [^{3}H] TdR for the last 4 h of incubation. In the case of IgM production, spleen cells were incubated with test samples for 72 h, and the content of IgM in the culture supernatants was measured by ELISA. The maximum activities of SF1 and SF2 for inducing blastogenesis and IgM production were comparable to those of LPS (Table 1). In terms of doses, SF1 was about 10-fold more effective than SF2.

Table 1. Induction of proliferation and IgM production on B cells by SF1 and SF2

	SF1		SF2		LPS		Control
	MA[a]	ED_{50}[b]	MA	ED_{50}[b]	MA[a]	ED_{50}[b]	
Blastogenesis (cpm)	136,547	0.97	112,785	3.64	115,675	0.52	2198
IgM (A415)	0.343	1.85	0.240	2.76	0.305	2.38	0.078

[a]maximum activity
[b]μg/ml

3.2 Effects of SF1 and SF2 on macrophages

We further examined the effects of SF1 and SF2 on the function of macrophages. Four soluble mediators IL-1, IL-6, TNF-α and NO produced by TG-elicited PEC in response to SF1, SF2 or LPS were measured (Table 2). SF1 induced the production of IL-1 as strong as LPS, while SF2 was inactive. In the case of IL-6, the equivalent amount of IL-6 was detected in the culture supernatants of SF1- and LPS-treated PEC. However, the induction of IL-6 by SF2 was relatively weak at the maximum level and the effective dose. The same tendency was observed in the TNF-α production. In the case of the NO production, SF1 and SF2 exhibited the comparable activity with LPS. However, higher doses of SF1 and SF2 were required to draw the maximum production. In this point, SF1 was more effective than SF2.

Table 2. Induction of soluble mediators on macrophages by SF1 and SF2

	SF1		SF2		LPS		Control
	MA[a]	ED50[b]	MA[a]	ED50[b]	MA[a]	ED50[b]	MA[a]
IL-1 (ng/ml)	6.72	39.96	0.30	$<3.9\times10^{-4}$	4.72	46.15	$<3.9\times10^{-4}$
IL-6 (ng/ml)	7.26	1.44	3.62	3.24	5.80	1.25	0.04
TNF (unit/ml)	553.3	7.39	353.3	12.97	732.3	5.86	N.D.[c]
NO (μM)	27.9	2.3	28.9	10.19	31.4	1.3	<0.39

[a]maximum activity
[b]μg/ml
[c]Not detected

4. Discussion

In this paper, we reported the preparation of high molecular weight substances from safflower extracts and their biological activities on B cells and macrophages. Dried safflower petals were subjected to boiling water extraction, ethanol precipitation, and Sepharose CL-2B column chromatography. High molecular weight fractions termed SF1 and SF2 were found to exhibit a specific blastogenic activity on B cells. SF1 and SF2 induced the proliferation and differentiation of B cells as strong as LPS.
Moreover, SF1 stimulated the production of various cytokines of PEC as comparable with LPS. SF2 showed some selectivity in the production of cytokines, especially, in that SF2 failed to induce IL-1 within the concentration examined. LPS activates the host immune responses. However, accompanied inflammatory responses results in the pathogenesis of sepsis and adult respiratory destress syndrome [5]. These fatal effects have precluded LPS from the clinical use for infectious diseases and cancer. In contrast, some polysaccharides such as lentinan [6], are currently used for the therapeutic treatments of cancer.
LPS can activate macrophages via at least two distinct pathways [7]. One is transmitted by CD14. CD14 is expressed strongly on monocytes and weakly on neutrophils. CD14 binds LPS complexed to LBP (LPS binding protein) [8]. The other pathway is independent of CD14 and LBP, and putative receptors have not been identified. In the presence of high concentrations of LPS (>100 ng/ml), CD14-independent pathway is shown to be operative [9]. In TG-elicited macrophages, SF1 and SF2 exhibited significant differences in their ability to induce the production of soluble mediators. SF1 induced the production of IL-1, IL-6, TNF-α, and NO to the equivalent level to those detected in LPS-treated cells. SF2 did not induce the production of IL-1, weakly induced that of IL-6 and TNF-α, but strongly that of NO. Thus, SF2, and possibly SF1 too, seem to stimulate macrophages via distinct pathways from that of LPS. Further studies on the mode of action of SF1 and SF2 will reveal novel pathways of activation of immune cells.

5. References

[1] Nakamura A, Nagai K, Suzuki S, Ando K & Tamura G (1986) Anovel method of screening for immunomodulating substances, establishment of an assay system and its application to culture broth of microorganisms, J. Antibiotics 39(8): 1148-1154
[2] Lim B. O, Yamada K, Yoshimura K, Watanabe T, Hung P, Taniguchi S,& Sugano M (1995) Free Bile Acids Inhibit IgE Production by Mouse Spleen Lymphocytes Stimulated by Lipopolysaccharide and Interleukins. Biosci. Biotech. Biochem. 59: 624-627
[3] Matsuda T, Hirano T,& Kishihara T (1988) Establishment of an interleukin 6 (IL-6)/B cell stimulatory factor 2-dependent cell line and preparation of anti-IL-6 monoclonal antibodies. Eur. J. Immunol. 18: 951-956
[4] Kalechman Y, Shani A, Dovrat S, Whisnant J. K, Mettiger K, Albeck M & Sredni B (1996) The Antitumoral Immunomodulator AS101 and Paclitaxel (Taxol) in Murine Model of Lung Adenocarcinoma. J. Immunol. 156: 1101-1109
[5] Jerry A. N, Natalie J. A, På}r G, Gary L. J and Worthen G. S (1996). Activation of a p38 Mitogen-Activated Protein Kinase in Human Neutrophils by Lypopolisaccharide. *J. Immuno.* 156, 4867-4875
[6] Chihara G, Hamoro J, Maeda Y. Y, Arai Y& Fukuoka F (1970) Fractionation and Purification of the Polysaccharides with Marked Antitumor Activity, Especally Lentinan, from ***Lentinus edodes*** (Berk.) Sing. (an Edible Mushroom). Cancer Research 30: 2776-2781
[7] Cohen L, Haziot A, Shen D. R, Lin X, Sia C, Harper R, Silver J & Goyert S. M(1995) CD14-Independent Responses to LPS Require a Serum Factor That Is Absent from Neonates. J. Immuno. 155: 5337-5342
[8] Ziegler-Heitbrock H. W. L and Ulevitch R. J (1993) CD14: Cell surface receptor and differentiation marker. Immunology Today 14: 121-125
[9]Haziot A, Ferreo E, Köntgen F, Hijima N, Yamamoto S, Silver J, Stewart C. L & Goyert S. M (1996) Resistance to Endotokin Shock and Reduced Dissemination of Gram-Negative Bacteria in CD14-Deficient Mice. Immunity 4: 407-414

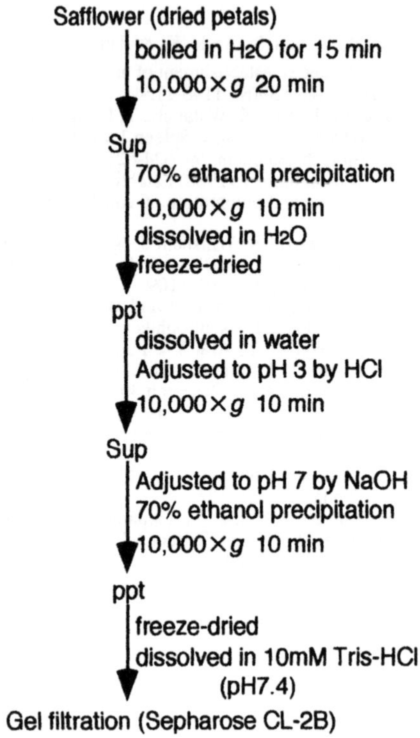

Figure 1 Purification methods of SF1 and SF2 from safflower dried petals

CHARACTERIZATION OF HUMAN T CELL CLONES ISOLATED FROM IMMORTALIZED T CELL LINES GENERATED BY ONCOGENE TRANSFECTION

Shahabuddin Alam, Yoshinori Katakura and Sanetaka Shirahata
Graduate School of Genetic Resources Technology, Kyushu University,6-10-1 Hakozaki, Higashi-ku, Fukuoka 812-81, Japan

ABSTRACT. Previously me established a method to reproducibly immortalize human T cells by oncogene transfection (2). Based on our findings which showed that these immortalized T cells demonstrated cytotoxicity against K562 cells in a MHC-nonrestricted manner, we attempted to obtain T cell clones which exhibit natural killer like activity. In this study, we tried to isolate immortalized T cell clones by two methods, i) direct limiting dilution, and ii) limiting dilution after stimulation with K562 cells. We succeeded in obtaining 35 T cell clones. Two clones among the 35 were $CD4^+$, but the rest of the clones were shown to contain a small population of $CD8^+$ cells together with highly dominating $CD4^+$ cells, suggesting that $CD4^+$ T cells were preferantially expanded from T cell lines immortalized by oncogene transfection. However, we can not yet assign the reason why this small population of $CD8^+$ T cells was retained in the 33 established T cell clones. Eleven out of 19 T cell clones established by method (i) and 11 out of 16 T cell clones established by method (ii) were shown to be cytotoxic against K562 cells, indicating that these T cell clones still retain their functional properties irrespective of their cloning methods.

1. Introduction

A subset of $CD3^+$ T cells exhibiting natural killer like activity are rarely found in the peripheral blood lymphocytes (1, 3, 5). Natural killer (NK) cells are a subpopulation of lymphocytes that have the ability to recognize and lyse a broad spectrum of tumor cell lines and virus-infected cells in the absence of prior antigenic sensitization or major histocompatibility complex (MHC)-restriction (3). NK cells and MHC-nonrestricted killer T cells appear to be the major cytotoxic populations which can be applied for use in antitumor therapy (4). Here, we report that T cell clones established from oncogene transfected immortalized T cell lines have MHC-nonrestricted killer activity.

2. Materials and methods

2.1. PREPARATION OF IMMORTALIZED T CELL LINES

The method to reproducibly immortalize human T cells by oncogene transfection has previously described by Alam *et al*, (2). These established immortalized T cell lines, which were derived from the peripheral blood of healthy donors, were used for this experiment.

2.2. CELL CULTURE

Immortalized T cell lines at a density of 1.7 cells/ml were directly inoculated into 96-well plate containing 20% FBS-ERDF medium (method -i), or immortalized T cell lines at a density of 1.0 x 10^6 cells were stimulated every 7 days with 1.0 x 10^4 mitomycin C-treated K562 cells (an erythroleukemia cell line used for the study of natural killer assay) for one month, then inoculated at a density of 1.7 cells/ml into a 96-well plate containing 20% FBS-ERDF medium (method-ii). These cloned T cells were maintained every 2-3 days for 2 months with ERDF medium containing 10% FBS, 1.5 µl/ml bone marrow extract (BME), 1.0 µl/ml egg yolk lipoprotein (YLP), 5 µg/ml insulin, 20 µg/ml transferrin, 20 mM ethenolamine, 25 nM sodium selenite, 20% spent medium of the T cell lines and 50 U/ml recombinant human interleukin-2 (rhIL-2, Genzyme, Cambridge, MA).

2.3. FLOWCYTOMETRIC ANALYSIS

Surface marker antigens on the T cell clones were analyzed by flowcytometer (FACS Calibur; Becton Dickinson, San Jose, CA). Antibodies (simultest CD4/CD8) were purchased from Becton Dickinson (San Jose, CA).

2.4. CYTOTOXICITY ASSAY

Cytotoxicity assays were performed by a modified lactate dehydrogenase (LDH)-release assay of natural cytotoxicity using a non-radioactive cytotoxicity assay kit (Promega Corp., Madison, WI) according to the manufacturer's protocol. Briefly, after optimizing the cell density of the effector T cells (T cell lines and clones) against K562 or other tumor cell lines, the effector cells together with the tumor target cells were transferred into ERDF medium without phenol red supplemented with 5% FBS and 20 U/ml rhIL-2. Cytotoxicity assays were carried out in round-bottomed 96-well microtiter plates. This assay is able to quantitatively measure the level of LDH that is released upon cell lysis in the same way as ^{51}Cr released in radioactive assays. The percentage of specific cytotoxicity was calculated using the formula: % cytotoxicity= 100 x (Experimental-Effector spontaneous LDH release-Target spontaneous LDH release)/(Target maximum LDH release-Target spontaneous LDH release).

3. Results

3.1. SURFACE MARKER ANTIGENS ON T CELL CLONES

Table 1 Surface marker antigens of T cell clones

oncogenes	T cell lines before cloning		T cell clones			
			direct L.D.		L.D. after stim. with K562	
	CD4	CD8	CD4	CD8	CD4	CD8
ras	+	+++	+++	+	+++	+
ras+myc	++	++	+++	+	+++	+
ras+fos	+	+++	+++	+	+++	-
ras+jun	+++	-	+++	+	+++	+
ras+myb	+	+++	+++	+	+++	+
myc+fos	+	+++	+++	+	+++	-
myc+jun	+++	+	+++	+	+++	+
myc+myb	+	+++	+++	+	+++	+
fos+myb	+++	+	+++	+	+++	+
fos+jun	++	++	+++	+	+++	+
jun+myb	+++	+	+++	+	+++	+

L.D.: Limiting Dilution

Table 1 shows flowcytometric analyses of immortalized human T cell lines and T cell clones. The ratios of $CD4^+$ and $CD8^+$ T cells in immortalized T cell lines were shown to be varied depending on which oncogene combination they were transfected with. Two out of 35 T cell clones were shown to be single $CD4^+$ but the rest were shown to contain as small population of $CD8^+$ cells together with highly dominating $CD4^+$ cells, indicating that $CD4^+$ T cell clones were preferantially expanded from T cell lines immortalized by oncogene transfection. These results also suggest that the coexpression of low $CD8^+$ T cells along with highly concentrated $CD4^+$ T cell clones may be a result of our immortalization procedure. Further work is needed to confirm the relationship of our cloning methods with cell surface expression.

3.2. CYTOTOXICITY ASSAY

Figure 1 shows the cytotoxic activities of immortalized T cell lines and the established immortalized T cell clones against K562 cells. Immortalized T cell lines were able to proliferate in presence of mitomycin C-treated K562 cells and were cytotoxic against K562 cells. Furthermore, immortalized T cell clones obtained by direct limiting dilution and immortalized T cell clones obtained by limiting dilution after stimulation with K562 cells were also shown to be cytotoxic against K562 cells. Among these clones, immortalized T cell clones obtained by limiting dilution after stimulation with K562 cells exerted the strongest cytotoxic activity against target cells. This demonstrates our success in generating T cells with natural killer activity.

4. Discussions

Our study suggests that functional T cell clones could be established from oncogene transfected immortalized human T cell lines in the presence of interleukin-2 without the need for feeder and/or antigen stimulation. The components of the complete medium, which is described in detail in Materials and Methods, were shown to be optimal for the generation of T cell clones and for maintenance in long-term culture. CD4 positive T cell clones with MHC-nonrestricted killer activity were preferantially expanded from T cell lines immortalized by oncogene transfection. Furthermore, K562 stimulation was shown to be effective for obtaining MHC-nonrestricted killer T cell clones, indicating that the frequency of cells reacting with antigen can be increased by sensitization *in vitro*, and therefore K562 sensitized T cell clones may represent a new approach for amplifying the population of highly cytotoxic T cell clones exhibiting MHC-nonrestricted killer activity. The generation of T cell clones with NK-like cytotoxic activities point toward a mechanism that may have excellent antitumor effects which may be applied to cancer therapy.

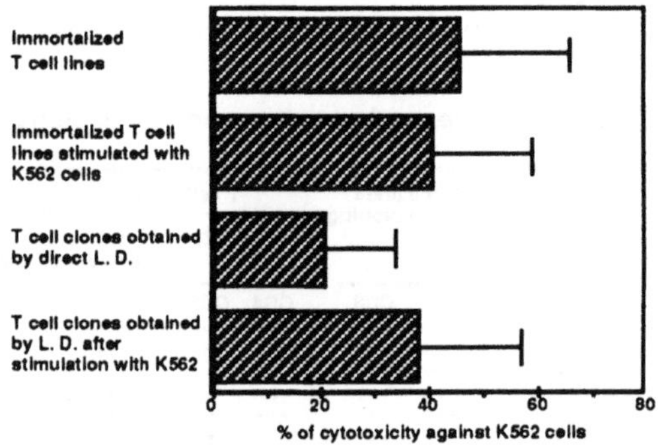

Fig. 1. Cytotoxic activities of immortalized T cell lines and established T cell clones. The figure represents data from oncogene(s)-transfected immortalized T cell lines and immortalized T cell clones against K562 cells. Results are presented as mean value ± standard error.

5. References

1. Acha-Orbea H., Groscurth P., Lang R., Stitz L and Hengartner H (1983). *J. Immunol.* 130: 2952-2959.

2. Alam S., Katakura Y., Yoshida H., Kim E-H and Shirahata S (1996). *Cytotechnology* (submitted).

3. Lanier L.L., Le A.M., Civin C.I., Loken M.R. and Phillips J.S (1986). *J. Immunol.* 136: 4480-4486.

4. Lotzova E (1992). *Leukemia and Lymphoma.* 7:15-28.

5. Young J D-E., Hengartner H., Podack E.R. and Cohn Z.A (1986). *Cell.* 44: 849-859.

Acquired Autoantibodies from Human Plasma B Cells by Light Chain Shifting

Siripen THONGPASSANO, Sanetaka SHIRAHATA and Hirofumi TACHIBANA*

*Graduate School of Genetic Resources Technology, *Department of Food Science and Technology, Faculty of Agriculture, Kyushu University, Hakozaki 6-10-1, Fukuoka, Japan*

ABSTRACT HTD8 is one of concanavalin A-resistant variant subclones of a human plasma B cell line. We have shown that HTD8 expresses various new light chains replacing the original light chain at a high rate. This process of replacing the original light chain in antibody secreting cells was called "light chain shifting". The subclones of transformed-HTD8 cells by introducing certain human antibody heavy chain gene, expressed new light chains which led to altered antigen binding in both specificity and affinity. These results raise the possibility that plasma B cell can be altered to be generate autoantibodies. We examined this possibility by assessing an autoantigens binding of hybrid antibody molecules from the transformed-HTD8 cells repertoire. The hybrid antibodies were obtained by panning against dsDNA and further examined with other autoantigens. Some of the antibodies produced from the transformed-HTD8 cells also express a different light chain and shown to be of high affinity with double-stranded DNA (dsDNA), human thyroglobulin, and Fc fragment of human IgG. These results indicated that light chain shifting can be utilized to generated autoantibodies with high affinity and polyspecificity to various autoantigens by an antibody having no significant affinity for autoantigens.

1. INTRODUCTION

Autoantibodies are antibodies that bind to antigens present in its own cells or extracellular proteins. It has been shown that high-affinity antibodies to dsDNA can be generated from an antibody having no significant affinity for DNA, solely by reconstruction of the heavy-chain complementarity determining region 3 [(1)]. Moreover, *in vitro* polyclonal stimulation with mitogenic lectins, or by the Epstein-Barr Virus, leads normal human B cells to produce rheumatoid factor, anti-DNA and other autoantibodies [(2)]. The objective of this work is to study the possibility to acquire anti-DNA autoantibodies by human plasma B cell line which express new light chain.

HTD8 is one of concanavalin A-resistant variant subclones of a human plasma B cell line. The HTD8 cells undergo active loss of production of the original λ light chain at a high rate followed by the production of a new λ chain. We call this process " light chain shifting" and is a possible mechanism by which expression of the original light chain is actively replaced with a new light chains in human plasma B cells[(3)]. It has been shown that replacing the original λ light chains with the new light chains leads to an alteration of the original antigen binding ability[(4)]. Therefore, to examine whether or not the new light chain producer can be altered to acquire an ability to bind autoantigens, we attempt to isolate autoantibodies from HTD8 cells which undergo light chain shifting. Some of the light chain shifted-antibodies had similar reactivity to the autoantigens although the original HTD8 antibody absolutely showed no reactivity to the autoantigens. Next, we tried to examine the possibility to acquire the autoantibodies by light chain shifting with another heavy chain. An anti-DNA autoantibody derived from the rheumatoid arthritis (RA) patient was also assessed for comparison. Some of the transformed-autoantibodies showed high affinity and polyreactivity to the panel of autoantigens.

2.MATERIALS AND METHODS

2.1 Cell Culture

The human plasma cell, HTD8, is one of the concanavalin A-resistant variants clones of HB4C5 cell [(3)]. The transformed-HTD8 cells are cultured in ERDF medium (Kyokuto Seiyaku, Tokyo, Japan) supplemented with 5 % fetal calf serum and maintained in a humidified, CO_2 gassed (5%) 37 °C incubator.

2.2 Cell transformation

New light chain secreting HTD8 cells were transfected by electroporation with plasmid encoding the BD9D12 heavy chain gene. For each transfection reaction, $5x10^6$ cells in 0.5 ml phosphate buffered saline (PBS) and 10μg of plasmid DNA were used An electric pulse (0.45kV/cm) was discharged across the cell suspension. After electroporation, cells were resuspended with 5% fetal calf serum, and plated into 96-well tissue culture plates (Becton Dickinson). After incubation at 37 °C for 48 hr, 2mg/ml of G418 (GibcoBRL, NY, USA) was added to the culture medium. Surviving clones were screened by ELISA for the transformant and then for autoantibodies.

2.3 ELISA assays for antigenic specificity

Antibodies were tested by ELISA for binding to human Thyroglobulin, Fc fragment of human IgG and dsDNA. *dsDNA.* Fifty microliter of dsDNA at 10 μg/ml was added to tissue culture 96-well plates (Nunc, Denmark), and dried overnight at 37 °C. After blocking with 0.05%Tween 20 in PBS, culture supernatants were incubated at 4 °C overnight followed by secondary antibody, peroxidase-conjugated anti-human IgG gamma chain at a 1:2000 dilution (Biosource, CA, USA) for 1 hr. After a final wash, the reaction was developed with horse radish peroxidase (HRP) and then were monitored at 405 nm in a Titertik ELISA reader. *Human thyroglobulin and human IgG Fc region.* 96-well microliter ELISA plates were coated with 50 μl of human thyroglobulin and human IgG/Fc region at 10 μg/ml at 4 °C overnight. The ELISA for these autoantigens binding was then performed as described above for dsDNA.

2.4 Western Blot Analysis

Culture supernatant of autoantibodies were boiled for 5 min in sample buffer, electrophoresed on SDS-polyacrylamide gels (10%), and transferred to a nitrocellulose membrane. The blotted membrane was blocked in Block Ace (Dainippon Pharmacy, Japan) then incubated in a 1:1000 dilution of peroxidase-conjugated goat anti-human λ light chain for 1 hr. The membranes were washed in PBS containing 0.05% Tween 20 and developed with 1.6 mM 4-chloro-1-naphthol, 0.01% H_2O_2 in PBS with 20% methanol.

3.RESULTS AND DISCUSSION

3.1 Acquired autoantibodies from the light chain-shifted antibodies produced by HTD8 cells transformed with certain heavy chain gene.

HTD8 cells continues expressing the new light chains replacing the original light chain at a high rate. After being expanded for 4 weeks, 70% of HTD8 cells continue to express the original light chain while 30% of those loss of production of the original light chain and continue to express the new light chain (data not shown). Secretion of the various new λ light chains was determined by their size different from original HTD8. In addition to secreting the original 32 kDa λ light chain species, the single cell variant clone of HTD8 cells also distinctly secreted the 26 kDa and 28 kDa λ light chain species. The reactivity of antibodies derived from HTD8 subclones to dsDNA, single stranded-DNA (ssDNA) and Poly I was examined to determine if the light chain-shifted antibodies were capable of binding to the autoantigens or not.

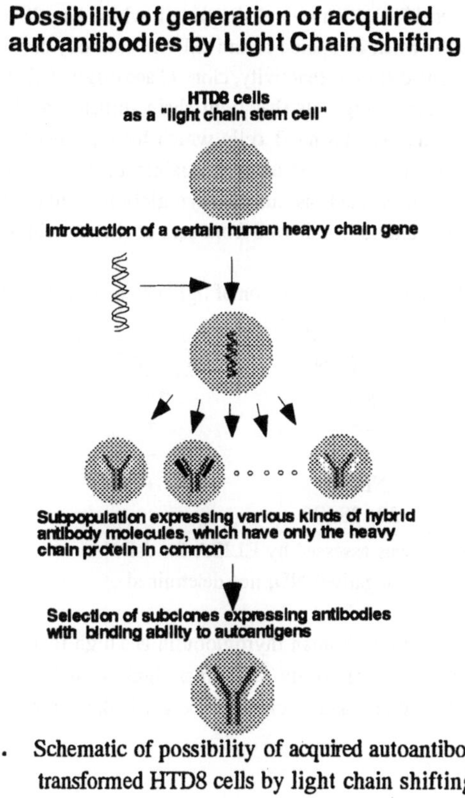

Figure 1. Schematic of possibility of acquired autoantibodies from transformed HTD8 cells by light chain shifting.

Some of the light chain shifted-antibodies had similar reactivity to the autoantigens when compared to anti-DNA antibody derived from autoimmune patient, although the original HTD8 antibody absolutely showed no reactivity to these autoantigens. This result makes a possibility to acquire autoantibodies from human plasma B cell by light chain shifting. Next, we try to examine the possibility to acquire the autoantibodies by light chain shifting with another heavy chain. We used the HTD8 cells as a "light chain stem cell" and transformed these cells by introducing certain human antibody heavy chain gene. For this purpose, we used human antibody heavy chain of anti-lung cancer antibody (BD9D12) as a sort of certain heavy chain gene. The subclones of the transformed-HTD8 cells can also produced various kind of hybrid antibody molecules, which have only the recombinant human immunoglobulin heavy chain protein in common (Figure 1).

To examine whether transformed-HTD8 cells able to generate autoantibodies or not, we screened about 100 cloned supernatants from the transformed-HTD8 cells repertoire to find anti-DNA antibody producing clones. Nearly 10% of anti-DNA positive clones were obtained only from panning of the antibodies against dsDNA. The anti-DNA autoantibodies also express new λ light chain with the new heavy chain. The reactivity of these light chain-shifted antibodies to dsDNA is shown in Table 1. Clone A antibody expresses BD9D12 heavy chain protein which has 28 kDa new light chain, showed higher affinity to dsDNA when compared to anti-dsDNA antibody derived from autoimmune disease patient while clone B showed similar reactivity, clone C and original HTD8 antibodies showed no reactivity to dsDNA. This result suggests that light chain shifting can be utilized to generate anti-DNA autoantibodies from human plasma B cells which have no reactivity to autoantigens. Furthermore, the anti-DNA autoantibodies were assessed against a panel of autoantigens, including those usually considered autoantigens, such as human thyroglobulin and human IgG/Fc fragment compared with the anti-DNA autoantibodies derived from sera of rheumatoid arthritis patient.

Table 1. Characteristic and reactivity determination of light chain-shifted antibodies produced from heavy chain-transformed HTD8 cells against dsDNA.

antibodies	λ chain (kDa)	Anti-dsDNA
Clone A	28	++
Clone B	28	+
Clone C	32	-
Anti-DNA antibody	ND	+
Original HTD8	32, 28, 26	-

Antigen binding of the antibodies was assessed by ELISA for dsDNA.
(++,strong positive; +, positive; -, negative) ND; not determined

The clone a showed similar reactivity to human thyroglobulin and high reactivity to human IgG/Fc fragment, while clone b had similar reactivity to human IgG/Fc and no reactivity to human thyroglobulin. In contrast, the clone c and original HTD8 antibodies showed no reactivity to the both autoantigens (Table 2).

Table 2. Reactivity determination of light chain-shifted HTD8 antibodies against human thyroglobulin and Fc fragment of human IgG.

antibodies	anti-human thyroglobulin	anti-human IgG/Fc fragment
Clone a	+	++
Clone b	-	+
Clone c	-	-
Anti-DNA antibody	+	+
Original HTD8	-	-

Antigen binding of the antibodies was assessed by ELISA.
(++, strong positive; +, positive; -, negative)

These results indicated that the light chain-shifted antibodies can demonstrated high affinity and marked cross reactivity to the panel of autoantigens. Although each of the antibodies in this study was monoclonal, the secreted autoantibodies generally were polyreactivity. Similarly, studies of monoclonal autoantibodies produced by hybridomas derived from lupus patients have demonstrated that anti-DNA autoantibodies are often polyspecific[5]. Includingly, human immunodeficiency virus type 1 (HIV-1) seropositive donors typically have high serum antibody titers to a range of autoantigens[6]. However, natural autoantibodies directed against DNA present in the serum of normal healthy individuals, are generally of low relative affinity for DNA, and exhibit polyreactivity[7]. Although polyreactivity appears to be a characteristic feature of natural autoantibodies, polyreactive anti-DNA autoantibodies can be derived both from patients with autoimmune disease and from normal individuals[8]. Eventhough, the mechanisms of light chain shifting to acquire autoantibodies are poorly understood, these results also suggest that light chain shifting can make a possibility to obtain autoantibodies by human plasma B cell line.

4. REFERENCES

1) Barbas, S.M., Ghazal, P., Barbas, C.F, III, and Burton, D.R. (1994) *J.Am.Chem.Soc.* **116**:2161.
2) Levinson, A.I., *et.al.* (1987) *J.Immunol.* **139**: 2237.
3) Tachibana, H., *et.al.*, (1996) *J. Biol. Chem.* **271**: 17404.
4) Tachibana, H., Kido, I. and Murakami, H. (1994) *J.Biol.Chem.* **269**:29061.
5) Yu, C.L., Huang, M.H., Tsai, C.Y., et.al. (1996) *Clin.Exp.Rheumatol* **14(2)**:137.
6) Ditzel, H.J., Barbas, S.M., Barbas III, C.F., and Burton, D.R. (1994) *Proc.Natl.Acad.Sci.USA* **91**:3710.
7) Naparstek, Y., Andre, S.T., Manser,T., et.al. (1986) *J.Exp.Med.* **164**: 614.
8) Rioux, J.D., Zdarsky, E., Newjurj, M.M., and Rauch, J. (1995) *Mol.Immunol.* **32(10)**:683.

CLINICAL APPLICATION OF ANTI-CARCINOMA HUMAN MONOCLONAL ANTIBODY

KEIJI KODA[1], MARK C. GLASSY[2], and NOBUYUKI NAKAJIMA[1].
[1]*First Department of Surgery, Chiba University School of Medicine, 1-8-1 Inohana, Chuo-Ku, 260 Japan.*
[2]*Novopharm Biotech, Inc., 6190 Cornerstone Court E., Ste. 106, San Diego, CA 92121, USA.*

A human monoclonal antibody SK1 (HuMAb SK1) recognizes a 42-46 KDa sialo-glycoprotein which is commonly expressed on carcinoma tissues. In order to examine the in vivo accumulation capacity of SK1, immunoscintigraphy was performed in colorectal cancer patients. Under informed consent, 4 patients with suspected recurrent colorectal cancers were entered into the study. On day 1, 1-2mg of biotinylated SK1 was administered intravenously to each patient. On day 2, 5mg of streptavidin was administered, followed one day later, biotinylated $^{111}InCl_3$ was administered. Either planar or SPECT (single photon emission tomography) images were obtained 6 to 24 hours later. Two cases out of three histologically-proven local pelvic recurrences were successfully imaged with this protocol. One case who had been diagnosed as local recurrence by a computed tomography (CT), and a magnetic resonance imaging (MRI) was true-negative in immunoscintigraphy. No side effects were observed. These findings suggested that HuMAb SK1 effectively accumulates to the tumor regions in vivo. In addition, it was also suggested that immunoscintigraphy using SK1 is a safe and an effective approach for the detection of local recurrence of rectal cancers.
Key Words: human monoclonal antibody, rectal cancer, immunoscintigraphy

1. Introduction

As reported previously, human monoclonal antibody SK1 recognizes a 42-46 KDa sialo-glycoprotein which is expressed in several types of cancer tissues[1]. This antibody may have the direct role to inhibit cancer cell invasiveness[2]. In addition, it was reported that this antibody may induce complement dependent cytolysis in vitro when sufficient antibody concentration be achieved[3]. In the current study, we examined in vivo accumulative capacity of HuMAb SK1 using immunoscintigraphy to suspected recurrent rectal carcinoma patients.

2. Materials and Methods

2. 1. HUMAN MONOCLONAL ANTIBODY SK1

A HuMAb SK1 was generated from in vitro immunized lymphocytes of a patient with Dukes B colon carcinoma[4]. Briefly, lymph node lymphocytes from a colon cancer patient were stimulated in vitro with a supernatant of pokeweed mitogen-stimulated T lymphocytes[5] for 7 days. Lymphocytes were then harvested and fused with a human lymphoblastoid cell line, SHFP-1, using PEG1500 as described[6]. Human-human hybridoma secreting immunoglobulin which binds to an immobilized colon cancer cell line (HT29) were selected with EIA. A human hybridoma secreting SK1 was thus selected and cultured in serum free media until sufficient amount of HuMAb was obtained. A pooled HuMAb was purified by three series of column work. The final product was tested and passed for sterility, pyrogen concentration, and general safety. SK1 was then biotinylated with N-hydroxysuccinimide-biotin, passed through a filter, and stored until use.

2. 2. PATIENTS

Four patients who had been pointed out a pelvic mass by CT or MRI, which suspected a local recurrence of rectal cancer, were entered into the study. All of the patients underwent positron emission tomography (PET) and CT-guided needle biopsy to confirm the diagnosis.

2. 3. THREE STEP IMMUNOSCINTIGRAPHY

The immunoscintigraphy was performed with three step administration method as described elsewhere[7] with some modification. Briefly, 1-2mg of biotinylated SK1 was administered i.v.. On the next day, 5mg of streptavidin was administered, followed one day later, 74MBq of biotinylated $^{111}InCl_3$ was administered. The scintigram was obtained 6-24 hours after the final injection.

3. Results

As shown in TABLE 1, two cases out of three histologically-proven local recurrences were successfully imaged by immunoscintigraphy using HuMAb SK1. One patient who had a local pelvic recurrence of 3 cm in diameter was unable to be imaged. One patient who had been suspected to have a local recurrence with CT, MRI, and PET, was true-negative with immunoscintigraphy.

TABLE 1 Comparison of immunoscintigraphy and other imaging modalities in the diagnosis of local pelvic recurrence.

patients	tumor size (cm)	diagnosis in other imaging modalities CT	MRI	PET	results
1	4.2	positive	positive	positive	true positive
2	5	positive	positive	positive	true positive
3	4	positive	positive	positive	true negative
4	3	positive	positive	positive	false negative

* In patient 1, the tumor was resected with total pelvic exenteration. In patient 2,3 and 4, approximate tumor size was measured with CT.

Figure 1 shows the typical SPECT (single photon emission tomography) images using HuMAb SK1. Because of the non-specific accumulation of biotinylated $^{111}InCl_3$ to kidneys and urinary tract, urinary bladder was always imaged in SPECT. In addition, urinary bladder irrigation was necessary to reduce the background in planar images. There were no side effects concomitant to the current protocol.

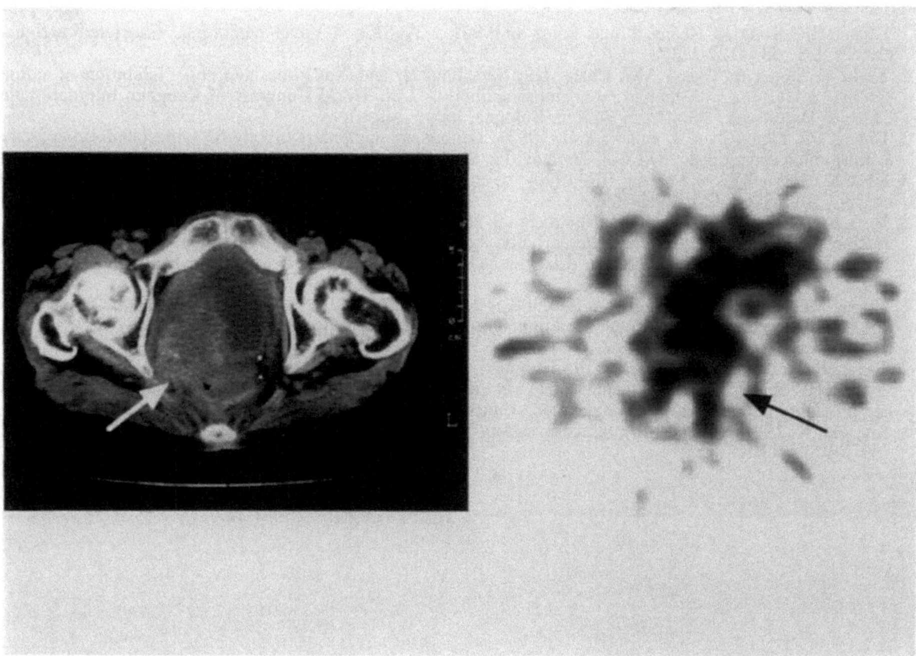

Figure 1 SPECT image of local pelvic recurrence. Accumulation of $^{111}InCl_3$ indicated local recurrece of rectal cancer (arrow) behind the urinary bladder

4. Discussion

Clinically, it is often difficult to make a definitive diagnosis of local pelvic recurrence in rectal cancers. The CT and MRI have been a useful approach to detect a pelvic mass, however, it is often insufficient in specificity to make a final diagnosis. The PET scan may reflect the enhanced metabolism in glycolysis which is often seen in malignant tissues. However, such enhancement may also be seen in active inflammatory tissues[8]. Immunoscintigraphy using cancer-specific monoclonal antibodies may overcome these problems. Although the immunoscintigraphy using murine monoclonal antibodies have been well documented, clinical importance of human monoclonal antibodies have not been discussed so far. In this report, we described that HuMAb SK1 effectively accumulates the colon carcinoma tissues in vivo. Of note, there was no side effect concomitant to the protocol. As reported previously[2], HuMAb SK1 exhibit a direct inhibition of cancer invasion in vitro. Considering all these data conjointly, we are now anticipating the successful biotherapy for the colon cancers using HuMAb SK1.

5. References

1. Chang HR, Koda K, Chang S, and Baird S.(1993) AgSK1, a novel carcinoma associated antigen. Cancer Res. 53: 1122-1127.
2. Koda K, Saitoh N, Glassy MC, Chang HR, Sarashina H, and Nakajima N.(1994) Inhibition of colon carcinoma invasion with human monoclonal antibody. 14th World Congress of Coligium Internationale Chirgiae Digestivae Monduzzi Editore, Bologna, Italy. (abst.).
3. Chang HR, Chavoshan B, and Park H.(1993) Human monoclonal antibody SK1-mediated cytotoxicity against colon cancer cells. Dis.Col. Rectum 36: 1152-1157.
4. Koda K, Glassy MC, and Chang HR.(1990) Generation of human monoclonal antibodies against colon cancer. Arch.Surg. 125: 1591-1597.
5. Borrebaeck CAK.(1987) Human monoclonal antibodies produced from primary in vitro immunized leucine methyl ester-treated peripheral blood lymphocytes. in "in vitro immunization in Hybridoma Technology" Borrebaeck CAK(ed.) Elsevier Science Publishers, Amsterdam, The Netherlands; pp 209-229.
6. Glassy MC, Handley HH, Hagiwara H, and Royston I.(1983) UC 729-6, a human lymphoblastoid B-cell line useful for generating antibody-secreting human-human hybridomas. Proc Natl Acad Sci USA 80: 6327-6331.
7. Paganelli G, Magnani P, Zito F, Villa E, Sudati F, Lopalco L, Rossetti C, Malcovati M, Chiolerio F, Seccamani E, Siccardi AG, and Fazio F(1991) Three-step monoclonal antibody targeting in carcinoembryonic antigen-positive patients. Cancer Res. 51: 5960-5966.
8. Imazeki, K.(1994) Study on the glucose metabolism in human leukocyte using PET. Chiba Med.J. 70: 169-179.

SOME PROPERTIES OF BISPECIFIC ANTIBODIES TO BOVINE LACTOFERRIN AND HORSERADISH PEROXIDASE SECRETED BY A HYBRID-HYBRIDOMA HH1-4-3

HIROSHI SHINMOTO*, MASUKO KOBORI*, TOJIRO TSUSHIDA* AND KAZUKI SHINOHARA**
**National Food Research Institute, Ministry of Agriculture, Forestry and Fisheries, 2-1-2 Kannondai, Tsukuba, Ibaraki 305, Japan, and **National Research Institute of Fisheries Science, Fisheries Agency, 2-12-4 Fukuura, Yokohama, Kanagawa 236, Japan*

Summary

A mouse hybrid-hybridoma HH1-4-3 secreting IgG1 class bispecific antibodies to bovine lactoferrin (bLF) and horseradish peroxidase (HRPO) was established. Using the supernatants as bispecific antibodies, a competitive ELISA was carried out to measure bLF concentration. The competitive ELISA gave a good standard curve at bLF concentrations of 2 to 100 μg/ml. However, the competitive ELISA using the HH1-4-3 culture supernatant was not sensitive enough to measure bLF concentration in biological fluids. To improve the sensitivity of the competitive ELISA, we fractionated bispecific antibodies by antigen affinity column chromatography. The competitive ELISA for bLF using the affinity purified bispecific antibodies through HRPO-immobilized column chromatography showed a good standard curve at bLF concentrations of 10 ng/ml to 100 μg/ml.

1. Introduction

Lactoferrin (LF) is an iron-binding glycoprotein found in many secretions, plasma and the secondary granules of neutrophils [1-4]. One of the major roles of LF is its effects on cellular functions such as suppressing the production of colony-stimulating activity [5], growth stimulation for certain cell lines in serum-free medium [6], and enhancing iron absorption in the intestinal tract [7]. We have also demonstrated that B-lymphocytes incorporate LF and then release it to the extracellular medium [8, 9]. In these studies, we measured the concentration of LF by a single radius immunodiffusion, an immunoelectrophoresis, or a multi-stepped enzyme-linked immunosorbent assay (ELISA). To improve measurement of the concentration of bovine lactoferrin (bLF), we have established a hybrid-hybridoma HH1-4-3 secreting bispecific antibodies specific to horseradish peroxidase (HRPO) and bLF [10].

Antibodies capable of binding two different antigens are termed bispecific (or bifunctional) antibodies [11-16]. Bispecific monoclonal antibodies are thought to be useful in detecting biological molecules by ELISA or immunostainings, and in treating cancer patients by lymphokine-activated killer cell targeting [17-21]. We have also established two hybrid-hybridomas secreting human IgM class bispecific monoclonal antibodies specific to ricin and diphtheria toxin, and neocarzinostatin and diphtheria toxin as the model experiments [22, 23].

In this study, we describe the fractionation of the bispecific antibodies by antigen affinity column chromatography and an improved competitive ELISA of bLF.

2. Materials and methods

2.1 HIGH DENSITY CULTURE OF A HYBRID-HYBRIDOMA HH1-4-3

We have established a hybrid-hybridoma HH1-4-3 secreting bispecific antibodies to horseradish peroxidase (HRPO) and bovine lactoferrin (bLF) [10]. HH1-4-3 cells were cultured in a CELLMAX artificial capillary system with a MPS module (molecular weight cut off of 30 kDa, Gibco BRL). HH1-4-3 cells (2.5 x 10^6 cells) were seeded into the extracapillary space (ECS) of the MPS module (ECS volume of 11.5 ml) and cultured in an ITES-ERDF medium (ERDF medium containing 5 μg insulin / ml, 10 μg transferrin / ml, 20 μM ethanolamine, and 25 nM sodium selenite) [11] at 37 °C for 17 days.

2.2 COMPETITIVE ELISA OF BLF WITH BISPECIFIC ANTIBODIES

The hybrid-hybridoma HH1-4-3 was cultured in a serum-free ITES-ERDF medium and the supernatants were used as bispecific antibodies for competitive ELISA of bLF. Standard samples of bLF (50 μl, n=3), the bispecific antibodies (50 μl) and HRPO solution (50 μl, 5 μg/ml) were mixed and incubated for 90 minutes at room temperature. Each reaction mixture (50 μl) was added to the bLF-immobilized ELISA plate and incubated for 1 h at room temperature. The plates were washed 6 times and HRPO activity immobilized via the bispecific antibodies was measured.

2.3 ANTIGEN-AFFINITY CHROMATOGRAPHY OF BISPECIFIC ANTIBODIES

One ml of HRPO solution (10 mg/ml) or bLF solution (2 mg/ml) was added to a HiTrap NHS-activated column (1 ml, Pharmacia) and incubated for 30 min. The column was washed with 0.5 M ethanolamine. The resulting HRPO-immobilized column and bLF-immobilized column were used to separate antigen-specific antibodies. Culture supernatant (1 ml for bLF, or 2 ml for HRPO) was added to each column and the column was washed with 0.1 M Tris-HCl buffer (pH 7.6) containing 0.5 M NaCl. The antibodies adsorbed on the column were eluted with 0.1 M Glycine-HCl buffer (pH 2.8) containing 0.5 M NaCl.

3. Results and discussion

3.1 COMPETITIVE ELISA OF BLF WITH BISPECIFIC ANTIBODIES

A mouse hybrid-hybridoma HH1-4-3 secreting IgG1 class bispecific antibodies to bovine lactoferrin (bLF) and horseradish peroxidase (HRPO) has been established [10]. The hybrid-hybridoma HH1-4-3 was cultured in a serum-free ITES-ERDF medium and the supernatants were used as bispecific antibodies for competitive ELISA of bLF.

Standard samples of bLF, the bispecific antibodies (culture supernatant, antibody concentration of 5 μg/ml), and HRPO solution were mixed and incubated for 90 minutes at room temperature. Each reaction mixture was added to the ELISA plate coated with bLF and HRPO activity immobilized via the bispecific antibodies was measured. Fig. 1 shows that the bispecific antibodies gave a good standard curve at bLF concentrations of 2 to 100 μg/ml. It was not necessary to prepare enzyme-labelled antibodies by a chemical method. We only mixed enzyme and antibodies to make enzyme-labelled antibodies for the competitive ELISA.

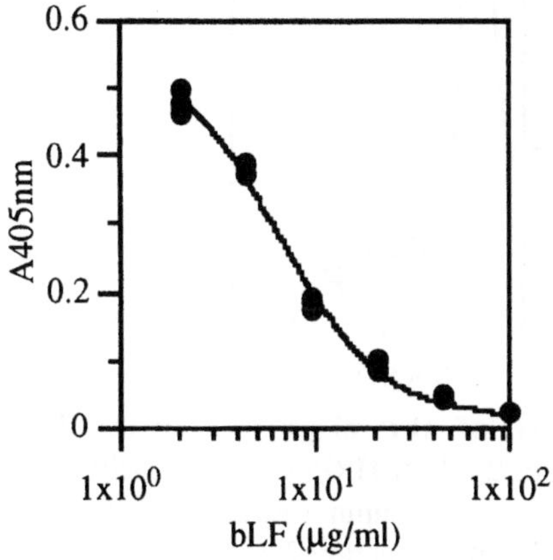

Fig. 1. Standard curve of bLF in competitive ELISA using HH1-4-3 supernatant. Standard samples of bLF, HH1-4-3 supernatant and HRPO solution were mixed and incubated. Each reaction mixture was added to the bLF-immobilized ELISA plate and incubated. The HRPO activity immobilized via the bispecific antibodies was measured.

However, the competitive ELISA using the HH1-4-3 culture supernatant was not sensitive enough to measure bLF concentration in biological fluids such as milk, saliva, and plasma. We postulated the low sensitivity of the competitive ELISA to be due to the presence of monospecific and nonspecific antibodies in the culture supernatant.

3.2 AFFINITY CHROMATOGRAPHY OF THE BISPECIFIC ANTIBODIES

Culture supernatant was separated by antigen-immobilized affinity column chromatography. Fig. 2 (left) shows the results of bLF-affinity chromatography of the bispecific antibodies. Approximately 60 % of the antibodies was adsorbed on the bLF-immobilized column.

Fig. 2 (right) shows the results of HRPO-affinity chromatography. We only obtained less than 5 % of the antibodies adsorbed on HRPO-immobilized column.

3.3 COMPETITIVE ELISA OF BLF WITH AFFINITY-PURIFIED ANTIBODIES

The fraction obtained by HRPO-immobilized affinity chromatography was used as bispecific antibodies for an improved competitive ELISA of bLF. Standard samples of bLF, the bispecific antibodies (fraction 8 in Fig. 2-HRPO, antibody concentration of 2 μg/ml), and HRPO solution were mixed and incubated. Each reaction mixture was added to the ELISA plate coated with bLF and HRPO activity immobilized via the bispecific antibodies was measured. Fig. 3 shows that the bispecific antibodies gave an improved standard curve at bLF concentrations of 10 ng/ml to 100 μg/ml. The competitive ELISA using the affinity-purified bispecific antibodies was sensitive enough to measure bLF concentration in biological fluids such as milk, saliva, and plasma.

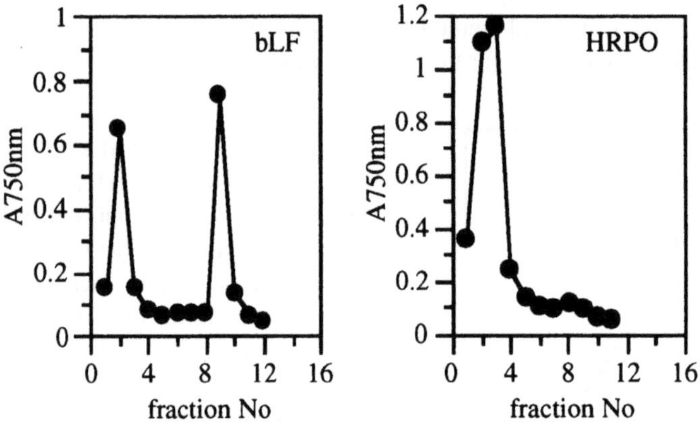

Fig. 2. Antigen-affinity chromatography of the bispecific antibodies. Culture supernatant (1 ml for bLF-column, left, or 2 ml for HRPO-column, right) was added to each column and the antibodies adsorbed on the column were eluted with 0.1 M Glycine-HCl buffer (pH 2.8) containing 0.5 M NaCl (Fraction 7-12).

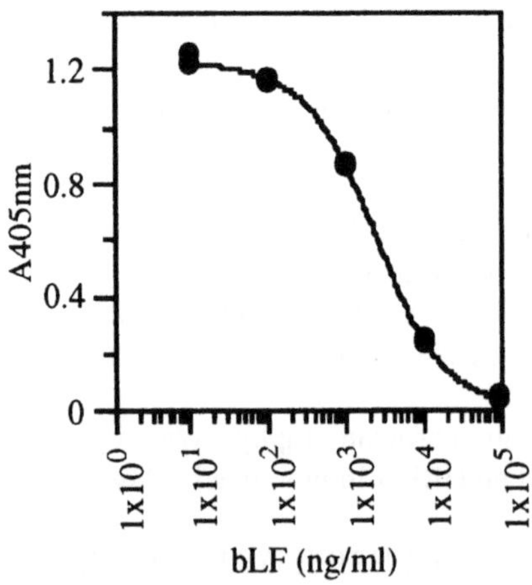

Fig. 3. Standard curve of bLF in competitive ELISA using affinity-purified bispecific antibodies. Standard samples of bLF, affinity-purified bispecific antibodies and HRPO solution were mixed and incubated. Each reaction mixture was added to the bLF-immobilized ELISA plate and incubated. The HRPO activity immobilized via the bispecific antibodies was measured.

4. Acknowledgement
This study was supported by a Grant BMP 96-V-4-3 from the Ministry of Agriculture, Forestry and Fisheries, Japan.

5. References

[1] Johansson, B.G. (1969) *Acta Chem. Scand.* **23**, 683-714.

[2] Wang, C-S., Chan, W-Y., and Kloer, H.U. (1984) *Comp. Biochem. Physiol.* **78B**, 575-580.
[3] Broxmeyer, H.E., Smithyman, A., Eger, R.R., Meyers, P.A., and de-Sousa, M. (1978) *J. Exp. Med.* **148**, 1052-1067.
[4] Finkelstein, M.B., and Finkelstein, R.A. (1982) *FEBS Lett.* **144**, 1-5.
[5]Bagby, G.C., McCall, E., and Layman, D.L. (1989) *J. Clin. Invest.* **71**, 340-344.
[6] Hashizume, S., Kuroda, K., and Murakami, H. (1983) *Biochim. Biophys. Acta.* **763**, 377-382.
[7] Kawakami, H., Hiratsuka, M., and Dosako, S. (1988) *Agric. Biol. Chem.* **52**, 903-908.
[8] Sato, K., Shinmoto, H., Tanimoto, M., Dosako, S., and Nakajima, I. (1990) *Agric. Biol. Chem.* **54**, 1275-1279.
[9] Sato, K., Shinmoto, H., Tanimoto, M., Kawasaki, Y., Dosako, S., and Taneya, S. (1991) in *Animal Cell Culture and Production of Biologicals* (Sasaki, R., and Ikura, K., Eds), pp. 367-372, Kluwere Academic Publishers, the Netherlands.
[10] Shinmoto, H., Kobori, M., Tsushida, T., and Shinohara, K. (1996) *Biotechnology Techniques* **10**, 661-666.
[11] Milstein, C., and Cuello, A.C. (1983) *Nature* **305**, 537-540.
[12] Reading, C.L. (1983) in *Hybridomas and Cellular Immortality* (Tom, B.H., and Allison, J.P., Eds), pp. 235-250, Plenum Press, New York.
[13] Ghetie, V., and Moraru, I. (1983) *Methods Enzymol.* **92**, 523-543.
[14] Milstein, C., and Cuello, A.C. (1984) *Immunology Today* **5**, 299-304
[15] Paulus, H. (1985) *Behring Inst. Mitt.*, **78**, 118-132.
[16] Suresh, M.R., Cuello, A.C., and Milstein, C. (1986) *Methods Enzymol.* **121**, 210-228.
[17] Allard, W.J., Moran, C.A., Nagel, E., Collins G., and Largen, M.T. (1992) *Molecular Immunol.* **29**, 1219-1227.
[18] Auriol, J., Guesdon, J-L. Mazie, J-C., and Nato, F. (1994) *J. Immunol. Methods* **169**, 123-133.
[19] Hakalahti, L, Vihko, P., Henttu, P., Autio-Harmainen, H., Soini, Y., and Vihko, R. (1993) *Int. J. Cancer* **55**, 590-597.
[20] Warnaar, S.O., De Paus, V., Lardenoije, R., Machielse, B.N.M., De Graaf, J., Bregonje, M., and Van Haarlem, H. (1994) *Hybridoma* **13**, 519-526.
[21] Knuth, A., Bernhard, H., Jager, E., Wolfel, T., Karbach, J., Jaggle, C., Strittmatter, W., and Meyer zum Buschenfelde, K-H. (1994) *European J. Cancer* **30A**, 1103-1107.
[22] Shinmoto, H., Dosako, S., Tachibana, H., Shirahata, S., Murakami, H. (1991) *Agric. Biol. Chem.* **55**, 2883-2885.
[23] Shinmoto, H., Dosako, S., Tachibana, H., Yamada, K., Shirahata, S., and Murakami, H. (1991) *Hum. Antibod. Hybridomas* **2**, 39-41.
[24] Murakami, H. (1989) *Advances in Biotechnological Processes* **11**, 107-141.

DEVELOPMENT OF AN ELISA BASED ASSAY FOR MEASURING GLYCOSYLTRANSFERASE ACTIVITY USING SYNTHETIC GLYCOPOLYMERS

MOHAMED OUBIHI[1], KEN KITAJIMA[1], KAZUKIYO KOBAYASHI[2], TAKAHIRO ADACHI[1], NAOHITO AOKI[1] and TSUKASA MATSUDA[1].
[1] *Department of Applied Biological Sciences, School of Agricultural Sciences, Nagoya University , Nagoya 464-01 , Japan*
[2] *Department of Molecular Design and Engineering, Graduate School of Engineering, Nagoya University, Nagoya 464-01, Japan*

1. Introduction

UDP-*N*-Acetylglucosamine β1,4-galactosyltransferase (GalT) belongs to the large family of glycosyltransferases, and is responsible for the transfer of galactose to *N*-acetylglucosamine terminated glycoproteins and glycolipids. Previous research works have shown the association of GalT with various physiological alterations such as cancer (Gaudushon et al 1983; Davey et al 1986; Uemura et al 1988), rheumathoid arthritis (Axford et al 1987; Wilson et al 1993) and hepatic diseases (Kim et al 1972; Stibler et al 1991; Xin et al 1995).

Scanning GalT activity in these pathologies has been performed by means of radiochemical assays by evaluating the radioactivity incorporated into acceptor sugar residues from radiolabelled sugar nucleotide donors, but this method has the drawbacks of being costly and time consuming. In that respect we designed an alternative method, which converges with the recently developed assays in that it is a non-radioactive ELISA based approach (Stults and Macher 1990; Taki et al 1990; Zatta et al 1991; Keshvara et al 1992 ; Keush et al 1995) but its particularity lies in the fact that it uses a synthetic glycopolymer as an acceptor substrate. This assay was applied to measure GalT activity in milk.

2. Materials and Methods

2.1. REAGENTS

Bovine milk β1,4-galactosyltransferase (GalT), UDP-Gal, *Ricinus communis* agglutinin 1 (RCA 1) lectin, anti-rabbit RCA 1 and peroxidase labelled anti-rabbit IgG antibodies were purchased from Sigma. The acceptor substrate Poly(*p*-acryloylaminophenyl-2-acetamido-2-deoxy-β-D-glucopyranoside) PAP(GlcNAcβ) was synthesized as described previously (Kobayashi et al 1994). ELISA plates were purchased from Nunc.

2.2. ASSAY FOR GalT ACTIVITY

The ELISA plates were coated with the glycopolymer (1 μg/ml in phosphate buffered saline: PBS) for 3 h at 4°C, washed 3 times with PBS containing 0.05 % Tween 20 (PBST) and then blocked with 1% bovine serum albumin (BSA) in PBST for 1 h at room temperature. The plates were washed and the enzyme reaction was started by adding to each well 100μl of enzyme-donor substrate mixture (1.5 mU/ml GalT, 10 mM $MnCl_2$, 0.1% BSA and 0.32 mM UDP-Gal) in 30 mM 3-morpholinopropanesulfonic acid (MOPS) (pH 7.4).

After 1 h incubation at 37°C, the reaction was stopped by either adding 20 μl of 200 mM EDTA or discarding the reaction mixture. The plates were washed and incubated with RCA 1 (1 μg /ml BSA in PBST) at 37°C for 90 min. After washing, the plates were incubated at 37°C for 1 h with rabbit anti-RCA 1 antibody, followed by the incubation with peroxidase-labelled anti-rabbit IgG antibody at 37°C for 1 h.

Color development by the enzyme reaction was performed by adding to each well 100 μl of *o*-phenyldiamine (1 mg/ml) in 0.1 M citrate-phosphate buffer pH 5.0 with 0.003% H_2O_2. After 15 min the reaction was stopped by adding 50 μl of 2.5 M H_2SO_4. The absorbance was measured at 492 nm with an ELISA reader .

2.3. ASSAY FOR GalT ACTIVITY IN CRUDE ENZYME FRACTIONS FROM MILK.

Fresh bovine milk provided by a local farm was centrifuged with 5000 g at 4°C for 20 min. Cream and precipitate fractions were removed, and skim milk was 2 times diluted with PBS followed by dialysis against PBS at 4°C for 6 h. The resulting fraction was 100 times diluted with PBS and used directly for measurement of GalT activity by the same procedures as described above, except an additional step: pronase digestion (100 μg/ml) overnight at 37°C after the enzyme reaction. For the inactivation of GalT in crude fractions, an aliquot of the milk sample was boiled for 5 min.

3. Results

Different glycopolymers were tested as acceptor substrates for GalT (Fig. 1), PAP(GlcNAcβ) was found to be highly specific for the enzyme compared with other glycopolymers, even though some are also GlcNAcβ-terminated such as PV[$(GlcNAc)_2$-GlcNAcβ] and PN(GlcNAcβ1,4-GlcNAcβ).

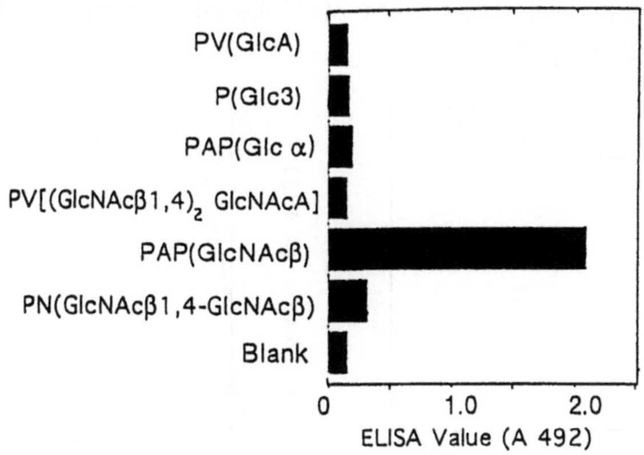

Figure 1: Acceptor substrate specificity of GalT.
GalT was incubated with various glycopolymers differing in backbone structures and bound sugar residues (Kobayash et al 1994A, 1994B) under the same conditions as described in the text.

In order to optimize conditions for GalT activity using PAP(GlcNAcβ) as an acceptor substrate, the enzyme was incubated in the wells coated with different concentrations of PAP(GlcNAcβ) (Fig. 2). The reaction rate was constant up to 1 nM and it became saturated above 4 nM. The Km for the acceptor substrate was estimated to be 0.03 nM.

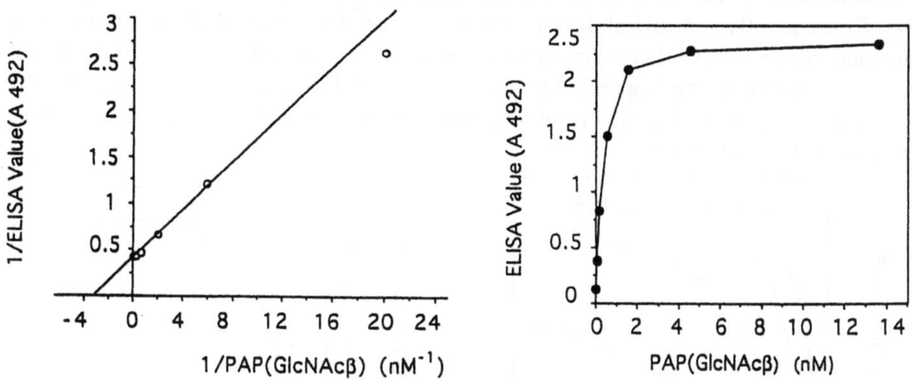

Figure 2: Effect of PAP(GlcNAcβ) concentration.
Right: The incubation conditions were the same as described in the text except for the increasing concentrations indicated.
Left: The corresponding Lineweaver-Burk plot

When GalT was incubated with increasing concentrations of UDP-Gal, the transfer of Gal residue reaches its maximum at about 200 μM of the sugar nucleotide as shown in Fig. 3. The Km value for UDP-Gal was about 60 μM. The effect of the divalent cation

Mn^{2+} was also examined, and at 2-10 mM Mn^{2+} the enzyme reaction was the most effective, as illustrated in Fig. 4.

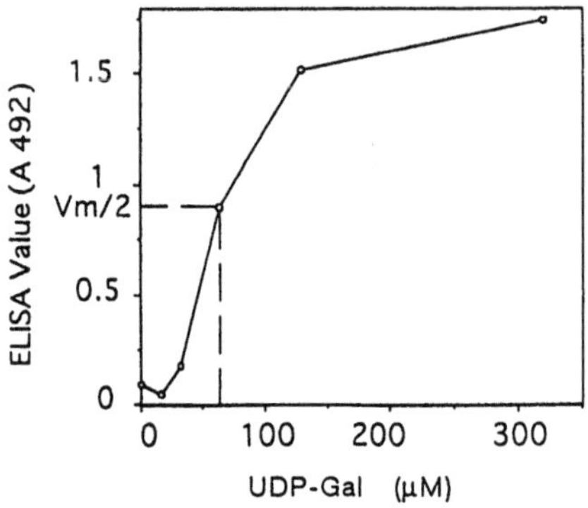

Figure 3: Effect of UDP-Gal concentration. The same conditions were used for the enzyme reaction as described in the text except for UDP-Gal concentrations .

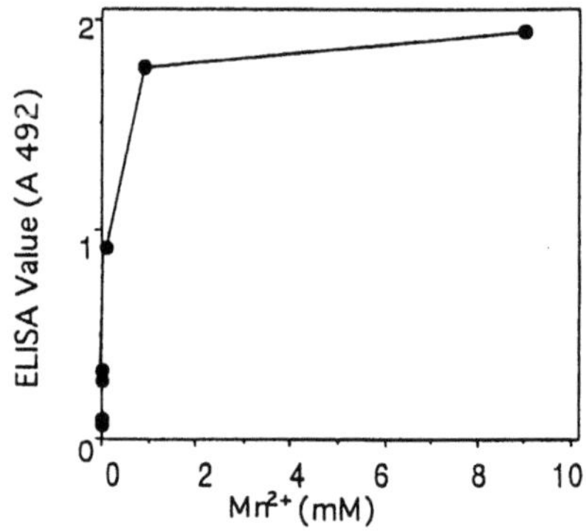

Figure 4: Effect of Mn^{2+} on GalT activity. The same incubation conditions were used with increasing Mn^{2+} concentrations

The effect of GalT concentration on the enzyme reaction was assessed on a time course experiment as shown in Fig.5. The reaction rate was constant up to 30 min for high concentrations of the enzyme (6 and 3 mU) with maximum activity, but for low concentrations (0.750 and 0.38 mU) it increased up to 2 h though the activities were low. Consequently, 1.5 mU of the enzyme was chosen for the standard conditions since the reaction rate remained constant up to 1 h with 70% of the maximum activity. GalT concentration was plotted against the corresponding ELISA values obtained for the 60 min incubation period (Fig. 6) in order to evaluate the enzymatic activity directly from the measured ELISA value.

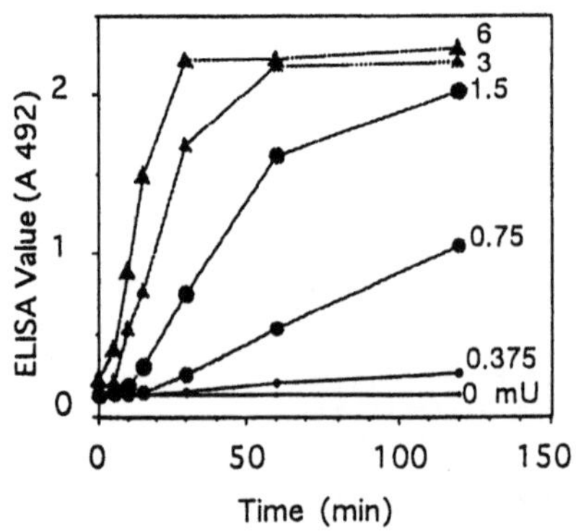

Figure 5: Time course reactions at different GalT concentrations.The incubation was undergone under the same conditions as described in the text . Several GalT concentrations were used as indicated.

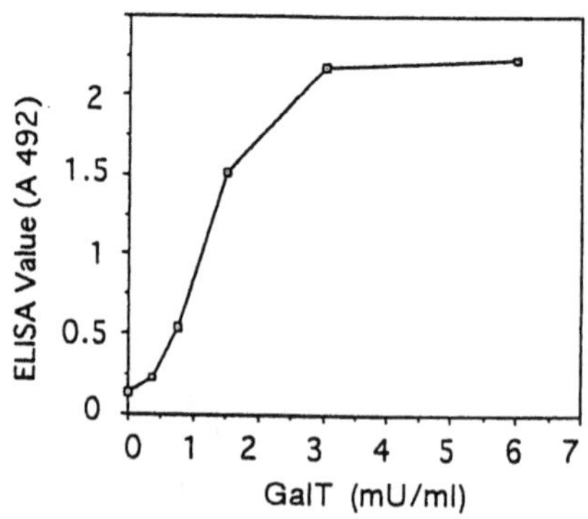

Figure 6: Effect of GalT concentration. The same incubation conditions were used as described in the text except for the enzyme concentrations indicated.

After establishing the standard method for measuring pure GalT activity, we tried to apply this assay to milk which is considered the major source for GalT. Dialyzed bovine skim milk was incubated in the presence and abscence of UDP-Gal. As shown in Fig.7, the activity of GalT in bovine milk was detected in the presence of UDP-Gal, but not in heat treated milk (inactivated GalT).

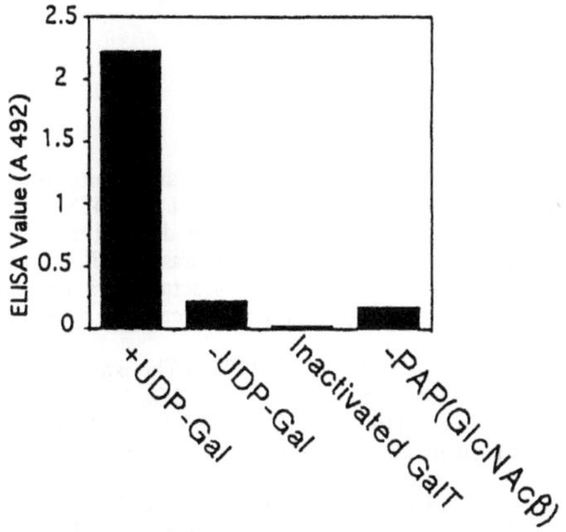

Figure 7: GalT activity in bovine milk.
Dialyzed skim milk was incubated in the presence(+) and abscence(-) of UDP-Gal and without the acceptor substrate [-PAP(GlcNAcβ)] under the conditions described in the text.

4. Discussion

The assay developed in this research is a new alternative for measuring the activity of soluble GalT without using radioisotopes. Our approach is an ELISA based assay using a synthetic glycopolymer as an acceptor substrate and RCA 1 lectin to detect the transfered Gal residue in a β1,4 linkage to terminal GlcNAc (Merkle and Cummnigs, 1987). When another GlcNAcβ1,4-GlcNAc-terminated polymer was used instead, the transfer of Gal was negligible, consistent with the previous results that GalT could not use di-*N*-acetylkitobiose as a substrate. The α–Glc terminated polymer was also assayed as a possible acceptor in the presence of α-Lactalbumin, but we could not observe any galactose transfer as already established for free Glc (Brodbeck et al, 1967; Brew et al, 1968).

The tight binding of the glycopolymer to the microtiter plate within a short period (3 h 4°C) and with as low as 1 μg/ml was attained. This concentration is 50 fold lower than that reported by Keusch et al 1995. This property of our glycopolymer is attributed to its highly hydrophobic backbone. Another property of the synthetic glycopolymer is a high sugar density which may increase the apparent affinity to GalT. The apparent Km value is 100000 times lower than the value reported by Stults et al (1990), which had been considered the lowest among other ELISA-based assays.

The sensitivity of this method is comparable to the recently developed non-radioactive ELISA-based assays. In addition, the synthetic glycopolymer has the advantage of being easy to prepare in large amounts. The chemical and enzymatic stability of the polymer would enable an accurate analysis of a large number of samples. This method was consequently applied to bovine milk in which GalT activity was found to be about three times higher than previously reported (Barker et al 1972). Such difference could be either due to the indirect estimation we adopted by using this method, or to the analysed sample itself since GalT activity varies with lactation stages as well as individuals.

5. References

Axford,J.S., Mackenzie,L., Lydyard,P.M., Hay,F.C., Isenberg,D.A and Roitt,I.M. (1987) Reduced B-cell galactosyltransferase activity in rheumathoid arthritis, *Lancet* **2**, 1486-1488.

Barker,R., Olsen,K.W., Shaper,J.H. and Hill,R.L. (1972) Agarose derivatives of uridine diphosphate and *N*-Acetylglucosamine for the purification of a Galactosytransferase, *J. Biol. Chem.* **247**, 7135-7147.

Brew, K., Vanaman, T.C. and Hill, R.L. (1968) The role of a-Lactalbumin and the A protein in Lactose synthetase: A unique mechanism for the control of a biological reaction, *Proc. Natl. Acad. Sci. USA* **59**, 491-497.

Brodbeck, U., Denton, W.L., Tanahashi, N. and Ebner, K. E. (1967) The isolation and identification of the B Protein of Lactose Synthetase as α-Lactalbumin, *J. Biol. Chem.* **242**, 1391-1397.

Keshvara,L.M., Newton,E.M., Good,A.H., Hindsgaul,O. and Palcic,M.M. (1992) Enzyme-linked immunosorbent assays for the measurment of blood group A and B glycosyltransferase activities, *Glycoconjugate J* **9**, 16-20.

Keusch ,J., Lydyard,P.M., Isenberg ,D.A. and Delves,P.J. (1995) β 1,4–Galactosyltransferase activity in B cells detected using a simple ELISA-based assay, *Glycobiology* **5**, 365-370.

Kim,Y.S., Perdomo,J., Whitehead,J.S. and Curtis,K.J. (1972) Glycosyltransferases in human blood : Study of serum galactosyltransferase and N-acetylgalactosaminyl-transferase in patients with liver diseases, *J. clin. invest.* **51**, 2033-2039.

Kobayashi,K., Akaike,T. and Usui,T. (1994) Synthesis of poly(N-Acetyl-β -lactosaminide-carrying Acrylamide):Chemical enzyme hybrid process. *Methods in Enzymology* **242**, *Neoglycoconjugates Part A :Synthesis*, Lee,Y.C. and Lee,R.T. Academic Press , San Diego, 226-235.

Kobayashi,K., Kobayashi,A. and Akaike,T. (1994) Culture of hepatocytes on a lactose-carrying polystyrene via asialoglycoprotein receptor mediated interactions, *Methods in Enzymology* **247**, *Neoglycoconjugates Part B, Biomedical applications*, Lee,Y.C. and Lee,R.T. Academic Press, San Diego, pp. 409-418.

Merkle,R.K and Cummings,R.D. (1987) Lectin affinity chromatography of glycopeptides. *Methods Enzymol.* **138**, 232 259.

Gauduchon,P., Tillie,C., Guyonnet ,C., Heron, J., Bar-Guilloux,E. and Le Talaer,J. (1983) Clinical Value of serum glycoprotein galactosyltransferase levels in different histological types of ovarian carcinoma, *Cancer Research* **43**, 4491-4496.

Davey,R.D., Milliken,S.M., Harvie,R.M., Cahill,E.J., Morgan,L.J., Kerestes,Z . and Levi,J.A.(1986) Serum galactosyltransferase as a prognostic marker in patients with solid tumors, *Cancer Research* **46**, 5973-5975.

Stibler,H.and Borg,S. (1991) Glycoprotein glycosyltransferase activities in serum in alcohol-abusing patients and healthy controls *Scand. J. Clin. Lab. Invest*, **51**, 43-51.

Stult ,C.L.M. and Macher,B.A. (1990) Measurement of β Galactosyltransferase activity in cell extacts with an ELISA-based assay, *Arch.Biochem.Biophys.* **280**, 20-26.

Taki,T., Nishikawa,S., Ishii,K.and Handa,S. (1990) A simple and specific assay of glycosyltransferase and glycosidase activities by an enzyme-linked immunosorbent assay method, and its application to assay of galactosyltransferase activity in sera from patients with cancer, *J. Biochem.* **107**, 493-498.

Uemura,M., Winant,R.C.and Brandt,A.E. (1988) Immunoassay of serum galactosyltransferase isoenzyme 2 in cancer patients and control subjects. *Cancer Research* **48**, 5335-5341.

Wilson,I.B.H., Platt,F.M., Isenberg,D.A. and Rademacher,T.W. (1993) Aberrant control of galactosyltransferase in peripheral B lymphocytes and Epstein-Barr virus transformed B lymphoblasts from patients with rheumatoid arthritis, *J. Rheumathol* **20**, 1282-1287.

Xin,Y., Lasker,J.M.and Lieber,C.S.(1995) Serum carbohydrate-deficient transferrin : Mechanism of increase after chronic alcohol intake, *Hepatology* **22**, 1462-1468.

Zatta,P.F., Nyame,K., Cormier,M.J., Mattox,S.A., Smith,D.F. and Cummings,R.D (1991) A solid phase assay for β1,4-galactosyltransferase activity in human serum using recombinant aequorin, *Anal. Biochem.* **194**, 185-191.

CONCANAMYCIN A, A VACUOLAR TYPE H^+-ATPASE INHIBITOR, SELECTIVELY INDUCES CELL DEATH IN ACTIVATED $CD8^+$ CTL

KEN-ICHI TOGASHI, TAKAO KATAOKA & KAZUO NAGAI
Department of Bioengineering, Tokyo Institute of Technology
4259 Nagatsuta-cho, Midori-ku, Yokohama, Kanagawa 226, Japan

Concanamycin A (CMA) and concanamycin B (CMB) are specific inhibitors of vacuolar type H^+-ATPase (V-ATPase). In our previous studies, it was shown that intraperitoneal injection of CMB suppressed the increase in $CD8^+$ CTL population without affecting $CD4^+$ and $B220^+$ populations in mice immunized with allogeneic tumors. To clarify the molecular basis of the selective decrease in the $CD8^+$ CTL population by CMB, we have performed a series of *in vitro* experiments with use of CMA. Cell viability of the $CD8^+$ population prepared from the immunized mice was preferentially decreased by CMA treatment for 4-8 h. Moreover, in the $CD8^+$ CTL clone, CMA induced a marked DNA fragmentation characteristic of apoptosis. Anti-CD3 or phorbol ester accelerated the CMA-induced reduction in cell viability of the $CD8^+$ CTL clone, but not $CD4^+$ T cell clones. However, this rapid cell death was not accompanied by DNA fragmentation. Thus, our data suggest that V-ATPase activity is essential for survival of $CD8^+$ CTL especially when activated.

1. Introduction

Vacuolar type H^+-ATPase (V-ATPase) supports several cellular functions through acidification of intracellular organelles such as lysosomes and Golgi apparatus [1, 2]. These functions are strongly perturbed by specific inhibitors of V-ATPase such as bafilomycins and concanamycins [3-6]. We have previously reported that the injection of concanamycin B (CMB) severely reduced the induction of $CD8^+$ CTL population and the perforin-dependent killing activity without affecting $CD4^+$ and $B220^+$ populations in mice immunized with allogeneic tumors [7]. In this paper, to address the mechanism of the selective decrease in $CD8^+$ population, we have performed a variety of *in vitro* experiments with use of CMA.

Abbreviations: CMA–concanamycin A, CMB–concanamycin B, CTL–cytotoxic T lymphocytes, FCS–fetal calf serum, KLH–keyhole limpet hemocyanin, OVA–ovalbumin, PBS–phosphate-buffered saline, PMA–phorbol 12-myristate 13-acetate, V-ATPase–vacuolar type H^+-ATPase.

2. Materials and Methods

2.1. MICE

C57BL/6, BALB/c and C3H/HeN mice (female, 6 wk-old) were purchased from Charles River Co. Ltd. (Yokohama, Japan). Six to eight wk-old mice were used for experiments.

2.2. CELLS

H-2^d-specific $CD8^+$ CTL clone OE4 [8], perforin-deficient H-2^k-specific $CD8^+$ CTL clone POK [9], KLH-specific I-E^d-restricted $CD4^+$ CTL clone BK-1 [10], OVA-specific I-A^d-restricted $CD4^+$ Th1 clone 42-6A [11], and I-A^b-specific $CD4^+$ Th2 clone D10.G4.1 [12] were cultured in the RPMI 1640 medium (Life Technologies, Grand Island, NY) supplemented with 10% (v/v) FCS (Bioserum, Victoria, Australia) and 5% (v/v) culture supernatant of rat spleen cells stimulated with 5 μg/ml of concanavalin A for 24 h. OE4, POK, and D10.G4.1 were stimulated with mitomycin C-treated spleen cells from BALB/c, C3H/HeN and C57BL/6 mice every two weeks, respectively. BK-1 and 42-6A were stimulated with mitomycin C-treated BALB/c spleen cells together with 10 μg/ml of KLH and OVA every two weeks, respectively. These cells were used for experiments at least over 3 wk after the last stimulation. A mastocytoma P815 (H-2^d) was maintained in the RPMI medium plus 10% (v/v) FCS.

2.3. REAGENTS

CMA was kindly provided by Dr. K. Mizoue, Taisho Pharmaceutical Company Ltd. (Tokyo, Japan). Phorbol 12-myristate 13-acetate (PMA) was the commercial product of Wako Pure Chemical Industries Ltd. (Osaka, Japan). Anti-CD4 and anti-CD8 monoclonal antibodies were purchased from Seikagaku Co. (Tokyo, Japan) and anti-mouse immunoglobulin polyclonal antibody was obtained from Cedarlane Laboratories Ltd. (Ontario, Canada).

2.4. PREPARATION OF LYMPHOCYTE POPULATIONS

C57BL/6 mice were injected intraperitoneally with P815 (2×10^7 cells/mice). Spleen cells were removed from the mice on day 11. After adherent cells were removed, spleen cells (1.5×10^7 cells) were incubated in antibody-coated culture dishes (35 mm diameter) for 30 min on ice. Non-adherent cells were removed by washing three times with PBS, and then antibody-bound cells were collected. Each population was shown to be >90% purity by flow cytometry.

2.5. MEASUREMENT OF CELL VIABILITY

Cells (2×10^4 cells/well) were cultured in microtiter plates for indicated periods. For the last 3h, the cells were incubated with 500 μg/ml of 3-(4,5-dimethlthazol-2-yl)-2,5-diphenyl tetrazolium bromide (MTT, Sigma Chemical Co., St. Louis, MO). One hundred microliters of 0.04 N HCl in isopropanol were added to each well, and mixed vigorously. Absorbance at 595 nm was measured. The experiments were performed in triplicate cultures.

2.6. DETECTION OF DNA FRAGMENTATION

[^{3}H]thymidine-labeled T cell clones (2×10^4 cells/well) were treated with drugs in microtiter plates as indicated, and then the cells were lysed by the addition of 0.01% Triton X-100. Intact DNA was precipitated by centrifugation (3,000 X *g*, 3 min), and then the supernatants were recovered. Radioactivity in the supernatants was measured. Percentage of DNA fragmentation was calculated by using following formula: (experimental release – spontaneous release)/(maximum release – spontaneous release) X 100. The experiments were performed in triplicate cultures.

3. Results

To clarify the molecular mechanism of the CMB-induced selective decrease in $CD8^+$ population *in vivo*, we prepared lymphocyte populations by panning method, and treated these cells with CMA, a close analog of CMB, *in vitro* (Table 1). All lymphocyte populations of both normal and the immunized mice retained their viability until 8 h, and then gradually lost their viability. The addition of CMA accelerated the decrease in cell viability of all populations. It should be noted, however that cell viability of $CD8^+$ population prepared from the immunized mice was preferentially decreased by CMA treatment. A certain group(s) of $CD8^+$ population in the immunized mice might be highly susceptible to CMA.

Above data prompted us to examine the effect of CMA on various $CD8^+$ and $CD4^+$ T cell clones (Fig. 1). CMA induced a significant DNA fragmentation in a $CD8^+$ CTL clone OE4 in 8 to 20 h (Fig. 1A). Moreover, nuclear condensation was also observed in CMA-treated OE4 (data not shown), suggesting that CMA induces apoptosis. However, CMA only marginally increased DNA fragmentation in $CD4^+$ T cell clones (Fig. 1B to 1D). Although CMA is reported to induce apoptotic cell death to the several cultured cell lines [13], CD8+ CTL clone seems to be most sensitive to CMA (see discussion).

A part of CD8+ population in mice immunized with allogeneic tumors might repeatedly receive activation signals from neighboring cells in the spleen and become more sensitive to CMB. To examine this possibility, CD8+ and CD4+ T cell clones were treated with CMA in the presence of stimuli such as immobilized anti-CD3 or PKC activator, PMA (Fig. 2). Although CMA itself gradually decreased the number of viable cells of OE4, CMA in combination with these stimuli dramatically decreased the viable cells within 4 h after the treatment (Fig. 2A). Similar results were obtained by dye exclusion assay using trypan blue (data not shown). The viability of perforin-deficient CD8+ CTL clone P0K was also decreased rapidly by the treatment with CMA plus PMA (Fig. 2B). On the other hand, PMA failed to affect the CMA-induced cell death of CD4+ T cell clones (Fig. 2C to 2E). These results suggest that the activation through T cell receptors or PKC activation increases the sensitivity of CD8+ CTL to CMA. To address whether CMA- and PMA-induced rapid cell death is caused by apoptosis, two typical characteristics usually associated with apoptosis were examined. PMA accelerated neither the DNA fragmentation (Fig. 2F) nor nuclear condensation (data not shown). These data imply that immediate cell death induced by CMA in the presence of PMA is not apoptosis.

Mice	CMA	Incubation (h)	Viability (%)		
			CD4+	CD8+	B220+
Normal	None	4	100.3	103.8	97.7
		8	92.5	94.7	94.5
		20	59.2	67.3	66.2
	100nM	4	85.5	86.7	80.1
		8	72.7	67.0	64.9
		20	27.8	22.5	31.7
P815-immunized	None	4	100.4	99.2	92.1
		8	97.1	93.2	94.7
		20	54.9	49.1	47.0
	100nM	4	83.1	69.0	83.1
		8	73.4	50.7	67.5
		20	20.2	12.1	20.1

Table 1. Effect of CMA on the viability of lymphocyte populations in spleen cells
Lymphocyte populations were obtained from P815-immunized or normal C57BL/6 mice by panning method. Each population was cultured in the presence or the absence of 100 nM CMA, then their viability at indicated periods was analyzed by MTT method.

4. Discussion

CD8+ population from P815-immunized mice showed higher sensitivity to CMA as compared with other lymphocyte populations from naive and the immunized mice. Moreover, CD8+ CTL clones, but not CD4+ T cell clones, were extremely sensitive to CMA, and exhibited two kinds of cell death: (1) apoptosis in the long-term culture and (2) rapid cell death in the short-term culture in the presence of PMA, which apparently lacked DNA fragmentation and nuclear condensation.

It is reported that CMA induces apoptotic cell death to several cultured cell lines [13]. However, CMA induces only about 30% of apoptosis in B lymphoma by 24 h treatment [13]. In our present data, CMA induced 90% of DNA fragmentation in OE4 by 20 h treatment, and CMA plus PMA induced rapid cell death within 4 h. Thus,

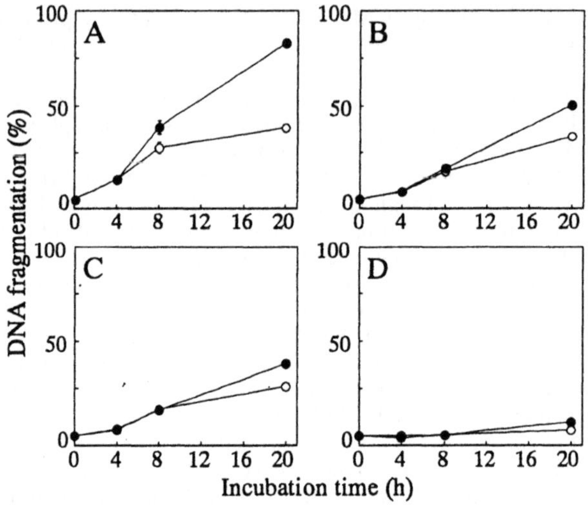

Figure 1. CMA induced DNA fragmentation in T cell clones
[^{3}H]thymidine-labeled OE4 (A), BK-1 (B), 42-6A (C), and D10.G4.1 (D) were incubated in the presence (closed circles) or the absence (open circles) of 100 nM CMA for indicated periods, and percentage of DNA fragmentation was estimated as described in Materials and Methods.

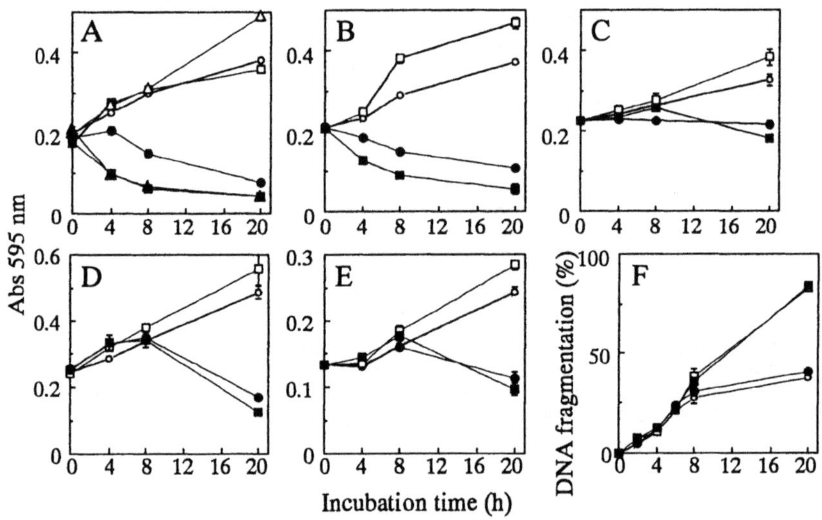

Figure 2. Effect of activation signals on CMA-induced cell death of T cell clones
OE4 (A), POK (B), BK-1 (C), 42-6A (D), and D10.G4.1 (E) were treated with anti-CD3 (triangles) or 100 nM PMA (squares), or without stimuli (circles) in the presence (closed symbols) or the absence (opened symbols) of 100 nM CMA. At indicated incubation periods, the viable cell numbers were tested by MTT method. (F) [^{3}H]thymidine-labeled OE4 was incubated in the presence of CMA and/or PMA for indicated periods and percentage of DNA fragmentation was estimated.

CD8+ CTL clones seem to be most sensitive to CMA among various cell types so far tested.

Why are CD8+ CTL clones most susceptible to CMA? As far as the CMA-induced apoptosis is concerned, one possibility is the involvement of apoptosis-inducing effector molecules (e.g. granzyme B) present in the lytic granules. Since CMA induced drastic morphologic changes of the lytic granules such as vacuolation [14], granular content(s) might leak from the granules into cytosol, thereby inducing apoptosis.

What mechanism(s) is involved in the rapid cell death of the CD8+ CTL clones in the presence of CMA and PMA (or anti-CD3 antibody)? Inhibition of acidification seems to be irrelevant, since monensin and chloroquine which neutralize the internal pH in acidic organelles had negligible effects (data not shown). CD8+ CTL clones might contain a large number of V-ATPase, in that these cells harbor a number of the lytic granules. If this is the case, CD8+ CTL clones might depend heavily on V-ATPase activity for cell survival. It is reported that PMA increases V-ATPase activity [15]. Thus, it is possible that CMA plus PMA treatment bring negative and positive effects on V-ATPase, leading to a severe unbalance on V-ATPase-regulating functions such as maintenance of the intracellular pH. At present, other mechanism(s), e.g. ecto-type ATPase sensitive to CMA and PMA-induced gene expression, are equally possible.

Which kind of the cell death, or both, are actually involved in the CMB-induced selective decrease in CD8+ population in the immunized mice has been unanswered. In the spleen of the immunized mice, CD8+ CTL might be repeatedly stimulated by cognate interaction with antigen-presenting cells or soluble factor(s). If this is the case, the rapid cell death might be dominantly involved in the elimination of activated CD8+ T cells.

5. References

1. Mellman I, Fuchs R & Helenius A (1986) Acidification of the endocytic and exocytic pathways. *Ann. Rev. Biochem.* **55**: 663-700.
2. Forgac M (1989) Structure and function of vacuolar class of ATP-driven proton pumps. *Physiol. Rev.* **69**: 765-796.
3. Woo JT, Shinohara C, Sakai K, Hasumi K & Endo A (1992) Isolation, characterization and biological activities of concanamycins as inhibitors of lysosomal acidification. *J. Antibiot.* **45**: 1108-1116.
4. Muroi M, Shiragami N, Nagao K, Yamasaki M & Takatsuki A (1993) Folimycin (concanamycin A), a specific inhibitor of V-ATPase, blocks intracellular translocation of the glycoprotein of vesicular stomatitis virus before arrival to the Golgi apparatus. *Cell Struct. Funct.* **18**: 139-149.
5. Dröse S, Bindseil KU, Bowman EJ, Siebers A, Zeeck A & Altendorf K (1993) Inhibitory effect of modified bafilomycins and concanamycins on P- and V-type adenosinetriphosphatases. *Biochemistry* **32**: 3902-3906.
6. Bowman EJ, Siebers A & Altendorf K (1988) Bafilomycins: a class of inhibitors of membrane ATPases from microorganisms, animal cells, and plant cells. *Proc. Natl. Acad. Sci. USA* **85**: 7972-7976.
7. Lee MH, Kataoka T, Magae J & Nagai K (1995) Prodigiosin 25-C suppression of cytotoxic T cells *in vitro* and *in vivo* similar to that of concanamycin B, a specific inhibitor of vacuolar type H^+-ATPase. *Biosci. Biotech. Biochem.* **59**: 1417-1421.
8. Staerz UD, Kanagawa O & Bevan MJ (1985) Hybrid antibodies can target sites for attack by T cells. *Nature* **314**: 628-631.
9. Kojima H, Shinohara N, Hanaoka S, Someya-shirota Y, Takagaki Y, Ohno H, Saito T, Katayama T, Yagita H, Okumura K, Shinkai Y, Alt FW, Matsuzawa A, Yonehara S & Takayama H (1994) Two distinct pathways of specific killing revealed by perforin mutant cytotoxic T lymphocytes. *Immunity* **1**: 357-365.
10. Shinohara N, Huang YY & Muroyama A (1991) Specific suppression of antibody responses by soluble protein-specific, class II-restricted cytolytic T lymphocyte clones. *Eur. J. Immunol.* **21**: 23-27.
11. Kakiuchi T, Watanabe M, Hozumi N & Nariuchi H (1990) Differential sensitivity of specific and nonspecific antigen-presentation by B cells to a protein synthesis inhibitor. *J. Immunol.* **145**: 1653-1658.
12. Kaye J, Gillis S, Mizel SB, Shevach EM, Malek TR, Dinarello CA, Lachman LB & Janeway Jr. CA (1984) Growth of a cloned helper T cell line induced by a monoclonal antibody specific for the antigen receptor: Interleukin 1 is required for the expression of receptors for interleukin 2. *J.Immunol.* **133**: 1339-1345.

13. Nishihara T, Akifusa S, Koseki T, Kato S, Muro M & Hanada N (1995) Specific inhibitors of vacuolar type H^+-ATPases induce apoptotic cell death. *Biochem. Biophys. Res. Commun.* **212**: 255-262.
14. Kataoka T, Takaku K, Magae J, Shinohara N, Takayama H, Kondo S & Nagai K (1994) Acidification is essential for maintaining the structure and function of lytic granules of CTL. *J. Immunol.* **153**: 3938-3947.
15. Nanda A, Gukovskaya A, Tseng J & Grinstein S (1992) Activation of Vacuolar-type proton pumps by protein kinase C. *J. Biol. Chem.* **267**: 22740-22746.

YIELD IMPROVEMENTS UTILISING FED-BATCH CULTIVATION AND OPTIMISATION STRATEGIES

C.S.SANDERSON , C.P.MARQUIS AND J.P.BARFORD
Department of Chemical Engineeering, University of Sydney 2006, N.S.W., Australia

In previous work, we have demonstrated the improvements that fed-batch systems can bring to the overall antibody productivity and antibody concentration obtained in batch culture.

It is now possible to transform these experimental findings into specilialsed feeding programs which are tailored to a particular cell line. This has the effect of generalising the approach for industrial purposes. It also allows for different strategies to be tested as deemed necessary.

The testing of various feeding options to find an optimum which can be tested experimentally relies on access to a sophisticated computer simulation and optimisation procedure as well as a series of carefully controlled laboratory experiments demonstrating its usefulness. To be effective commercially, there needs to be a demonstration of improvement with a number of cell lines in industry and industry access to the procedure needs to be cost effective in terms of increased productivity.

1. Introduction

The use of fed-batch cultivation of animal cells has the advantage of extended culture time with high cell viability, the ability to control the accumulation of toxic end products (or their rate of accumulation) as well as , in theory, allowing rational feeding strategies to be developed.

The basis for developing rational feeding strategies requires one to develop a mathematical description of the batch and fed-batch process and to evaluate various feeding regimes on some defined objective basis. In our laboratories, we have developed a extensive simulation of animal cells and have developed some "first-guess" objective functions by which various feeding strategies can be tested on a profit /

reactor volume / time basis. In addition to testing a number of defined feeding strategies which deliver feed at particular times, we have developed continuous feeding optimisation which allows the feeding pattern to be continuously adjusted during the fed-batch. In principle, this is a powerful technique for the increase in productivity of antibody production. However to date, the application of such advanced feeding strategies has not been widely accepted.

In this paper, we present the advantages of such an approach and discuss the steps necessary for this to be more commonly applied in industry. In accompanying works, (Black et al.,1997; Renneberg et al.,1997) we look at regulatory aspects surrounding fed-batch systems and the role that biosensors may play in accelerating the application of on-line process control in the animal cell technology industry.

2. Background

The basis of the simulation developed by our group and the dynamic optimisation program have been previously described (Sanderson et al.,1997 (In Press)).

3. Results and Discussion

A fed-batch culture of the mouse hybridoma AFP - 27 (Marquis, 1994) was performed with a period of constant dilution rate to determine if increased antibody secretion could be maintained under controlled conditions. A batch or initial medium and a feeding medium were prepared as follows:

Component	Batch (1.5 l)	Fed-Batch (2.5l)
Glucose (mM)	50	50
Glutamine (mM)	0.5	10
MEM Amino Acids	1 X	2 X
MEM Vitamins	0	1 X
FCS	2%	2%

During the batch phase, 6 mmol of glutamine was fed at a constant rate from a 60 mM solution over approximately 60 hours, to avoid excessive loss of glutamine due to spontaneous breakdown. During the main feeding phase, the feed pump was per-programmed to deliver at an exponentially increasing rate, in an attempt to achieve a constant dilution rate of 0.018 h^{-1}. The main feed was commenced at 63.5 hours and ceased at 117 hours, post inoculation. The feed profile and calculated dilution rate are described in Figure 1. This experiment is an example of a fixed , defined feeding regime based of previous knowledge of the batch culture growth of the hybridoma and does not try to identify an optimum feeding strategy.

The biomass profile for this experiment is described in Figure 2 and the secreted antibody is shown in Figure 3. Biomass increased to a level of 1.64 +/- 0.2 X 10^6 per ml at 83 hours and this level was more or less maintained until 130 hours, from which viable cell decline began. During this period, the non-viable cell population consistently rose. Antibody increased during the batch growth, increased slightly over the feeding period and increased considerably (from 50 to 80 mg per ml) during the decline phase, once feeding had ceased. The ammonia profile shown in Figure 4, shows that a level of approximately 3 mM was reached during the batch phase, rising to 3.8 to 4 mM which was maintained during the feeding stage then increasing steadily during the post-feeding stage to 6 mM. Glucose was consumed continuously from a high initial level of 50 mM to 20 mM. Lactate rose consistently during the same period, reaching a level of nearly 30 mM at the end of the culture (Figure 5). This experiment shows that yield improvements can be gained by fed-batch operation but the strategy if far from optimal . The optimal feeding strategy may be obtained by a series of these experiments in a "trial and error" manner - however this is both time-consuming and costly.

As an alternative to a "trial and error" approach, we have investigated the use of optimisation algorithms to determine the optimal initial concentrations of glucose and glutamine and the volume of and time at which shots of feed should be added This has been described in detail elsewhere (Sanderson et al.,1997(In Press)). The following optimal results were obtained when the composition of the feed stream and the initial concentrations of the other components were held constant (to limit the dimensionality of the problem):

Initial Glutamine Conc'n (μM)	**Initial Glucose Added Conc'n (μM)**	**Volume (ml / l)**		**Time of Addition (h)**		**$/h / l**
Base Case - Optimised Medium						
4,831	24,399					383
2 x 10 ml / l - Glutamine						
9,952	21,421	10	10	0.2	34	384
2 x 10 ml / l - Mixed						
5,730	27,509	9.5	10	97.5	130	531

In the first strategy, only a 200 mM glutamine solution was added to the culture whilst, in the second feeding strategy, a medium supplement was added . The concentrations of the substrates was determined in the following way :

(i) The concentration of the substrates at the beginning and the end of the batch were compared.

(ii) The volume of a feed stream containing 200 mM glutamine that would be required to restore the glutamine concentration to its initial level was calculated

(iii) The concentration of the other components that would be required in this shot to restore those components to their inoculation levels was found.

This was compared to the composition of the commercially available media and the lesser of the two values used.

The resultant medium composition (mM) was :
Ala : 0; Arg : 10; Asp : 6; Asp: 1; Cys : 6; Glu : 4; Gly : 0; His : 5; Ile : 20; Leu : 20; Lys : 20; Met : 4; Phe : 6; Pro : 0; Ser : 3; Thr : 10; Trp : 2; Tyr : 6; Val : 20; Gl : 100;

It can be seen that this strategy, in principle, allows the development of a rational feeding program .

However, there still remain some barriers to the full-scale industrial implementation of this scheme. An industry must be prepared to either :

(i) Understand the knowledge base contained in the simulation and optimisation package. This would require a commitment to computerisation of the production process.

or

(ii) Provide experimental batch data to a “servicing” company who would then provide “tailored” feeding strategies to the industry on a fee paying basis .

The latter option is the cheaper but it takes the control of the operation of the process away from the industry client. The former requires more expenditure (for packages, licences etc) and requires a higher level of training and participation. It does, however, have the advantage of keeping the expertise “in house” and allows the application of the methodology to a range of products and cell lines.

CONCLUSIONS

Fed-batch operation offers increased productivity. Optimal feeding strategies calculated using a computer simulation of animal cell growth and optimisation routines offer increased productivity and a rational feeding policy. Prior to industrial implementation, it is necessary to demonstrate the applicability of this approach for a large number of cell lines. Once established as a useful methodology, industry will need to decide whether to use this approach "in house" or have it applied for them on a "contracted out " basis.

REFERENCES

Black, DJ, JP Barford & C.Harbour (1997) Quality aspects of batch animal cell cultivation - relevance to fed-batch implementation in industry, Proceedings of 9th Annual Meeting of the Japanese Association For Animal Cell Technology, Yokohama, Japan.

Marquis CJ (1994) Batch and fed-batch cultivation of murine hybridomas and human lymphoblastoid cell lines, Ph.D. thesis, University of Sydney.

Renneberg, R, JP Barford & S.M. Tse (1997) The use of biosensors in control of fed-batch processes - opportunities and limitations, Proceedings of 9th Annual Meeting of the Japanese Association For Animal Cell Technology, Yokohama, Japan.

Sanderson CS, JP Barford and GW Barton (1997) Simulation and dynamic optimisation of animal cell culture, Cytechnology (In Press).

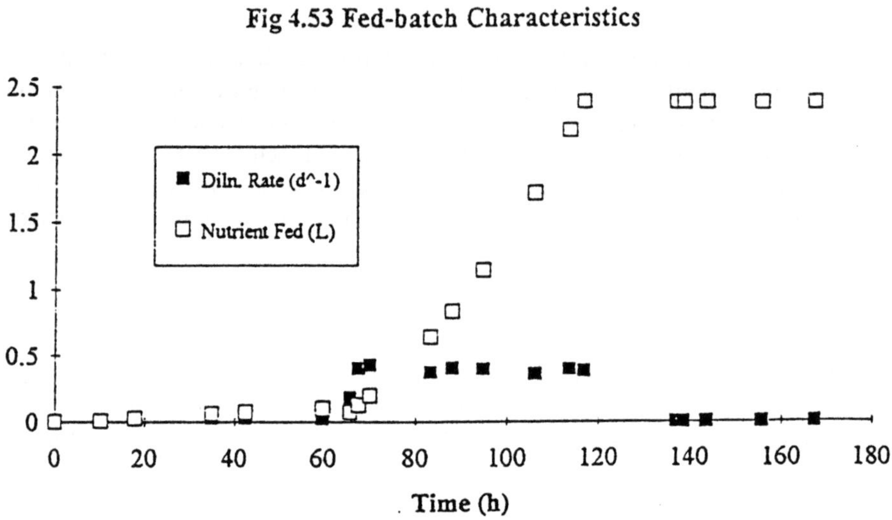

Figure 1 Fed-Batch Feed Characteristics

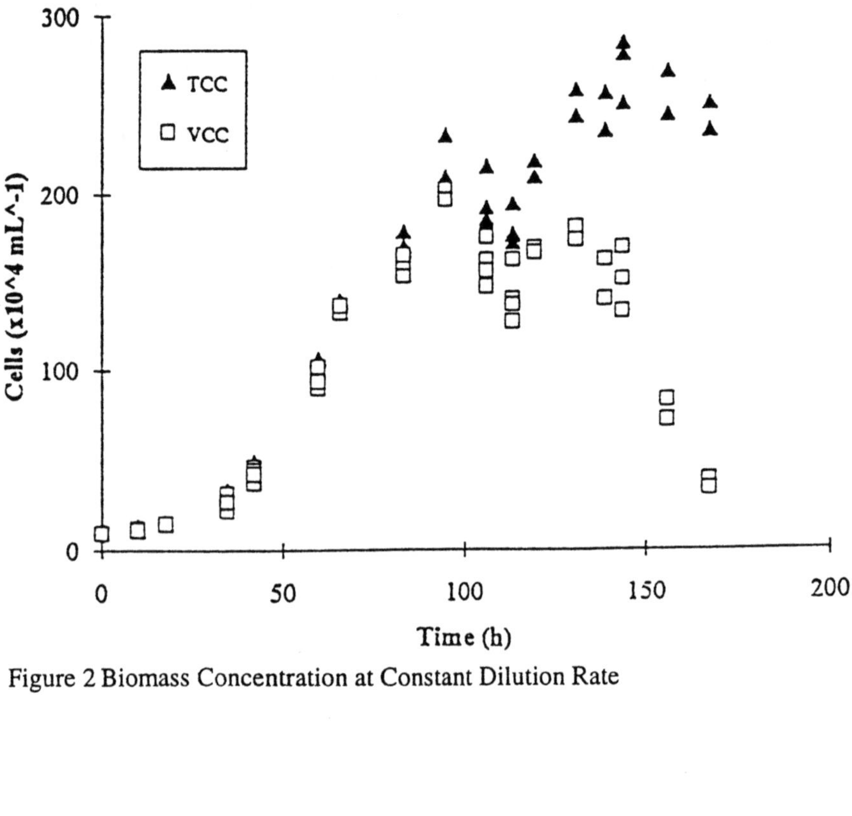

Figure 2 Biomass Concentration at Constant Dilution Rate

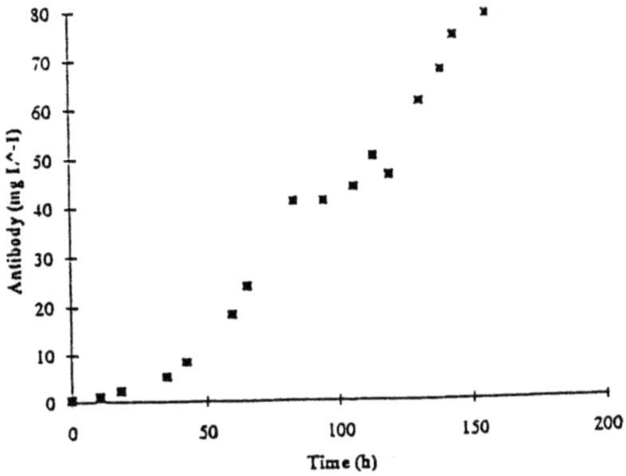

Figure 3 Antibody Concentration at Constant Dilution Rate

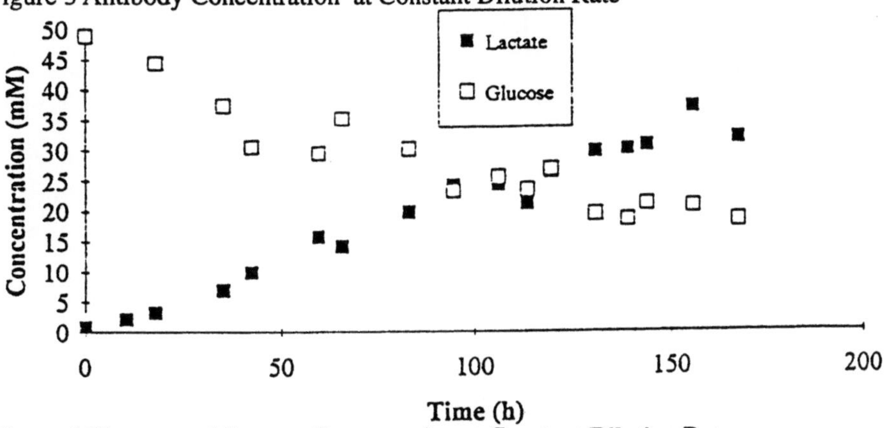

Figure 4 Glucose and Lactate Concentration at Constant Dilution Rate

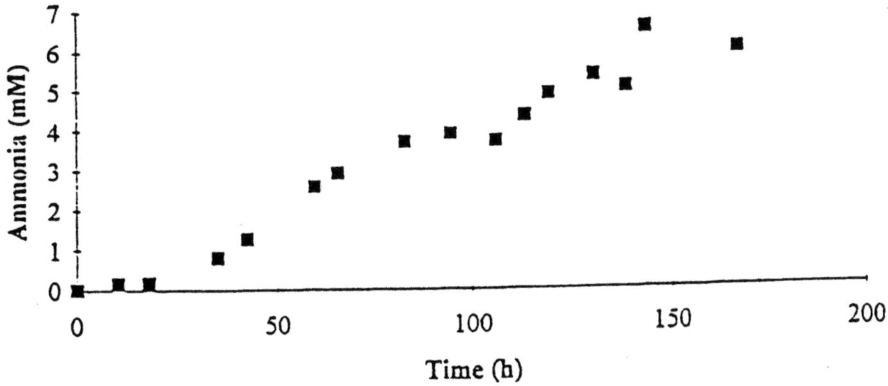

Figure 5 Ammonia Concentration at Constant Dilution Rate

QUALITY ASPECTS OF BATCH ANIMAL CELL CULTIVATION - RELEVANCE TO FED-BATCH IMPLEMENTATION IN INDUSTRY

D.J.Black [1], *J.P.Barford*[1] *and C.Harbour*[2]
Departments of Chemical Engineering[1] *and Infectious Diseases*[2] ,
University of Sydney 2006, NSW, Australia.

There is often the implicit criticism of fed-batch cultivation techniques due to their transient nature. Just how variations in media composition and product quality are considered by regulatory authorities, present a perception which is important to address.

In this work, the effect of growth phase in batch cultivation on functional activity and glycosylation of an Anti-D secreted by BTSN4 lymphoblastoid cell lines is examined. Rosetting assays, sugar(s) composition and relative distributions of glycoforms were shown to vary depending on the growth phase.

From this work, it is argued that significant variations in quality exist in batch culture and that fed-batch should not be considered to pose further issues of quality variations of regulatory concern.

1. Introduction

The issue of product quality addresses variations in the composition of products obtained from animal cell cultivation. Variations may have serious implications in terms of the product's ultimate therapeutic potential as well as causing problems for regulatory authorities. Fed-batch, because of inherent transient nature, is often considered to be a risky fermentation strategy (from a regulatory viewpoint) given these concerns. In this paper, we examine the variation in product quality during batch fermentation and demonstrate that significant variations appear during this type of cultivation as well. As such, it is questionable whether product variation should be the only criteria by which a regulatory authority considers a product. Whether process repeatability might be a better monitor of product consistency and effectiveness is raised by the results of this work.

2. Materials and Methods

A human lymphoblastoid cell line producing anti-D was used .Glycosylation analysis consisted of overall monosaccharide composition of the sugar chains and detection of different glycoforms present. Neutral monosaccharide and sialic quantification were achieved using HPLC. Two methods were used to determine the diversity of glycoforms present at the asparagine-297 glycosylation site of purified anti-D. Each method involved an initial step of enzymic cleavage and isolation of intact oligosaccharide chains from the protein backbone. In one protocol, the individual sugar species were subsequently separated by anion exchange chromatography and in the other method a electrophoretic separation on a highly cross-linked agarose gel was used. These methods are fully described elsewhere (Black, 1996).

BTSN4 cells were grown in batch mode. Samples taken at the end of the lag phase, in mid-exponential and mid-stationary phase and late in the death phase were purified and used to determine the effect of growth phase on the structure and functional activity of the anti-D produced. Supernatant samples of 150 mls were taken from each of four 250 ml spinner flasks at various points in the batch curve. Each flask contained IMDM medium supplemented with 10% heat-activated, bovine IgG-free F.C.S. and 2 mM glutamine. Each culture was inoculated at an initial viable cell density of 5 x 10^5 cell / ml from a common source produced by pooling several T-flask cultures. Incubation was at 35° C, agitation at 35 rpm using a magnetic stirring plate. the cultures were oxygenated and the pH controlled by discontinuous gassing with sterile filtered 9.5% CO_2 in air. As seen in Figures 1 and 2, there was no significant difference in the viable cell density of antibody profiles of these flasks.

3. Results and Discussion

Table 1 compares the functional (i.e Fc-mediated) activity of anti-D produced during each of the growth phases studied. As can be seen from these results, there was no significant difference in functional activity of anti-D between that produced during the exponential and stationary phases of growth, as determined by rosetting assay with U937 cells.

The heavy chain glycosylation on anti-D isolated during each phase of growth in batch culture was analysed both in terms of overall content of constituent monosaccharides and in terms of relative number and proportion of different glycoforms present. These results are presented in Table 2.

Each anti-D yielded the monosaccharides N-acetlyglucosamine, galactose, mannose, fucose and sialic acid in varying proportions when hydrolysed with trifluoroacetic acid. The highest sugar content was seen in the anti-D isolated from the spinner harvested during the stationary phase. It may be surmised that the observed differences in anti-D sugar content between mid-exponential and stationary phase cultures was linked to the sugar content between mid-exponential and stationary phase cultures was linked to the rate of antibody secretion, higher secretion rates may result in lower degree of post-transitional processing than that experienced by antibody secreted at a lower rate in the culture. The monosaccharide ratios calculated for each of the samples showed that anti-D harvested in the stationary phase contained significantly higher N-acetylglucosamine and galactose than the anti-D from the mid-exponential phase, whereas the fucose and sialic acid contents were significantly lower. Finally, the glycosoylation structure of the anti-D secreted during the lag phase in batch is considered. It can be seen for Table 2 that this antibody has a significantly lower mannose content and a significantly higher sialic acid content than the anti-D from the mid-exponential phase culture.

Figure 3 shows the glycoforms detected by the FACE system for the anti-D derived from batch cultures harvested during the mid-exponential, stationary and death phases. As seen in Table 3, the highest sugar content was seen in the stationary phase culture. This agrees with the results of the monsaccharide analysis in Table 2.

4. Conclusions

The results presented here have a number of implications for the culture of these cells cells commercially. Firstly, there is significant variations in carbohydrate composition during the various phases of batch culture. Hence, the harvested antibody will be heterogeneous. The implications of this for a regulatory body (and the likely extension of these findings to fed-batch culture) requires one to question the basis of regulations requiring product homogeneity. Rather acceptable variations in product heterogeneity and batch to batch (or fed-batch to fed-batch) consistency is likely a more realistic demand. For this cell line, it is concluded that the best time to harvest as late in the death phase as possible (if cell lysis can be controlled) or , otherwise, in the late stationary phase. This results in maximum antibody productivity and ensures batch to batch consistency of the product. The product heterogeneity present should also be consistent from batch to batch using this strategy.

REFERENCES

Black DJ, Glycoslation Studies on Human Lymphoblastoid Cell Lines, PhD Thesis, University of Sydney.

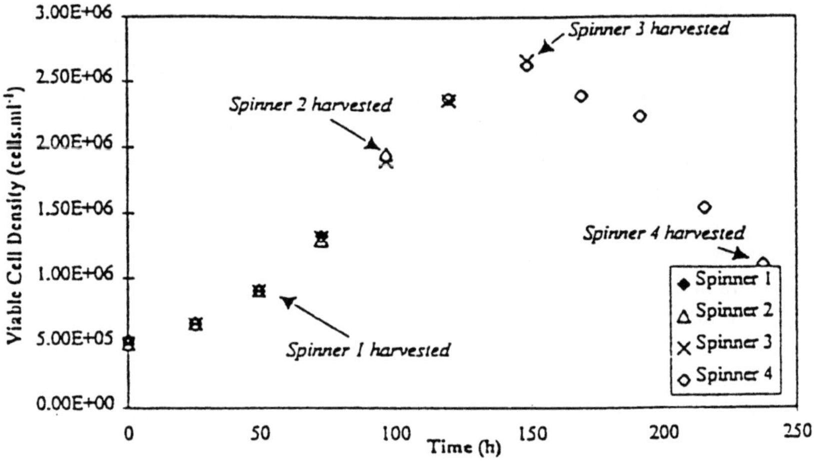

Figure 1 Comparison of BTSN4 Cell Growth in Each Batch Culture

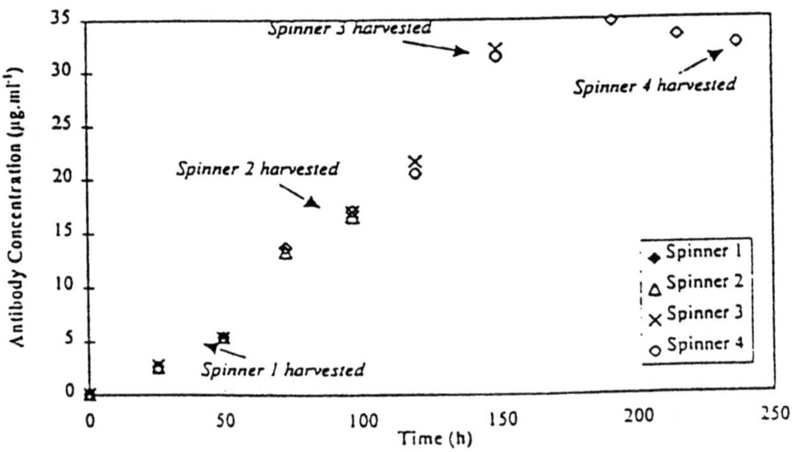

Figure 2 Comparison of Anti-D Production by BTSN4 Cells in Batch Spinner

Spinner No.	*Growth Phase When Harvested*	*Monocyte ADCC (%RBC Lysis)*	*Rosetting (% U937 Cells Forming Rosettes)*
1	Late Lag/Early Exponential	N. D.	14.5 ± 0.8
2	Mid-Exponential	N. D.	17.6 ± 1.0
3	Mid-Stationary	N. D.	18.2 ± 1.1
4	Late Death	N. D.	26.6 ± 0.7

Table 1 Functional Activity of Anti-D Produced by BTSN4 Cells During Various Phases Of Batch Growth (N.D. = Not Done)

Growth Phase	*Units*	*N-Acetylglucosamine*	*Galactose*	*Mannose*	*Fucose*	*Sialic Acid*
Lag Phase	ng Sugar.(μg Antibody)$^{-1}$	9.27 ± 0.23	3.47 ± 0.15	8.21 ± 0.23	N. Q.	1.93 ± 0.08
	Ratio to 3 Mannose	3.39 ± 0.13	1.27 ± 0.07	3		0.70 ± 0.03
Mid-Exponential Phase	ng Sugar.(μg Antibody)$^{-1}$	9.61 ± 0.16	3.45 ± 0.09	6.28 ± 0.14	1.86	1.33 ± 0.11
	Ratio to 3 Mannose	4.59 ± 0.13	1.65 ± 0.06	3	0.89 ± 0.02	0.64 ± 0.05
Stationary Phase	ng Sugar.(μg Antibody)$^{-1}$	16.68 ± 0.89	5.85 ± 0.28	9.55 ± 0.11	2.53 ± 0.03	1.74 ± 0.05
	Ratio to 3 Mannose	5.24 ± 0.29	1.84 ± 0.09	3	0.79 ± 0.01	0.55 ± 0.02
Death Phase	ng Sugar.(μg Antibody)$^{-1}$	11.12 ± 0.24	4.01 ± 0.23	5.70 ± 0.05	1.55 ± 0.19	1.46 ± 0.17
	Ratio to 3 Mannose	5.85 ± 0.13	2.11 ± 0.12	3	0.81 ± 0.10	0.77 ± 0.09

Table 2 Comparison of Monosaccharide Composition of BSTN-4 Derived Anti-D Produced During Different Growth Phases (N.Q. - not quantifiable)

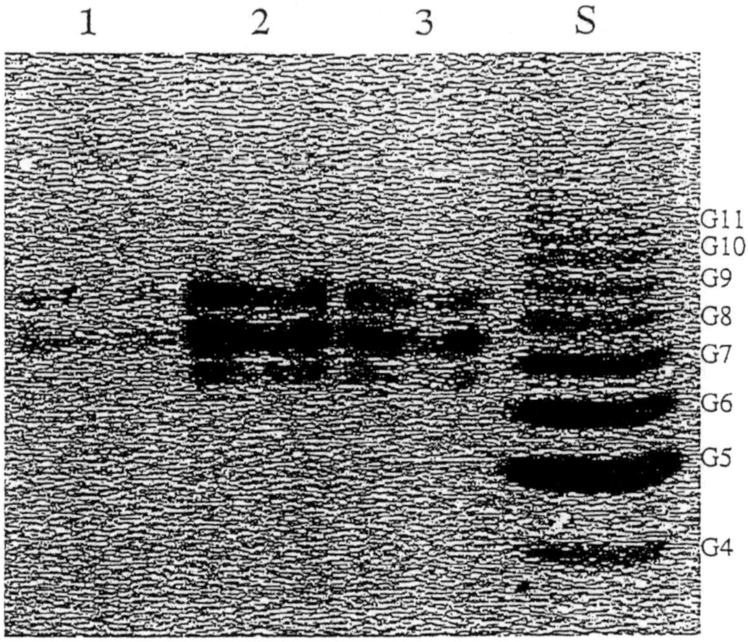

Figure 3 Effect of Growth Phase on Glycoforms Present in BSTN-4 Derived Anti-D. Lane S - FACE Oligo Ladder Standard; Lane 1 - Mid-Exponential; Lane 2 - Stationary Phase; Lane 3 - Death Phase

Growth Phase		*Glycoforms with G7 Equivalent Chain Length*	*Glycoforms with G8 Equivalent Chain Length*	*Glycoforms with G9 Equivalent Chain Length*	*Totals*
Mid-Exponential Phase	pmol	5.48	17.89	14.69	38.06
(25 µg Anti-D)	%	14.4	47.0	38.6	100
Stationary Phase	pmol	21.37	76.98	58.66	157.01
(50 µg Anti-D)	%	13.6	49.0	37.4	100
Death Phase	pmol	18.07	37.53	28.11	83.71
(50 µg Anti-D)	%	21.6	44.8	33.6	100

Table 3 Glycoforms Present in Each Phase of Batch Growth

TECHNICAL APPROACHES TO MINIMIZE REGULATORY HURDLES TO FED-BATCH AND CONTINUOUS MEDIUM SUPPLEMENTATION.

David W. Jayme, Richard M. Fike, Stephen F. Gorfien and Glenn P. Godwin
Life Technologies, Inc., 3175 Staley Road, Grand Island, NY 14072 USA

Abstract

Mammalian cell culture biological production applications have traditionally implemented batch production techniques to minimize real or perceived obstacles to product approval by regulatory agencies. Analysis of the bioreactor environment during the course of a batch fermentation run demonstrates significant fluctuations in levels of critically-limiting nutrients, accumulation of catabolic products and increased medium osmolality. These factors can be demonstrated to impact specific productivity and to result in product microheterogeneity. Extended bioreactor campaigns may exhibit superior productivity and longevity by controlled nutrient delivery via fed batch or continuous perfusion of concentrated supplements. Regulations defining criteria for "well-characterized products" and international efforts to harmonize standards may minimize barriers to extended production campaigns and process modifications. This paper examines the evolution of these regulatory guidelines and their application to process changes which permit productivity enhancement through supplementation with salt-free nutrient concentrates.

Historical Regulatory Emphasis

Regulatory agencies have traditionally safeguarded introduction of new therapeutic products (e.g., drugs, devices, biologics) by demanding thorough documentation of safety, therapeutic efficacy, and biochemical identity [1]. Many criteria useful for small, synthetic molecules have been difficult to extrapolate to biological macromolecules. Similarly, process controls relevant to batch manufacturing processes have been challenging to extend to fed-batch and continuous biological production processes. Anticipatory anxiety regarding regulatory agency requirements caused some reticence by prospective biologics and biopharmaceuticals manufacturers to invest in developmental research and clinical evaluation [2]. Concerns were heightened by proposed legislation to limit the potential return on drug development investments. Apparent jurisdictional conflicts between independent centers of the U.S. Food and Drug Administration (FDA) led to confusion over which types of products would be designated as "devices", "drugs" or "biologics", particularly since there existed minimal harmonization between centers regarding approval requirements or the application process.

Evolving Regulatory Trends

Bipartisan legislation to reform the drug evaluation process and global efforts to harmonize regulatory requirements promise to resolve these historical impediments and to expedite review and approval of quality biopharmaceuticals based upon a scientific approach [1-2].

Three key elements remain fundamental to regulatory authorization: (1) documentation for the master and working cell banks; (2) demonstration of controlled manufacturing processes, minimally covering culture maintenance and expansion and product harvest and purification; and (3) biochemical and performance characterization of the target biomolecule, including safety and efficacy criteria, and defining acceptable parameters for microheterogeneity.

However, anxiety remains regarding the impact of changes during the product development and clinical evaluation processes on regulatory approval, as exemplified by the following generic case study: A promising biological compound is developed by the research and development group of a biotechnology company. As this molecule exhibits favorable activity and specificity during pre-clinical trials, using both *in vitro* and animal models, it enters a fast track program for clinical evaluation. Pilot production of phase 1 clinical trial materials are rapidly produced and clinical evaluation is initiated. In due time, favorable results emerge from the phase 1 study and the company is anxious to enter pivotal phase 2/3 clinical trials. But, interim efforts by process development have recommended significant changes in the manufacturing protocol to improve productivity. Such recommended alterations by process development might impact:

Cell line (e.g., modifications to enhance product expression, improve robustness within the production bioreactor, reduce feed chain times or enhance maximal cell density, or other alterations of the master working cell bank)

Nutrient medium (e.g., conversion from a serum-supplemented to a serum-free cultivation system, addition of concentrated supplement to replenish exhausted nutrients, other changes in the composition or mode of preparation of nutrient formulation)

Bioreactor process cycle (e.g., conversion from batch bioreactor process to a fed-batch or continuous mode of medium replenishment and product harvest)

Product purification (e.g., steps in the overall purification scheme to eliminate adventitious contaminants which impact product recovery and biochemical purity)

Product assay (e.g., alterations in assay protocol or instrumentation which impact sensitivity or specificity of analyte detection or which evolve from the analytical requirements preferred for product development to the robustness and scalability demanded of product qualification)

Equipment and Facility (e.g., qualitative alteration in the type of culture bioreactor or purification process, or scaling adjustments during evolution from pilot to production requirements, relocation of manufacturing to an alternate site for increased production scale, or implementation of a contract manufacturer)

The hypothetical company in this generic case history is now faced with a dilemma which pits its objective of capturing market share by being "first to market" against its objective of ensuring a sustainable competitive advantage by delivering the most efficiently-manufactured product. Among the answers it must seek are regulatory issues, such as: What process changes will invalidate previous clinical trial data and require starting over from the beginning? and If this change is implemented, what additional data must be produced to demonstrate product equivalence?

The best interim guidance while regulatory guidelines are evolving and converging is that regulatory agencies will continue to evaluate individual circumstances on a "case-by-case" basis. The best assurance of a positive experience is to discuss proposed changes in advance with appropriate regulatory administrators, documenting objectively potential benefits and risks on a sound scientific basis. This option is advised over waiting until after completion of clinical evaluation using materials produced using varying protocols, and then attempting to sequester evidence, to minimize justification for process changes, or to coerce regulatory authorities into product approval. The U.S. government has mandated "re-engineering of government" (REGO II) initiatives to streamline and normalize evaluative efforts by FDA centers [1-2]. An International Conference on Harmonization (ICH) is moving aggressively to develop uniform regulatory requirements for 18 major biotechnology markets. Evolving from these efforts is a transition in emphasis from process to product characterization. This concept proposes that controlled modifications in manufacturing process elements resulting in minimal, reproducible alterations in a "well-characterized" biological molecule may be implemented with reduced levels of qualification [1-2].

Controlled process elements [1] associated with chemistry, manufacturing and controls impact all aspects of biological production, including expression, fermentation, purification, formulation, characterization and quality control. Validation of structural and biological equivalence of the end product must be correlated with reference materials prior to implementing proposed procedural changes identified by process development. However, it may be unnecessary and costly in terms of developmental time and financial investment to reproduce the full battery of pre-clinical or early clinical investigations, particularly for a biological molecule which is reasonably well-characterized.

The fundamental requirement is to establish equivalent identity, purity, potency, safety and efficacy of the target biological manufactured using the proposed manufacturing process change, compared with reference materials produced by the former process. However, interpretation of "equivalent" appears to be evolving, and industry and regulatory scientists agree that the molecules do not have to be identical [1-2]. In fact, more sensitive analytical tests and careful sampling within the course of a production run clearly demonstrate that even within a single batch fermentation, depending upon various culture parameters, there is a defined, acceptable level of product microheterogeneity. The challenge becomes, either for an existing manufacturing process or for a proposed process modification, to demonstrate reproducible appearance and proportion of isoforms and to document their stability and consistency. Similarly, quality control analysis may evolve from the highly-sensitive, specialized, labor-intensive methods useful during product development to more robust assays suitable for routine performance by less technically-skilled individuals for high-throughput evaluation of product stability and release testing.

Approaches to Nutrient Optimization

Identical cell populations will perform differently in different nutritional environments [3]. Control mechanisms have been demonstrated at the levels of intermediary metabolism, transcriptional events, and sub-population selection. The negative impacts of nutrient deprivation are more obvious [3-5]. Years of experience reinforce failure to sustain high viable cell density and specific productivity once critical nutrients, such as glucose or glutamine, have become exhausted. Less well-documented but also critically impacting bioreactor productivity are maintenance of sufficient levels of other amino acids, lipid constituents and nucleic acid precursors [3,6]. Vitamins, trace metals and other nutritional factors may also become limiting under high density cultivation conditions.

Excessive levels of desirable nutrients may also be detrimental to bioreactor productivity [3,7]. Supra-optimal glucose may result, under conditions of transient oxygen limitation, in accumulation of lactate. While direct cytotoxicity of lactate accumulation remains controversial and, at best, cell type specific, localized and mass acidification of the culture environment can lead to necrotic or apoptotic events and to elevated medium osmolality due to automated pH adjustment by base addition. Fluctuations in availability of glucose and other intermediary carbohydrates can also impact the type, complexity and homogeneity of protein glycosylation. Glutamine readily deamidates to yield ammonia, which can accumulate to relatively high levels, particularly with non-purged, extended culture bioreactor systems. Elevated ammonia has been demonstrated to inhibit cell growth and biological productivity. Generally, excessive levels of any intermediary metabolite creates competitive demand for metabolic energy and for transport and processing enzymes.

The objective of nutrient optimization is to create a homeostatic environment which minimizes "peaks" and "valleys" in exogenous metabolites. In this objective, there appears to be a convergence between cellular preferences and regulatory requirements: to sustain long-term, consistent culture productivity with minimal fluctuation in composition of bathing nutrient fluids or in resultant biological product. Fed-batch and continuous bioreactor alimentation regimens have inherent potential to improve biological product yield as a percentage of invested time, facility space, technical labor and consumable cost [8-9]. The question becomes: how best to implement such bioreactor process changes to enhance productivity while minimizing regulatory exposure? There are no clear answers, but several remedies appear more problematic, both from technical and from regulatory perspectives:

Major qualitative change in media composition: Particularly during advanced clinical trials, such alterations cause significant concerns. The burden resides with the manufacturer to demonstrate that the change has not materially impacted either the harvested biological product or its purification process. Deletions or reductions may be regarded as less critical than addition of a new constituent.

Animal-derived additives: With improved awareness of adventitious contaminants and detection methods, concerns regarding residual effects of animal-derived constituents on biological products intended for human or animal therapy have increased. There is increasing pressure to qualify biochemically-defined, protein-free nutrient formulations, to genetically-engineer production cells to eliminate requirement for peptide factors, or to ensure process validation and terminal sterilization of any animal-derived materials.

Enhanced medium replenishment rate: Unwarranted increases in medium delivery negatively impact profitability, both resulting from increased medium cost and from costs associated with process harvesting, purification of the desired product and disposal of waste fluids.

Supplementation with complete medium: Published reports and technical service inquiries have revealed some investigators utilizing complete medium (including salt components) formulated as a 10X bioreactor supplement. The negative impact of this practice on medium osmolality, cell viability and bioreactor productivity becomes readily apparent.

Remedies to supplying adequate quantities of critical nutrients to the bioreactor, while minimizing technical or regulatory anxieties, include the following two approaches:

Basal medium enrichment: A customized modification of the basal nutrient medium may be formulated which selectively enriches (by 2X-5X) non-salt constituents. This approach is particularly useful for batch production processes to enhance productivity, but may also be applied to continuous production processes.

Concentrated nutrient supplement: A protein-free, salt-free additive, containing 5X-10X levels of key metabolites may be prepared to replenish exhausted nutrients. Concentrated nutrient cocktails may be implemented to elevate maximal cell density viability and to improve bioreactor productivity in both fed-batch and continuous perfusion systems. Supplement composition may either be calculated as a multiple of the base formulation or analytically-derived from HPLC determination of spent fluids and estimation of individual nutrient consumption kinetics.

Summary

Regulatory guidelines continue to evolve with improved understanding of the benefits and risks of biotechnology-derived products, with significant efforts toward global harmonization. Transition of emphasis from process to product characterization suggests that manufacturing process improvements may be considered at advanced developmental stages, based upon emerging "well-characterized molecule" criteria. Nutrient enrichment by adding salt-free, protein-free concentrate of exhausted metabolites can improve productivity and minimize regulatory impact.

References

1. Lubiniecki, AS (1996) *"Monoclonal antibodies -- an industry perspective"*, presented at The Waterside Monoclonal Conference, April 22-25, 1996, Norfolk, VA.
2. Zoon, K (1996) *"A perspective on FDA goals, objectives and policies for 1996: Drugs & Biologics"*, presented at BioEast '96, January 15-18, 1996, Washington, D.C.
3. Jayme DW, Cytotechnology (1991) 5: 15-30, *"Nutrient optimization for high density biological production applications."*
4. Bibila TA, Ranucci CS, Glazomitsky K, Buckland BC, and Aunins JG, Biotechnol. Prog. (1994) 10: 87-96, *"Monoclonal antibody process development using medium concentrates."*
5. Mahadevan MD, Klimkowsky JA and Deo YM, Cytotechnology (1994) 14: 89-96, *"Media replenishment: a tool for the analysis of high-cell density perfusion systems."*
6. Jayme DW, Kubiak JM and Fike RM (1995) in Animal Cell Technology: Developments toward the 21st Century, eds. EC Beuvery, JB Griffiths and WP Zeijlemaker, 199-204 (Kluwer), *"Productivity enhancement using liquid medium concentrates."*
7. Fike R, Kubiak J, Price P and Jayme D, BioPharm (1993) 6: 49-54, *"Feeding strategies for enhanced hybridoma productivity: Automated concentrate supplementation."*
8. Roth G, Kubiak JM, Long JF, and Schoofs GM, BioPharm (1995) 8: 31-35, *"An automated system to produce cell culture media from liquid medium concentrates."*
9. Jayme DW, Kubiak JM, Battistoni TA, Cady DJ and Fike RM, Cytotechnology (in press) *"High capacity method for producing nutrient medium from concentrated intermediates for cell culture bioproduction applications."*

THE USE OF BIOSENSORS IN CONTROL OF FED-BATCH PROCESSES - OPPORTUNITIES AND LIMITATIONS

R.RENNEBERG [1] , J.P.BARFORD[2] AND S.M.(ANITA) TSE[1]
Department of Chemical Engineering, The Hong Kong University of Science and Technology[1] , Clear Water Bay, Hong Kong and University of Sydney[2] ,Sydney 2006, NSW, Australia

The introduction of biosensors in an industrial environment is only possible if they present a well adapted solution to an important analytical problem. In the past few years, this has been demonstrated with enzyme sensors for glucose, lactate and glutamine measurement. These biosensors have been connected via flow-injection analysis (FIA) devices and applied to batch cultures of hybridoma cells and also perfusion systems using recombinant BHK cells for the production of IL-2. It has been shown that these biosensors allow direct feedback control of the concentrations of the main nutrients.

Furthermore, the on-line determination of the biotechnological product (e.g. monoclonal antibodies, interleukin-2) is a prerequisite for the development of reliable process control strategies. Novel amperometric immunosensors and automated immunocartridge systems allow the analysis of protein products in minutes (compared to hours with ELISA techniques).

It has been clearly demonstrated that glucose, lactate, glutamate and high molecular weight product determinations with associated process control is possible. Renneberg et al., (1991), for example, showed that on-line control was possible for the process control of perfused bioreactor systems containing mammalian cell lines producing interleukin - 2. Glucose, lactate and glutamate were determined using enzyme sensors. Interluekin - 2 was detected using immunosensors. Figure 1 illustrates the experimental set-up of the flow injection analysis (F.I.A.) biosensor-bioreactor coupling and compares on-line and off-line data. Figure 2 illustrates the principle of the immunosensor and the experimental results obtained using it (Warsinke, 1992;

Renneberg et al., 1992). Such systems, however, have not been utilised extensively by the biotechnological industry. A possible reason for this is that early expectations that biosensors could be directly inserted (like pH and DO probes) into the bioreactor have clearly underestimated the problems of fouling and calibration of the biosensor and activity loss of the biocomponent.

We feel, therefore, that in-situ biosensors are not yet a practical industrial solution but rather by-pass systems integrating sampling, sample transport, sample pretreatment, separation and data handling are more appropriate. These biosensor systems, used as "measuring fingers", are the appropriate strategy for control of fed-batch processes and the optimisation of productivity.

1. Introduction

In the previous papers (Sanderson et al.,1997; Black et al.,1997) , we have shown that fed-batch systems offer many productivity advantages over batch cultivation and that regulatory difficulties often considered related to fed-batch systems are also evidenced in batch culture and should, therefore, not be a real issue in industrial scale production. In addition, we have shown that computerised fed-batch operation, where computer simulations are used in conjunction with optimisation routines, increases the productivity even further over simple "trial and error" methods.

Biosensors offer us the prospect of on-line control - particularly when linked to computer simulations and optimisation routines.

In this paper, we investigate the potential benefits and barriers to the implementation of animal cell fermentation systems employing biosensors in conjunction with computerised fed-batch strategies.

In fed-batch systems, the primary object of the feeding policy is to :

(i) Regulate the growth rate to a desired value (less than the maximum) to maximise product productivity

(ii) Keep the concentration of inhibitory products to a minimum

(iii) Satisfy the nutritional requirements of the cell without providing excess nutrients which may lead to osmolarity problems

(iv) Balance the ratio of glucose to glutamate to achieve an optimum catabolic and anabolic activity.

There are now commercially available biosensors for glucose, glutamate, lactacte, and sensors for ammonia. The comparison of the measured concentrations with those obtained from biosensor measurement is shown in Figure 3. It can be seen that the the agreement is excellent. The biosensor for glucose consists of buffer pumps, standard and rinse solution bottles, valves, a dialysis module, pumps, the immobilised enzyme electrode. A personal computer and a printer are also necessary. Other biosensors have similar equipment requirements. Ammonia is measured not by a biosensor but by an off-line sensor. The basis of this sensor is that the ammonia ion in solution is changed to ammonia gas when the pH is raised to pH 12. This ammonia will transmit through an ammonia gas filter module (Teflon) and be absorbed on a carrier solution. The pH will then be changed and the pH change may be converted to the ion density. Again the agreement between sensor measured and the experimental estimate is excellent (Figure 4).

In theory, these sensors should be able to be used to control fed-batch cultivation on-line since the output from the computer feeding and optimisation can be compared to the actual culture measurement in real time (the analyses take a few minutes)

The question then remains : what are the restrictions which have led to this methodology not being extensively used. These may be summarised:

Biosensor :

(i) Cost
Recent commercial biosensors (Aota et al., 1997) for on-line and off-line measurements are priced as follows:

(a)	Glucose biosensor	on-line	Yen 1,800,000 (approx $US 15,000)
(b)	Glucose biosensor	off-line	Yen 50,000 (approx $US 417)
(c)	Ammonium sensor	off-line	Yen 2,000,000 (approx $US 16,667)
(d)	Glutamine biosensor	on-line	Yen 1,800,000 (approx $US 15,000)
(e)	Lactate biosensor	off-line	Yen 1,800,000 (approx $US 15,000)

Other manufacturers (Yellow Springs Instruments, Universal Sensors (both U.S.A.) and E.K.F. Industrie Electronics (Germany) have similar costs ranging from $HK 40,000 - 70,000 (approx. $US 5,000 - $10,000) (E.K.F.) to $HK 89,000 (approx $US $11,500) (Yellow Springs). Consequently, they represent a major capital cost.

(ii) Reliability
The sensor life is quoted at about 3 months for the glucose biosensor, 2 months for the lactate biosensor and 1 month for glutamine biosensor. The reliability is quoted as +/- 5%. There is thus an on-going operating expense although this is only a fraction of the initial capital cost (less that 1 %).

(iii) Other Factors

There are still some potential difficulties which may be encountered in using biosensors. Amongst these are :

(a) Direct sterilisation of the biosensors is not possible
(b) In most cases, the analyte concentrations exceed the linear range of the sensor (Scheller and Schubert,1992)
(c) The presence of interfering compounds (Scheper et al.,1989; White et al.,1995)
(d) Mechanical and thermal stress on the biosensor
(e) Obtaining representative samples for off-line biosensors
(f) The use of amperometric biosensors for lactate using lactate oxidase and oxygen and hydrogen peroxide sensing electrodes are affected by fluctuations in the sample oxygen concentration. For hydrogen peroxide sensing electrodes, a high potential is required to operate the electrode. They also are prone to interference form other electroactive species such as uric acid, ascorbate or glutathione that are oxidised at high potential
(g) Biosensors involving the substrate of dehydrogenases are limited by the supply and regeneration of the cofactor NAD(P)

Many of these limitations are slowly becoming overcome by on-going research.

Simulation and Optimisation :

(i) High Technical Skill Required

To run the complex simulation and optimisation procedures following biosensor measurement requires a high level of technical skill. As a result, an industry must decide between a fixed feeding strategy (which is provided for a particular cell line on a "contract out" basis) or whether real on-line control is desirable / warranted. The latter clearly has better productivity but must be weighed against the increased complexity and cost.

(ii) Cost

The minimum requirement for on-line control is a PC with the associated simulation and optimisation software. Excluding the simulation software, this cost would be less than $US 10,000. In addition, personnel of sufficient technical skill would need to be involved in the fermentation runs. These issue are discussed elsewhere (Sanderson et al., 1997)

CONCLUSIONS:

Technology exists now for the on-line control of fed-batch animal cell cultivation which will lead to large increase in productivity over batch systems. The challenge is to make these cost effective and user friendly so that industry can benefit from their application.

REFERENCES

Aota, T, M Kaneko, S Nankai, & M Miyahara (1996) Glucose control by biosensor, 9th Annual Meeting of the Japanese Association For Animal Cell Technology, Yokohama, Japan, P-14.

Black, DJ, JP Barford & C.Harbour (1997) Quality aspects of batch animal cell cultivation - relevance to fed-batch implementation in industry, Proceedings of 9th Annual Meeting of the Japanese Association For Animal Cell Technology, Yokohama, Japan.

Renneberg, R., G Trott-Kriegeskorte, M Lietz, V. Jaeger, M Pawlowa, G Kaiser, U Wollenberger, F Schubert, R Wagner, RD Scmidt & FW Scheller (1991) Enzyme sensor-FIA-system for on-line monitoring of glucose, lactate and glutamine in animal cell cultures, J Bacteriol 21:173-186.

Renneberg, R., A Warsinke & G. Kaiser (1992) Amperometric immunosensors for bioprocess control, G.B.F. Monographs, 17:25-30.

Sanderson CS, CJ Marquis & JP Barford (1997) Yield Improvements Utilising Fed-Batch Cultivation and Optimisation Strategies, Proceedings of 9th Annual Meeting of the Japanese Association For Animal Cell Technology, Yokohama, Japan.

Scheper F & F Schubert (1992) Biosensors in Techniques and Instumentaion in Analytical Chemistry, vol 1, Elsevier

Warsinke, A. (1992) Dissertation , Humboldt University, Berlin.

White SF, APF Turner, U Bilitewski, J Bradley & RD Schmid, On-line monitoring of glucose, glutamate and glutamine during mammalian cell cultivations, Biosensors and Bioelectronics, 10:543-551

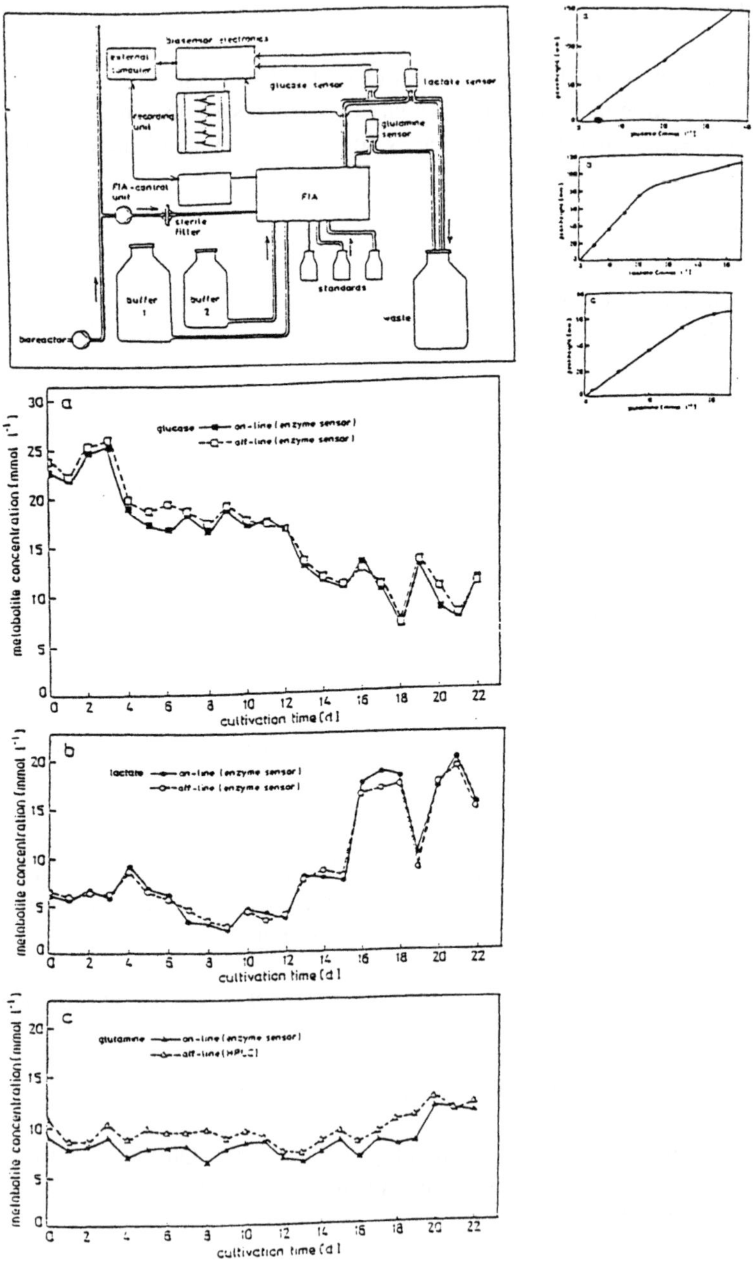

Figure 1 Enzyme Sensors - Biosensor - FIA - Bioreactor System (Top Left), Calibration Curves (Top Right), Comparison On-Line and Off-Line Data (Bottom Three Figures)

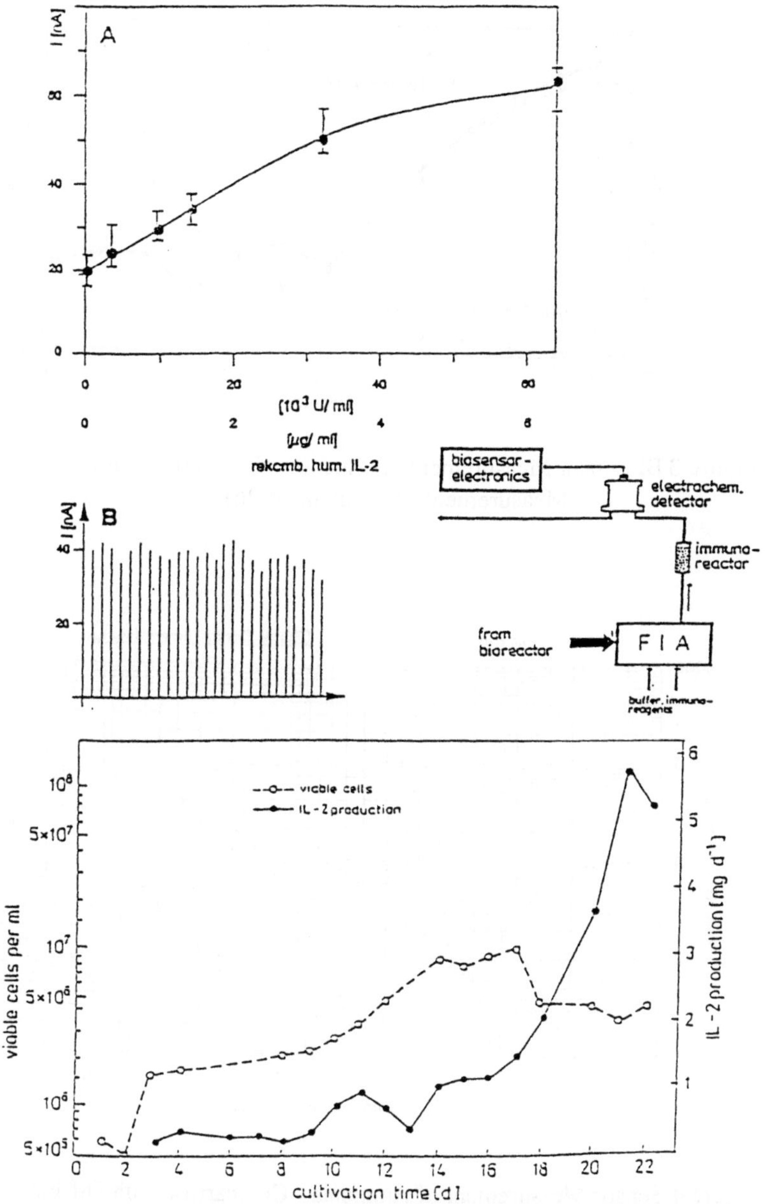

Figure 2 Immunosensors - Calibration Curve For the Detection of Interluekin - 2 (Top); Reusability of the Immunosensor (Middle Left), Coupling With F.I.A. (Middle Right); Data For the Production Process (Bottom) (Modified after Warsinke, 1992; Renneberg, 1991; Renneberg, 1992)

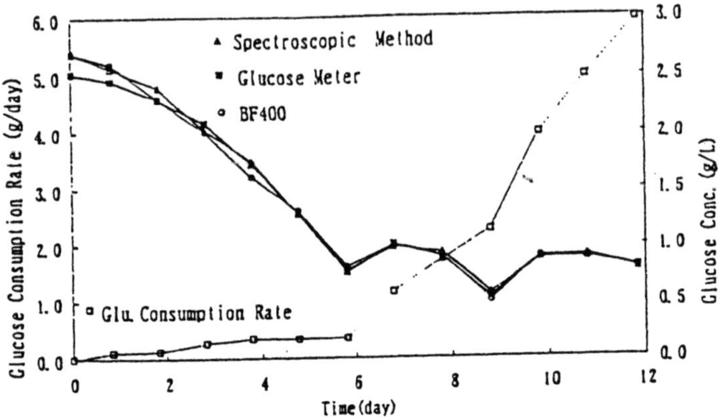

Figure 3 Biosensor Measurement of Glucose - Comparison with Off-line Measurement (Aota et al., 1996)

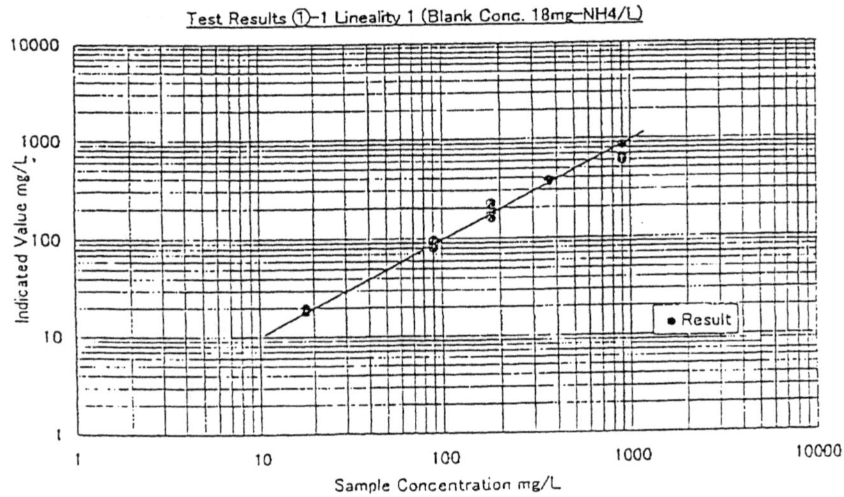

Figure 4 Sensor Measurement of Ammonia - Comparison with Off-line Measurement

Metabolic System Engineering Approach for *Saccharomyces cerevisiae* and Hybridoma Cell Culture

Kaiming YE, Huidong SHI, and Kazuyuki SHIMIZU
Department of Biochemical Engineering and Science
Kyushu Institute of Technology, Iizuka, Fukuoka 820, Japan

1. Introduction

There has been increasing interest in quantitative analysis of intracellular fluxes since it is a useful gauge of the physiological state of cells in culture, which can serve for process control, optimization and design of medium formulation. Currently, a variety of methods have been developed to combine stoichiometric model with extracellular measurements to estimate intracellular fluxes and thereafter validating using NMR techniques (Craig and Stephanopoulos, 1995). The idea behind this method is that once a stoichiometric model is constructed and validated with *in vivo* measurements, the model can be used to analyze the fluxes with some extracellular measurements.

On the other hand, the bioreactions network existing inside cells can be organized as a signal flow diagram (Endo and Inoue, 1976). Using this method, the relationship of cell growth, product synthesis, evolution of carbon dioxide and the consumption of carbon source, uptake of oxygen or response of cell metabolism to external stimulation can be examined (Jin et. al., 1995).

In this work, an integrated intracellular fluxes analysis method was developed, which was applied to distinguish the changes in metabolic flux branches in response to the culture conditions in *Saccharomyces cerevisiae* cultivation. It was aimed to develop a method that can be used on-line evaluation of cell metabolism. Furthermore, hybridoma cell metabolism was discussed using intracellular fluxes analysis technique focusing on change in cell metabolism due to the change in ratio of glucose and glutamine.

2. Integrated Intracellular Flux Model

We have developed a metabolic signal flow diagram model for analysis of the flow direction of flux and activity of metabolic loops in the metabolic pathway of intereste (Jin et. al. 1996), where a metabolic pathway network was converted into a directed signal flow diagram as shown in Fig. 1. In that model, a bioreaction of metabolite A to metabolite B is treated as the transformation of signal A^* to signal B^*. The catalytic function of an enzyme was referred as a transfer function. There are two input nodes: v (specific glucose consumption rate), q_{O2} (specific oxygen uptake rate); three output nodes: μ (specific cell growth rate), q_e (specific ethanol production rate), q_{CO2} (specific carbon dioxide production rate). Parameters a, b, c, … ...y, v_1, v_2, v_3 are defined as metabolic transfer coefficients. Using the equivalent transformation technique according to graph theory, we have (Jin, et. al., 1996)

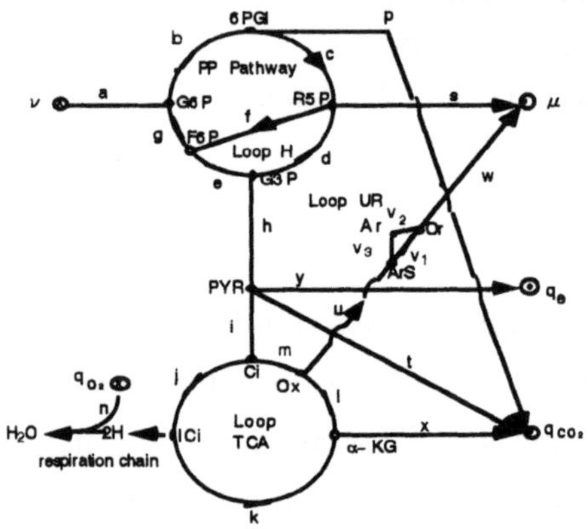

Fig. 1 metabolic signal diagram for yeast cell culture.

$$\boldsymbol{Y}=\boldsymbol{C}\boldsymbol{X} \tag{1}$$

where $\boldsymbol{Y}=(\mu, q_{CO2}, q_e)^T, \boldsymbol{X}=(\nu, q_{O2})^T,$

$$\boldsymbol{C}=\begin{pmatrix} c_{11} & c_{12} \\ c_{21} & c_{22} \\ c_{31} & 0 \end{pmatrix}.$$

By assuming that cell has its objective as maximizing ATP production (Ko, et. al., 1993; 1994), we have

$$\mu = 0.2\nu - \frac{0.328\beta}{(0.11+2.94z)}q_e - \frac{0.2\beta}{(0.11+2.94z)}q_{O_2} \tag{2}$$

$$q_{CO_2} = (1-1.68z)\nu + 1.31q_e + q_{O_2} \tag{3}$$

where **z** is the fraction of total carbon flux in the branch of glucose-6-phosphate flowing into the glycolysis pathway. β is the fraction of carbon dissimulated.

Substituting Eqs. 2 and 3 into Eq. 1, leads to

$$C_{11} = 0.2 - \frac{0.328\beta}{(0.11+2.94z)} C_{31} \tag{4}$$

$$C_{12} = -\frac{0.2\beta}{(0.11+2.94z)} \tag{5}$$

$$C_{21} = 1 - 1.68z + 1.31C_{31} \tag{6}$$

$$C_{22} = 1 \tag{7}$$

$$q_e = C_{31}\,\nu \tag{8}$$

3. Intracellular Fluxes Analysis of *Saccharomyces cerevisiae*

The integrated intracellular fluxes model developed in this work was applied to analyze the fluxes in the continuous cultivation of *S. cerevisiae*. The calculation results are presented in Table 1.

Table 1. The metabolic transfer coefficients in the continuous fermentation of *Saccharomyces cerevisiae.*

D(1/h)	C_{11}	C_{12}	C_{21}	C_{22}	C_{31}
0.20	0.20	-0.04	-0.39	1.00	0.00
0.29	0.20	-0.05	-0.28	1.00	0.00
0.32	0.13	-0.05	0.64	1.00	0.89
0.36	0.09	-0.06	1.01	1.00	1.16
0.41	0.07	-0.06	1.23	1.00	1.31

There are tow patterns for continuous fermentation of *S. cerevisiae*. Namely, when dilution rate was smaller than 0.29 h^{-1}, C_{22} was positive ,while C_{21} was negative, which suggests that

$$L_{TCA}<1 \tag{9}$$

$$L_H>1 \tag{10}$$

$$\frac{(bcf+g)ehijkx}{(1-L_H)(1-L_{TCA})} > abp \tag{11}$$

They inplies that most CO_2 was produced from TCA cycle and EMP pathway rather than from PP pathway.

In the similar way, the following inequalities can be given from the signs of C_{11} and C_{12}.

$$(bc + egf)s < \frac{(bcf + g)ehijkluv_1 w}{(1 - L_{\mathrm{TCA}})(1 - L_{\mathrm{UR}})} \qquad (12)$$

This suggests that most biomaterials for cell growth were synthesized through TCA cycle rather than through HMP pathway.

However, when dilution rate was greater than 0.29 h^{-1}, C_{21} became positive, while the other parameters did not changed. Therefore,

$$(bc + egf)s > \frac{(bcf + g)ehijkluv_1 w}{(1 - L_{\mathrm{TCA}})(1 - L_{\mathrm{UR}})} \qquad (13)$$

This implies that most biomaterials for cell growth were synthesize through HMP pathway. It can be clearly seen that a very complex metabolic pathway network can be analyzed by the determination of a transfer function matrix C.

4. Intracellular Flux Analysis of Hybridoma Cell Grown at Different Ratio of Glucose to Glutamine

4.1 Problem Definition

In this section, hybridoma cell is used as a model to study how the intracellular fluxes change with a change in culture environment, i.e., a change in the ratio of glucose to glutamine. According to the knowledge on mammalian cells metabolism as far as we know, it is quite difficult to analyze the whole metabolic pathway network existing in living mammalian cells. An alternative way is to analyze a central metabolic pathway, where most carbon source is dissimulated and most of energy is

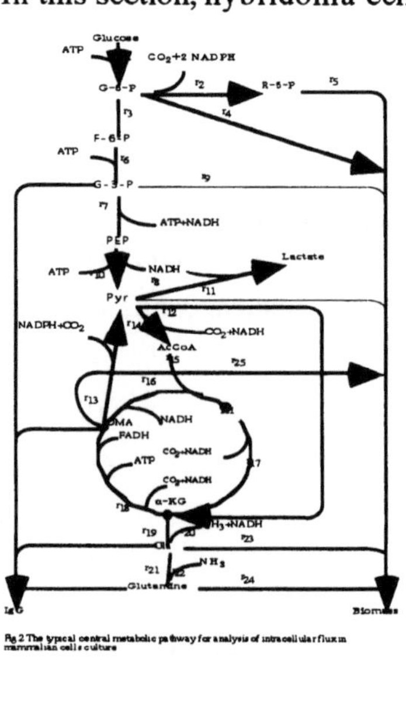

Fig 2 The typical central metabolic pathway for analysis of intracellular flux in mammalian cells culture

produced. A typical central metabolic pathway occurred in mammalian cells is presented in Fig. 2.

4.2.1 Cell Line and Assay

A murine hybridoma cell line, ATCC CRL 1606 was used in this work. Viable cells were counted by trypanblue dye exclusion method. The specific glucose and glutamine consumption rates, lactate and ammonia production rates, were estimated using the data collected during the exponential growth phase, respectively.

4.3. Effect of Glucose and Glutamine on The Metabolism of Cells

4.3.1. Effect of Glucose on Metabolism of Cell

The specific consumption and production rates of metabolites at different glucose concentrations are presented in Table 2.

Table 2 Specific metabolites rates at different glucose concentration

s	Q_{glu}	Q_{gln}	Q_L	Q_N	$Y_{lac/glu}$	$Y_{N/gln}$	$Y_{x/gln}$
1	3.17	0.99	6.39	1.04	2.13	0.58	4.27
4	3.30	0.97	5.55	0.84	1.55	0.84	4.11
9		0.80	6.97	0.82	1.75	0.99	2.58

Note: s, glucose concentration; Q_{glu},Q_{gln}, Q_L,Q_N are specific glucose and glutamine consumption rates, lactate and ammonium production rates, respectively. The unites for these rates are 10^{-10} mmol/h//cell; $Y_{lac/glu}$, yield of lactic acid for glucose; $Y_{N/gln}$, yield of ammonia for glutamine; $Y_{x/glu}$, yield of biomass for glutamine. The units for these yields are 10^8.

Although the glucose consumption rate increased with an increase in glucose concentration, glutamine consumption rate decreased suddenly at high glucose concentration. Sudden decrease in glutamine consumption rate at higher glucose concentration means that much more glucose entered TCA cycle to produce energy for bioreactions and biosynthesis. Ammonia production rate seemed to be irrelevant to the glucose concentration, while lactate production is strongly relevant to the glucose concentration.

4.3.2. Effect of Glutamine on Metabolism of Cell

Table 3 shows the specific metabolites rates at different initial glutamine concentration. From this table, we can see that ammonia production rate has a good linear relationship with glutamine consumption rate, while the yield of

Table 3 Specific metabolites rates at different glutamine concentration

g	Q_{glu}	Q_{gln}	Q_L	Q_N	$Y_{lac/glu}$	$Y_{N/gln}$	$Y_{x/glu}$
0.6	3.17	0.99	6.39	1.04	2.13	0.58	4.27
1.8	2.60	1.16	5.44	1.11	2.29	0.61	2.18
5.8	2.88	2.36	5.60	2.26	1.99	0.68	1.97

Note: g, glutamine concentration; $Y_{x/glu}$, biomass yield for glucose. Others are the same as those in Table 2.

lactate for glutamine seemed to be independent of initial glutamine concentration. The lactate yield for glucose was almost constant at different glutamine concentrations, which suggesting that increase of glutamine concentration does not affect the glucose dissimulation. The biomass yield for glucose at a 5.8 g/L concentration of glutamine decreased about three-fold compared with that at 0.6 g/L of glutamine concentration.

4.4. Metabolic Model of Hybridoma Cell lines Grown at Different Ratio of Glucose to Glutamine

A metabolic model for hybridoma cells was developed to study the effects of glucose and glutamine on the intracellular fluxes distribution, based on the consideration of bioreactions occurred in the central metabolic pathway as shown in Fig. 3. A stoichiometric bioreaction matrix derived from stooichiometry is, therefore, constructed such that

$$\mathbf{\Gamma R = B} \tag{14}$$

where **R** is a vector of the fluxes, while **B** is a vector of metabolites rates in which the actual measured extracellular metabolites are contained and the intracellular intermediates are assumed to be zero. **Γ** is a stoichiometric coefficients of the bioreactions. To find solution of Eq. (14), an optimization procedure was desirable. This is undergoing.

5. REFERENCES

Craig, Z. and Stephanopoulos, G. 1995. Intracellular flux analysis in hybridomas using mass balance and *in vitro* ^{13}C NMR. *Biotechnol. Bioeng.* **45**, 292-303.

Endo, I. and Inoue, I. 1976. Metabolic activities of yeast cells in batch culture. *Kagaki Kogaku*, **2**, 416-421. (in Japanese).

Jin S., Ye, K., and Shimizu, K. 1995. Metabolic pathway analysis of recombinant *Saccharomyces cerevisiae* with a galactose-inducible promoter based on a signal flow modeling approach. *J. Ferment. Bioeng.* **80**, 541-551.

Ko, Y.F., Bently, W.E., and Weigand, W.A. 1994. A metabolic model of cellular energetic and carbon flux during aerobic *E. coli.* fermentation. *Biotechnolo. Bioeng.* **43**, 847-855.

ADAPTIVE FUZZY CONTROL OF MAMMALIAN CELL CULTURE IN FED-BATCH REACTOR FOR PRODUCTION OF AN ANTITHROMBIN III VARIANT

Petros Lenas[1], Tamotsu Kitade[1], Hideki Watanabe[2], Hiroyuki Honda[1] and Takeshi Kobayashi[1] [1]*Department of Biotechnology, Graduate School of Engineering, Nagoya University, Furo-cho, Chikusa-ku, Nagoya 464-01, Japan*; [2]*Eisai Co. Ltd., Tsukuba Research Laboratories, 1-3 Tokodai 5-chome, Tsukuba, Ibaraki 300-26, Japan*

Abstract An adaptive fuzzy controller was developed to control the glucose and glutamine concentrations in the reactor constant at the desired level. It was applied to antithrombin III production from a recombinant BHK anchorage-dependent cell line performed in a fed-batch reactor on-line connected with an HPLC system. From six experiments with various set points, it was found that ammonia had a determining effect on productivity and its production was increased by an increase of glutamine or decrease of glucose set point concentrations. A combination of low glutamine with high glucose concentration was more efficient in maximizing the cell concentration and protein productivity.

1. Introduction

One of the methods to avoid the accumulation of inhibitory waste products, lactate and ammonia, is the controlled supply of nutrients glucose and glutamine in order to maintain their concentrations at low levels. Although it is well known that maintaining the glucose concentration low results in a decrease of the lactate concentration, an increase of glutamine consumption and ammonia production can be observed under the condition (Hu *et al.*, 1987; Gaertner and Dhurjati, 1993). An optimal glucose to glutamine ratio that depends on the metabolic characteristics of the specific cell line and its relative sensitivity to lactate and ammonia seems to be a more promising control strategy (Wang *et al.*, 1995).

In our previous study (Kurokawa *et al.*, 1994), the performance characteristics of different control strategies (adaptive, fuzzy and PI) for the control of glucose and glutamine concentrations in the reactor have been analyzed. Since each one of the above strategies had some advantageous characteristics, we tried in the present paper to retain them in a combined strategy with more general applicability designing a fuzzy PI-like controller with continuous adaptation of the parameters through another fuzzy strategy.

2. Materials and methods

A BHK21 cell line producing a recombinant variant of antithrombin III was used in this study. The basal medium was glucose-, glutamine- and serum-free Dulbecco's modified Eagles medium (DMEM) supplemented with 10 mgl^{-1} insulin, 5 mgl^{-1} transferrin, 30 mM ethanolamine, 40 nM sodium selenite and 100 mgl^{-1} LDL. Static attachment with intermitted shaking to Cytodex-3 microccariers (Pharmacia Co., Uppsala, Sweden) was performed in dishes in the medium supplemented with 5% FCS.

Figure 1a shows the flow diagram of the culturing system we previously developed (detailed description in Kurokawa *et al.*, 1994). Cells attached to microcarriers (3.5 gl^{-1}) were inoculated in a spinner flask (culture volume 750 ml). Culture broth was drawn out from the reactor every 2 hours automatically by an autosampler (LC-ROBO, Able Co., Tokyo) . The autosampler injected the sample into the on-line HPLC measuring system. Analyzed data of glucose, glutamine and lactate were sent to a computer for glucose-glutamine control.

The viable cell concentration was determined by hemocytometer with the dye exclusion method using 0.3 % trypan blue solution. Ammonia concentration was enzymatically measured by an ammonia assay kit (170-A, Sigma Chemical Co., St. Louis, MO). Antithrombin III was quantified by an enzyme-linked immunosorbent assay (ELISA).

3. Control strategy analysis

In a static fuzzy control, the process state variables are the glucose (or glutamine) error (E) and the change of the error (DE). The control output variable is the change of the feeding time (DF) of the glucose (or glutamine) feeding pump. Since the variables are treated as fuzzy variables, they take linguistic instead of numerical values such as negative small (NS) etc. The range of the DF membership function from zero to FTMAX or -FTMIN determines how much is any positive or negative change of the feeding time.

The influence of the parameters FTMAX and FTMIN on the controller performance was checked in simulations using the glucose (or glutamine) balance equation in the reactor every two hours (data not shown). An optimal value of FTMAX(MIN) around 20 gave satisfactory performance. However, this value must be increasing gradually during the cultivation following the increase of nutrients consumption caused by the increasing cell concentration. Therefore, the adaptation schema was considered.

In adaptive controller, both of the lactate concentration (LAC) and the sum of the four previous errors (SE) are used as an indicator of culture phase and a controller performance index, because the exact values of lactate when cell death starts are different for different experiments and not apriori known. The fuzzy adaptation strategy decides the change of the FTMAX(MIN) parameter value (DM) as an output. The combination of the static fuzzy controller with the above adaptation strategy in an hierarchical structure gives an adaptive fuzzy controller (Fig. 1b).

The adaptive controller exhibited satisfactory performance in simulations performed with the consumption function and the lactate function. The under(over)shoot of the early(late) phase observed with the static fuzzy controller was avoided with the adaptation of the parameters.

4. Results from controlled cultivations and discussion

Results of six cultivations with different glucose-glutamine set point concentrations (3.0-0.2, 2.0-0.2, 0.5-0.2, 3.0-0.6, 2.0-0.6, 0.5-0.6 gl^{-1}) are presented in Fig. 2. The maximum cell concentration depended on the ammonia concentration. In any experiments of 0.2 gl^{-1} glutamine in which the concentration of ammonia was lower (Fig. 2b), the maximum cell concentration was higher than those of the 0.6 gl^{-1} glutamine experiment. When the ammonia reached the concentration of 2.5 mM, cell growth inhibition was observed in the experiments with 0.6 gl^{-1} glutamine set point, while no inhibitory phase was evident in the experiment of 0.2 gl^{-1} glutamine. Lactate seems not to play a determining role in cell viability since cell death started at very different lactate concentrations (Fig. 2c).

In the condition of the decrease of glutamine or the increase of glucose set point, less ammonia production was observed. In a study with hybridoma cells cultured in flasks (Gaertner and Dhurjati, 1993), high glucose concentration above 8mM (1.5 gl^{-1}) and glutamine concentration of 1 mM (0.146 gl^{-1}) were suggested as the nutrients concentrations that optimize the antiboby productivity. The recommended nutrient concentrations are similar to that we found.

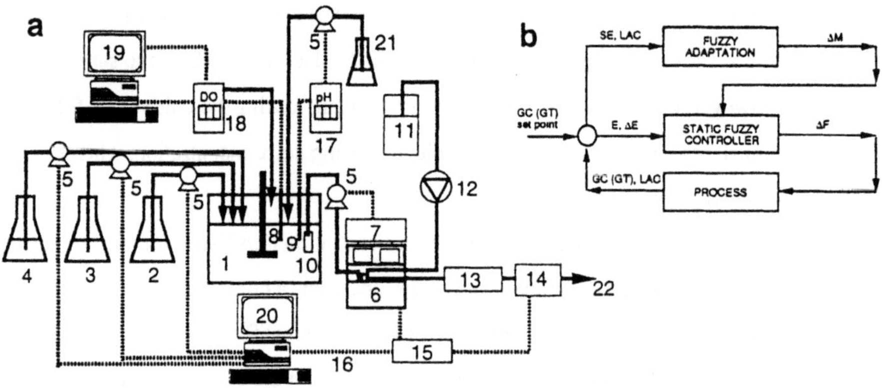

Figure 1. (a) Flow diagram of the culturing system. 1. reactor, 2. glucose feeding medium, 3. glutamine feeding medium, 4. volume control feeding medium, 5. pumps, 6. autosampler, 7. timer, 8. DO electrode, 9. pH electrode, 10. stainless-steel filter, 11. mobile phase solution, 12. HPLC pump, 13. HPLC column, 14. RI detector, 15. integrator, 16. RS-232C cable, 17. pH on-off controller, 18. DO meter-gas flow controller, 19. computer for DO control, 20. computer for gluc.-glut. control, 21. base, 22. HPLC waste. (—— material flow, ········ information flow)
(b) Structure of the adaptive fuzzy controller.

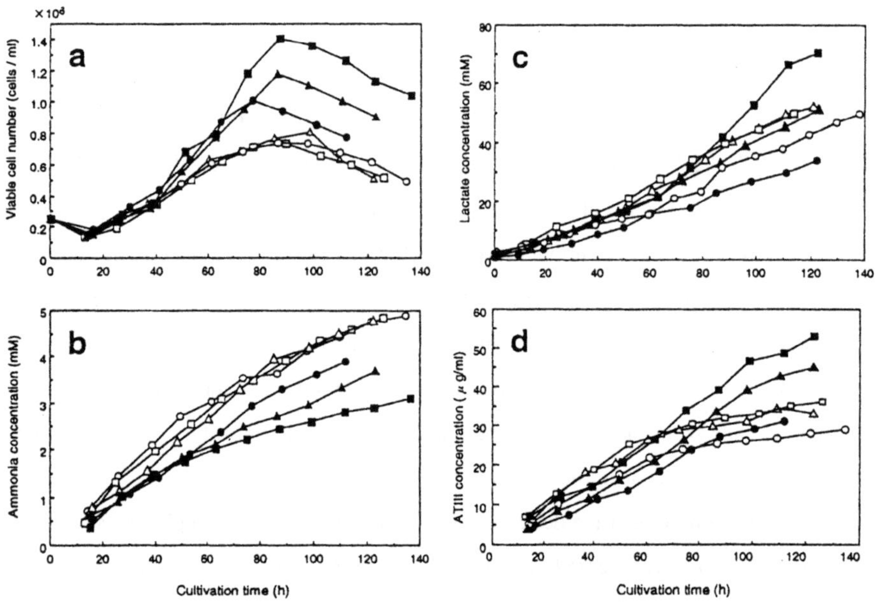

Figure 2. Results from controlled cultivations. (a) Viable cell concentration. (b) Ammonia concentration. (c) Lactate concentration. (d) Recombinant protein (Antithrombin III) concentration. (Glucose-glutamine set point concentrations in gl^{-1} : ● 0.5-0.2, ▲ 2-0.2, ■ 3-0.2, ○ 0.5-0.6, △ 2-0.6, □3-0.6)

However for cells that are sensitive to both, lactate and ammonia, low glucose and glutamine concentrations are the optimal (Kurokawa *et al.* 1993).

The highest final protein concentration was achieved in the cultivation of 3.0-0.2 gl^{-1} gluc.-glut. set points (Fig. 2d) which had also the higher maximum cell concentration. In the 3.0-0.2 cultivation, the final protein concentration was two times higher than that of the 0.5-0.2 culture which had the lowest final protein concentration.

5. Conclusions

In the present work using the minimum information for the behavior of the system, an adaptive fuzzy controller that efficiently controls the glucose and glutamine concentrations in the culture broth was designed. It was found that ammonia had a determining effect on productivity and its production was increased by an increase of glutamine or decrease of glucose concentrations. A combination of low glutamine with high glucose for cells with higher sensitivity to ammonia than lactate was more efficient in maximizing the cell concentration and protein productivity.

Acknowledgment

The fellowship from European Union (P. L.) is gratefully acknowledged.

References

Gaertner J and Dhurjati P (1993) Fractional factorial study of hybridoma behavior. 2. Kinetics of nutrient uptake and waste production. *Biotechnol. Prog.* **9**: 309-316.

Hu WS, Dodge TC, Frame KK and Himes VB (1987) Effect of glucose on the cultivation of mammalian cells. *Develop. Biol. Standard.* **66**: 279-290.

Kurokawa H, Ogawa T, Kamihira M, Park YS, Iijima S and Kobayashi T (1993) Kinetic study of hybridoma metabolism and antibody production in continuous culture using serum-free medium. *J. Ferment. Bioeng.* **76**: 128-133.

Kurokawa H, Park YS, Iijima S and Kobayashi T (1994) Growth characteristics in fed-batch culture of hybridoma cells with control of glucose and glutamine concentrations. *Biotechnol. Bioeng.* **44**: 95-103.

Wang J, Honda H, Lenas P, Watanabe H and Kobayashi T (1995) Effective tPA production by BHK cells in nutrients-controlled culture using an on-line HPLC measuring system. *J. Ferment. Bioeng.* **80**: 107-110.

COST-SAVING DESIGN AND OPERATIONAL OPTIONS FOR LARGE-SCALE PRODUCTION OF NUTRIENT MEDIUM AND BUFFERS.

David Jayme, James Kubiak, Richard Fike, Stephan Rashbaum and Shawn Smith,
Life Technologies, Inc., 3175 Staley Road, Grand Island, NY 14072 USA

Abstract

Technical and economic advantages may derive from use of liquid concentrates to formulate nutrient medium and buffers. Batch reconstitution may be performed in formulation tanks, within customized bulk packaging, or directly within the bioreactor. Improvements in operating convenience, utilization efficiency of facility, equipment and technical personnel, and manufacturing cost were observed when concentrate reconstitution was performed using a mixing device for continuous delivery of diluted fluids directly into a bioreactor or into interim storage vessels. The preferred alternative may depend upon bioreactor culture conditions, volumetric consumption requirement, and buffer application and composition.

Introduction

In his remarks for a JAACT '95 symposium on Modeling and Control of Animal Cell Culture, David Naveh [1] noted several emerging trends within the maturing biotechnology industry in the category of process economics. He observed that companies were evolving away from being "first to market" in favor of being an early market entrant with the most economical manufacturing cost. While cost reduction efforts have traditionally emphasized variable cost factors (e.g., amplified expression, improved recovery), he cautioned that volume-sensitive fixed costs (e.g., labor, utilities, waste treatment, utilities, maintenance) represented about 50% of cost of goods sold and might be a more fruitful target for reduction in product cost. Suggested areas included optimization of process time/space and relationship to overall product yield, outsourcing of "auxiliary unit" operations, and multipurpose plant utilization with generic technologies.

Outsourcing of auxiliary unit operations was perceived as superior to vertical integration because it permitted the biotechnology company to focus its resources (e.g., facility space, capital investments, technical personnel) on its core competencies. Candidate operations for outsourcing included cell bank cryopreservation and maintenance, and preparation of nutrient medium and buffered salt solutions.

The intervening year has confirmed Dr. Naveh's observations, at least from the perspectives of media and buffer supply. Companies have consciously elected not to construct a media kitchen in their new manufacturing facility, preferring to purchase ready-to-use liquid medium in bulk containers. Other firms with established media kitchens for hydration of homogeneous powders into liquid medium have converted either to purchased liquid medium or to liquid concentrated configurations and have confirmed an overall cost reduction, superior facility space utilization, and alleviation of manufacturing complexity. In addition to nutrient media, there has been increased interest in outsourcing production of buffers and other solutions formerly manufactured in-house. Typically, the primary decision drivers for outsourcing have focused upon manufacturing convenience, utilization efficiency of manufacturing space, equipment and personnel, and overall cost reduction [2]

This paper consolidates public information regarding novel alternatives for large-scale preparation of nutrient media and buffers from concentrated intermediates and discusses facility and equipment design and operational implications. It also notes trends in partial nutrient supplement alimentation of high density bioreactors to improve productivity in fed batch and perfusion culture modes.

Media production from concentrates

Several modes for reconstitution of liquid media concentrates are routinely employed or under qualification for preparation of large-scale nutrient medium. While most current users of concentrated media intermediates perform reconstitution in batch mode, interesting studies are in progress to qualify continuous modes of medium reconstitution utilizing an automated mixing device [3-4].

Sanitary mode reconstitution: The majority of liquid medium concentrate users reconstitute medium in an open formulation tank previously used for hydration of powdered medium. A measured volume of production water is added to the formulation tank, followed by rapid, sequential addition of concentrated components according to a specified protocol. Bicarbonate is added and final adjustment of pH and osmolality is performed similar to traditional powdered medium preparation. Final filtration of reconstituted medium is performed prior to aseptic delivery either directly to the bioreactor or to intermediate storage vessels.

Aseptic batch mode reconstitution: Enhanced methods for formulation and configuration of liquid medium concentrates permit reconstitution within a closed formulation vessel without manual titration. One option requires a closed formulation tank capable of sterilization-in-place. Production water is sterilized within the formulation tank and concentrated components are metered directly into the formulation tank. Due to technical considerations made in formulation of the concentrated sub-groups, no subsequent adjustment of pH or osmolality is required to meet final specifications. Depending upon the batch volume and perceived risks of contamination, some users perform aseptic connection of concentrated intermediates and dispense with a requirement for final filtration post-reconstitution.

A second mode for aseptic batch reconstitution eliminates the need for an intermediate formulation tank and performs reconstitution of media concentrates directly within a sterilized bioreactor prior to cell inoculation. This method combines automatic pH and osmolality adjustment of concentrated intermediates with aseptic processing of concentrated intermediates to facilitate reconstitution. Risk of microbial contamination is minimized through use of bacterial-retentive filters and the bacteriostatic biochemical composition of concentrated sub-groups.

Continuous reconstitution: The design features and process flow features for an integrated medium preparation system, combining concentrated intermediates using positive displacement pumps and an on-line static mixing device and computerized monitoring of reconstitution, have been described previously [4-5]. Mixing devices of similar design and function are under active evaluation by several prospective users for rapid preparation of large batches (10-50,000 liters) of liquid medium per day. Less complex devices for extended continuous delivery of nutrient medium prepared from concentrated intermediates to perfusion bioreactors have been reported [6] and are in use for preparation of materials for advanced clinical trials by several companies.

Buffer production options

Buffer concentrates may be prepared using criteria analogous to those previously-described for nutrient media concentrates, except that buffers differ from typical media in complexity and ionic strength. Buffer concentrate configurations are generally less complex, owing to the reduced number of biochemical constituents. However, the widely-ranging conductivity and pH specifications required by typical buffer applications complicate configurational design and stability.

Large scale: Large volume requirements, exceeding several hundred thousand to over one million liters per week, may be prepared from buffer concentrates using an automated mixing device of high throughput and reduced complexity. Buffer concentrates (1-2 components per buffer formulation) are provided in bulk packaging with appropriate connectors for closed delivery to the mixing device for continuous reconstitution. As with large-scale media preparation, this option improves the utilization efficiency of facility space, equipment and technical staff and provides convenient delivery of reconstituted buffers to chromatography column, storage vessel or final containers with a significant reduction in facility space requirement and in overall buffer manufacturing cost.

Intermediate scale: Two alternatives are under evaluation for intermediate scale (several hundred to a few thousand liters per week) volume buffer requirements. The first example is analogous to batch reconstitution of liquid medium. Buffer concentrates are designed to produce buffer with specified pH and conductivity upon reconstitution without manual titration. Post-formulation filtration may be performed, if required. Key features of this option are the rapid buffer preparation time and the ability to utilize existing buffer kitchen equipment and facilities.

A second option delivers single-component buffer concentrates as a partial fill within bulk packaging. An integral inlet filter is provided for delivery of sterile diluent and the pre-filled concentrate is diluted within the bulk container. Recirculation tubing is included to ensure homogeneous mixing. Advantages include minimal facility footprint for buffer preparation, elimination of requirement for formulation tanks or post-dilution filtration, and convenient storage and delivery of diluted buffer.

Small scale: Frequently, buffer kitchens commit a significant percentage of their total resources to manufacturing a relatively high number of small volume buffers (less than a few hundred liters per week). To satisfy these requirements, 1X buffers may be provided ready-to-use in bulk or unitized packaging (10-200 liter containers), complete with application-specific connectors for closed delivery. Outsourcing of small volume buffer preparation permits the user to focus its resources on high value core competencies. It may also result in more consistent buffer quality and performance and actually reduce the overall buffer manufacturing cost.

Concentrated nutrient feed options

Various studies have documented the rapid exhaustion of key metabolites from high density cell culture bioreactors [7-9]. Enrichment of the basal formulation to augment levels of critical nutrients has been particularly critical with the evolution of serum-free and protein-free bioproduction requirements [7-10]. Analytical techniques may be utilized to determine the pseudo first order consumption kinetics of key nutrients and to design a customized nutrient cocktail which is stoichiometrically balanced for the specific cell culture system [8,11]. However, a simple and effective approach to achieve improved productivity is to selectively augment certain nutrients within the basal formulation (e.g., amino acids, carbohydrates, vitamins, nucleosides) by two-to-five-fold, while maintaining constant levels of inorganic salts and buffering constituents. Such enriched modifications may even be specified for proprietary media, working in collaboration with media suppliers to identify key nutrients and to generate a customized formulation.

Elevated viable cell densities and biological production may be sustained by supplementation of culture medium with concentrated nutrient additives, either in fed-batch or continuous delivery modes [9,12]. Unsuccessful attempts at nutrient supplementation have utilized "complete" 10X formulations, with the inevitable consequences of hyperosmotic culture conditions. "Salt-free" and "protein-free" supplemental nutrient concentrates may be designed by eliminating sodium chloride and other inorganic salts, as well as costly peptides, from the feed medium. Such nutrient supplements may derive from the basal formulation, or they may be designed following HPLC analysis of spent fluids. This practice also reduces the overall cost of nutrient medium by decreasing the volume of waste materials and by facilitating downstream product recovery and purification processes.

Discussion

Key features of batch and continuous reconstitution of buffers are similar to previously-reported advantages associated with preparation of liquid medium from concentrated intermediates [2-3]. Batch processes emphasize reduction in manufacturing space and time, improved utilization of personnel and equipment, and improved quality and consistency of formulated fluids. Significant reductions in overall fluids manufacturing costs have also been achieved through "campaigned" production of similar formulations, using constant and variable concentrated additives [3,12].

Continuous production of media and buffers from concentrated intermediates utilizing a on-line mixing device augments these advantages by a further reduction in facility footprint and other capital investment costs, improved manufacturing control systems, and significant reduction in all batch-related production costs [2,4].

The traditional approach to nutrient depletion has been to accelerate the medium exchange rate within the bioreactor. This approach is costly in terms of media consumption, accumulation of waste fluids, and recovery of diluted biological product. Concentrated, salt-free nutrient additives may be an economical option to improve bioreactor productivity.

Summary

The visionary prospect suggested at last year's JAACT meeting that economic realities would force biotechnology to focus on core competencies and to outsource auxiliary unit operations is beginning to be realized, particularly in the areas of large-scale nutrient medium and buffer preparation. Moreover, novel modes for manufacture and delivery of these large volume fluids can favorably impact facility design and overall biological production economics. Bioreactor productivity may be economically enhanced by nutrient optimization, either through optimization of the basal formulation or addition of salt-free, protein-free concentrated additives.

References

1. Naveh D (1995) Proceedings of the Eighth Annual Meeting of the Japanese Association for Animal Cell Technology, *"An industrial perspective to development of animal cell culture processes for production of therapeutics."*
2. Jayme DW, Kubiak JM, Battistoni TA, Cady DJ and Fike RM (1995) Proceedings of the Eighth Annual Meeting of the Japanese Association for Animal Cell Technology, *"High capacity method for producing nutrient medium from concentrated intermediates for cell culture bioproduction applications."*
3. Jayme DW, Fike RM, Kubiak JM, Nash CR and Price PJ (1993) in Animal Cell Technology: Basic & Applied Aspects, eds. S. Kaminogawa, A. Ametani and S. Hachimura, volume 5, pp. 215-222, *"Use of liquid medium concentrates to enhance biological productivity."*
4. Jayme DW, Kubiak JM, Battistoni TA and Cady DJ, Cytotechnology (1996) 22: 255-261, *"Continuous, high capacity reconstitution of nutrient media from concentrated intermediates."*
5. Jayme DW, Kubiak JM and Price PJ (1994) in Animal Cell Technology: Basic & Applied Aspects, eds. T. Kobayashi, Y. Kitagawa and K. Okumura, volume 6, pp. 383-388, *"Continuous, automated reconstitution of liquid media concentrates."*
6. Roth G, Kubiak JM, Long JF, and Schoofs GM, BioPharm (1995) 8: 31-35, *"An automated system to produce cell culture media from liquid medium concentrates."*
7. Jayme DW, Cytotechnology (1991) 5: 15-30, *"Nutrient optimization for high density biological production applications."*
8. Xie L and Wang DIC, Biotechnology and Bioengineering (1994) 43: 1164-1174, *"Stoichiometric analysis of animal cell growth and its application in medium design."*
9. Xie L and Wang DIC, Biotechnology and Bioengineering (1994) 43: 1175-1189, *"Fed-batch cultivation of animal cells using different medium design concepts and feeding strategies."*
10. Jayme D, Watanabe T and Shimada T, Cytotechnology (in press), *"Basal medium development for serum-free culture: A historical perspective."*
11. Fike R, Kubiak J, Price P and Jayme D, BioPharm (1993) 6: 49-54, *"Feeding strategies for enhanced hybridoma productivity: Automated concentrate supplementation."*
12. Jayme DW, Kubiak JM and Fike RM (1995) in Animal Cell Technology: Developments toward the 21st Century, eds. EC Beuvery, JB Griffiths and WP Zeijlemaker, 199-204 (Kluwer), *"Productivity enhancement using liquid medium concentrates."*

EFFECT OF SHEAR STRESS ON THE rCHO CELLS

M.MOTOBU, P.C. WANG and M.MATSUMURA
Institute of Applied biochemistry, University of Tsukuba
1-1-1 Tennodai, Tsukuba, Ibaraki, 305, Japan

1. Introduction

Effect of shear stress on the endothelial cell has been studied by many workers, but few studies on the effect of shear stress for other cell lines have been reported. We have investigated the effect of shear stress on the rCHO using flat-bed type reactor with non-woven fabric mat and found that the renin productivity was enhanced and the cells attached to the non-woven fabric were aggregated by increasing shear stress. It is therefore considered that shear stress or cell aggregation may affect productivity(1).

The growth of eucaryotic cells has been described as a cell cycle consisting of four distinct phases occurring in a continuous succession : mitotic phase (M), gap phase (G1) between mitosis and DNA synthesis, DNA synthesis phase (S), and gap phase (G2) between completion of DNA synthesis and the following mitosis. Synthesis and secretion of immunoglobulins by hybridoma have been reported to be restricted to the late G1 and early S phases (2). The experiments studying on the phase which enhances the renin production by the r CHO cells have not been investigated yet.

In order to clarify the effect of shear stress on the cell productivity and cell cycle , the culture of rCHO using flat-bed type reactor, which does not occur the cell aggregation, was investigated. The r CHO cells were cultivated for 1-3 days at 0.05N/m^2 in this reactor, the results were compared with that of static culture.

2. Material and methods

2.1. CELL LINE

The cell line used in this study was the CHO dhfr- (DXB-11) transfected with human renin expression vector containing mouse dhfr gene. The medium used was DMEM supplemented with 5% FBS.

2.2. FLAT-BED TYPE REACTOR

Figure 1 shows the schematic diagram of flat-bed type reactor . The r CHO cells were inoculated in a 1.8 cm x 5 cm glass plate coated with collagen (inoculum size was 3 x 10^4 cells/cm^2), and the glass plate was fixed to the flat-bed type reactor after the cells reached confluent. The flat-bed type reactor system with a total volume 40 ml was settled in the CO_2 incubator. The cultivations were started by circulating medium at a shear stress of 0.05 N/m^2.

2.3. CONTROL EXPERIMENT

To compare with the results of the effect of shear stress on r CHO cells, the cells were cultivated without shear stress. The r CHO cells were inoculated in 4.9 cm x 4.7 cm glass plate coated with collagen (inoculum size was 3 x 10^4 cells/cm^2), and the glass plate wassettled in dish with fresh medium after the cells reached confluent. The cells on the glass plate were cultivated without being subject to shear stress.

2.4. ANALYSIS

After the cells were exposed to shear stress for 1-3 days, the renin activity in the supernatant was measured by RIA. The cells were treated with BrdU for 30 min at the end of the cultivation, then the cells were harvested. The cell cycle was analyzed by Flow cytometry after labeling with anti BrdU-FITC and PI.

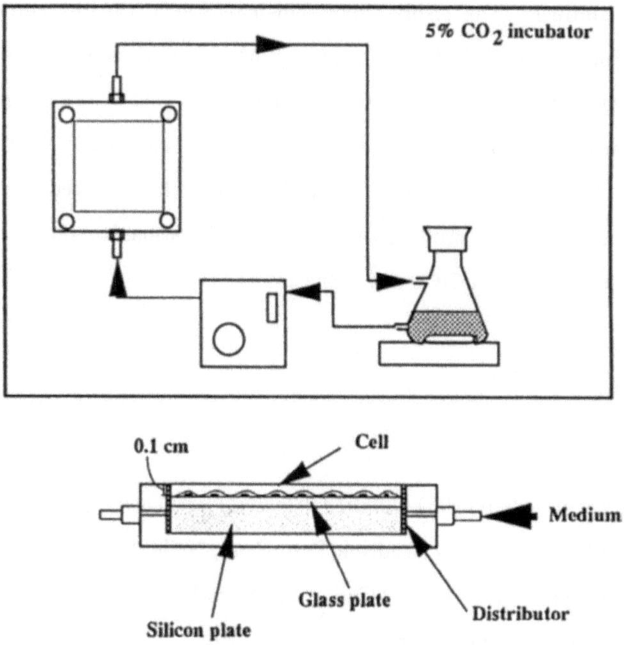

Fig. 1. Schematic diagram of flat-bed type reactor .

3. Results and discussion

To investigate the relationship between the cell cycle and the specific renin production rate of the rCHO cells, the static culture was performed in the 25 cm^2-flask and the results were shown in Fig.2. The cell cycle and specific renin production rate were analyzed for 3 days after inoculation. It is observed that the more the cell number in the G2/M phase increases, the more the cell productivity decreases. It may suggest that the renin production by the r CHO cells occurs under G1/G0 and S phase.

On the basis of the above result, the effect of shear stress on the cell cycle and specific renin production rate using flat-bed type reactor was further investigated. The cultivations were performed for 1-3 days at 0.05 N/m^2 after the cells on the glass plate became confluent. The results were shown in Fig.3. In the culture without shear stress, the ratios of G1/G0 , S, and G2/M phases were almost the same. On the other hand, most r CHO cells stayed in G1/G0 and S phases in the culture with shear stress and the specific renin production rate was about 4 times higher than that in the culture without shear stress. Taken together, shear stress enhances the renin productivity and this fact is probably due to its depression on the cells remaining in G1/G0 and S phases.

4. References

1. E.C.Beuvery et al. (eds.) (1995) *Animal cell technology* : Developments towards the 21 st Century, 811-815, Kluwer Academic Publishers, Netherlands.
2. Suzuki, E. and Ollis, D. F.(1989) Cell cycle model for antibody production kinetics, *Biotechnology and Bioengineering* **34**, 1398-1402.

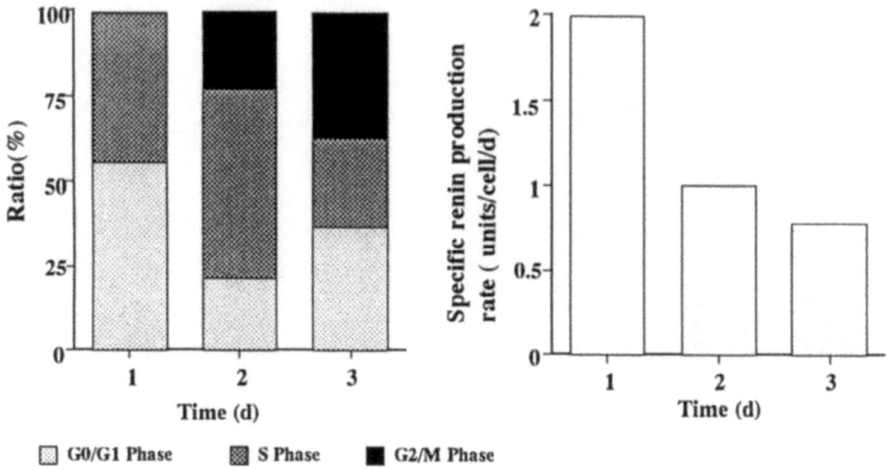

Fig.2. The relationship between the cell cycle and the specific renin production rate of the r CHO cells in the static culture.

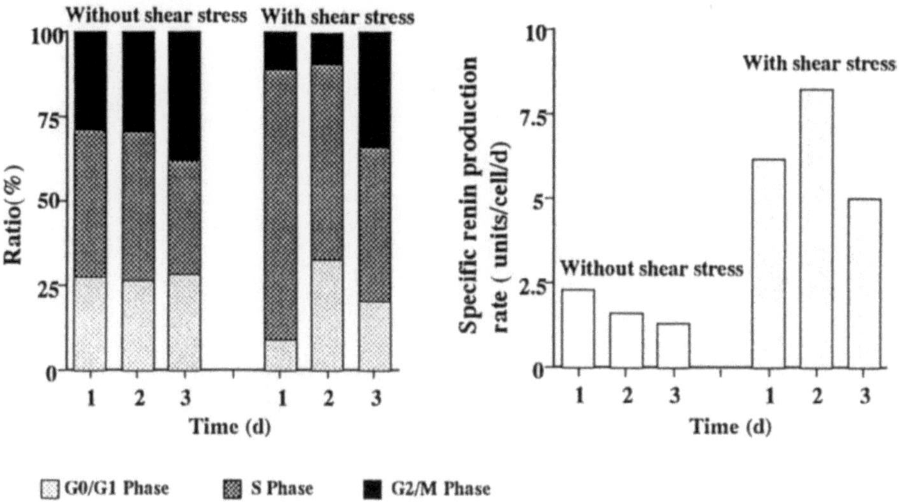

Fig.3. The effect of shear stress on the cell cycle and the specific renin prodution rate of r CHO cells in the culture using flat-bed type reactor.

QUAIL EMBRYO CULTURE USING ARTIFICIAL VESSELS

SATOSHI OGUCHI, MASAMICHI KAMIHIRA AND
SHINJI IIJIMA
Department of Biotechnology,
Graduate School of Engineering, Nagoya University,
Chikusa-ku, Nagoya, 464-01, Japan

ABSTRACT. The establishment of avian embryonic culture is important for both analysis of developmental process and establishment of transgenic chickens which produce useful biological materials in egg contents. However, hatchability of cultured embryos was at most 50% by now. We identified that the decrease in hatchability for cultured embryos was caused by limitation of oxygen and calcium supplies. In quail embryo culture using a chicken egg shell as a culture vessel, viability in middle phase of culture was improved and 30% of embryos hatched by oxygen enrichment. Furthermore, hatchability was improved very much (80%) by the addition of Ca lactate in addition to oxygen aeration.

We designed a fully artificial vessel for quail embryo culture in which gas permeable teflon membrane was used. By the addition of fine egg shell powder and Ca lactate, quail embryos grew and developed normally, and 45% of embryos was led to hatch. Although the hatchability is lower than that of cultures using a chicken egg shell as a culture vessel, we succeeded in avian embryonic hatching using a fully artificial vessel.

1. Introduction

Transgenic animal is expected to be a powerful tool for genetic improvement of species and production of useful biological substances in addition to basic scientific study. In the case of avian embryonic manipulation using chickens and quails, an egg shell has been holed and resealed after the manipulation. Therefore, hatchability falls down to less than 10% during the manipulation. Many researchers have tried to culture avian embryos without egg shell, but all of them were unsuccessful[1)]. In recent years, Perry reported *in vitro* culture method for a chicken embryo from single cell stage, where a host egg shell was used as a culture vessel[2)]. Now, hatchability of this method increased up to 50%[3,4,5)], but it is still low compared with that of normal incubation. Furthermore, no one have succeeded in embryonic hatching using a fully artificial culture vessel.

We have reported that the decreased hatchability for cultured embryo was caused by limitation of oxygen and calcium supply[6,7,8)]. In the present study, we examined culture conditions to improve hatchability of quail embryo culture using an egg shell as a culture vessel and developed a fully artificial vessel for quail embryo culture.

2. Materials and Methods

Materials

Fertilized quail eggs (WE strain) were obtained from Nisseiken (Yamanashi). Fine egg shell powder was prepared by breaking up quail egg shell with a mill. Calcium lactate was purchased from Sigma (St. louis, MO, USA).

Culture vessels

Chicken white leghorn egg shells or glass cups with teflon membrane were used as a culture vessel for quail embryo cultures. In the case of a chicken egg shell, blunt end of an egg shell was cut to make a hole of 40 mm in diameter with an electric minirouter.

Figure 1 shows an artificial culture vessel we developed in this study. A round-bottom shaped gas-permeable porous teflon membrane (Millipore, Tokyo) was put into a glass cup which has several vent holes. The glass cup was autoclaved before use.

Embryo culture

Quail embryos were cultivated in an incubator equipped with a rocking device (Showa Furanki, Tokyo). Laid quail eggs were incubated at 37.8°C and relative humidity of 65% while being rocked round at a 90 degree angle every 15 minutes for two days. The egg content was then transferred into a culture vessel, and they were sealed with plastic film. The culture vessels containing the egg content were incubated at 37.8°C and 65% relative humidity with a 30 degree angle rocking every one hour. One day before the expected hatching day, the rocking of the culture vessels was stopped and the plastic film was holed to facilitate embryonic respiration.

3. Results and Discussion

In the previous study, we examined quail embryo culture using a chicken egg shell as a culture vessel according to Perry's procedure[3,4,6,7]. However, we could not achieve embryonic hatching, although some embryos were alive until just before hatching (Fig. 2). In order to identify the reasons of this unsuccessfulness, egg contents such as yolk, albumen, amniotic and allantoic fluids during the embryonic development were collected separately to analyse concentrations

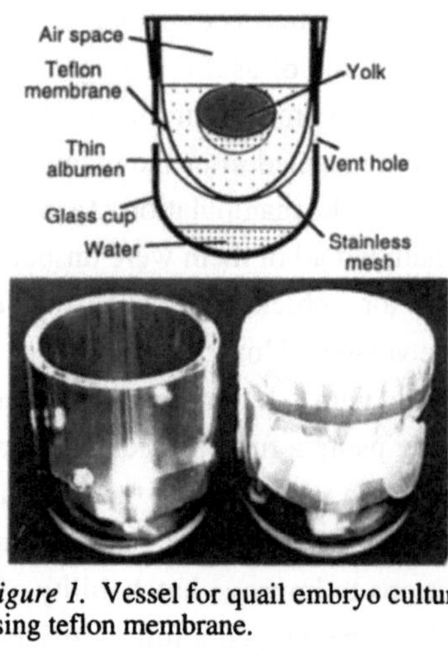

Figure 1. Vessel for quail embryo culture using teflon membrane.

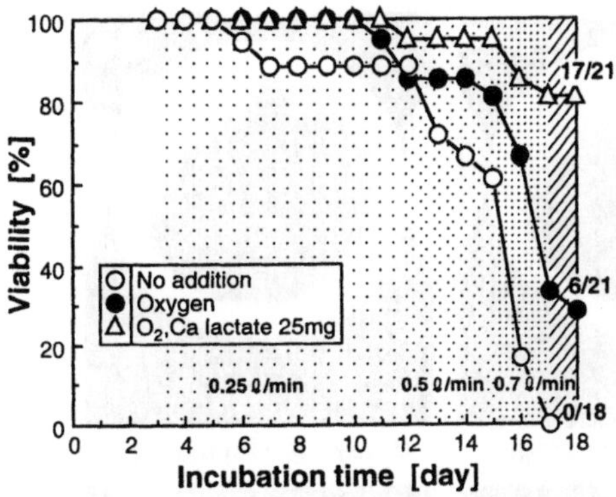

Figure 2. Effect of oxygen aeration and egg shell powder addition on viability and hatchability of quail embryo in egg shell culture. ○: no addition, ●: oxygen aeration, △: oxygen aeration and calcium lactate addition. Oxygen aeration was begun at a rate of 0.25 ℓ/min from 3 days to 12 days stage, and was then raised up to 0.5 ℓ/min to 15 days stage and 0.75 ℓ/min to hatching. Embryos hatched on day 17 or 18.

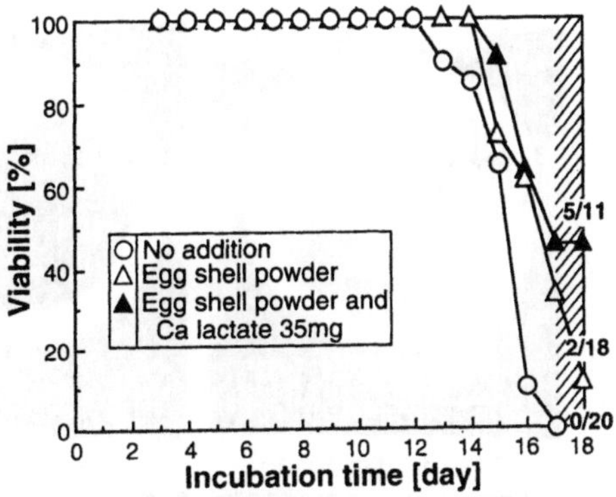

Figure 3. The time course of viability in teflon membrane culture. ○: no addition, △: egg shell powder addition, ▲: egg shell powder and calcium lactate addition. Embryos hatched on day 17 or 18.

of metal ions such as Na, K, Mg and Ca, and metabolites such as lactate, ammonia, urea and uric acid. We found that amniotic and allantoic fluids of the cultured embryos contained high concentration of lactate, suggesting the limitation of oxygen supply for the cultured embryos. Furthermore, calcium concentration in yolk of the cultured embryos drastically decreased at 4-6 days in culture period. In order to improve oxygen limitation, oxygen gas was introduced into an incubator (Fig. 2). By this oxygen enrichment in gasneous phase, viability of the embryos in middle phase of the culture was improved and 29% of embryos hatched.

For the next step, we examined Ca-containing materials such as egg shell fine powder, hydroxyapatite, $CaCl_2$, Ca lactate and Ca phosphopeptide to be added to albumen. Among these materials, Ca lactate showed good effect for embryonic growth. When 25 mg Ca lactate was added per an embryo in addition to oxygen aeration, the viability during the culture was improved very much and about 80% of embryos hatched (Fig. 2). The hatchability was comparable to normal incubation of embryo without manipulation.

We then designed a fully artificial vessel for quail embryo culture (Fig. 1). An egg content was enclosed by a gas permeable teflon membrane so that gas exchange can be conducted from all directions like an egg shell. Lactate concentrations in amniotic and allantoic fluids for embryos cultured in the teflon membrane were maintained low even without oxygen aeration, therefore oxygen was not limiting factor in this culture. Figure 3 shows time course of viability in embryo culture using the artificial vessel with teflon membrane. The viability was kept 100% until 12 days, but it decreased drastically afterward and no embryo could hatch. An egg shell has function as mineral source during embryonic

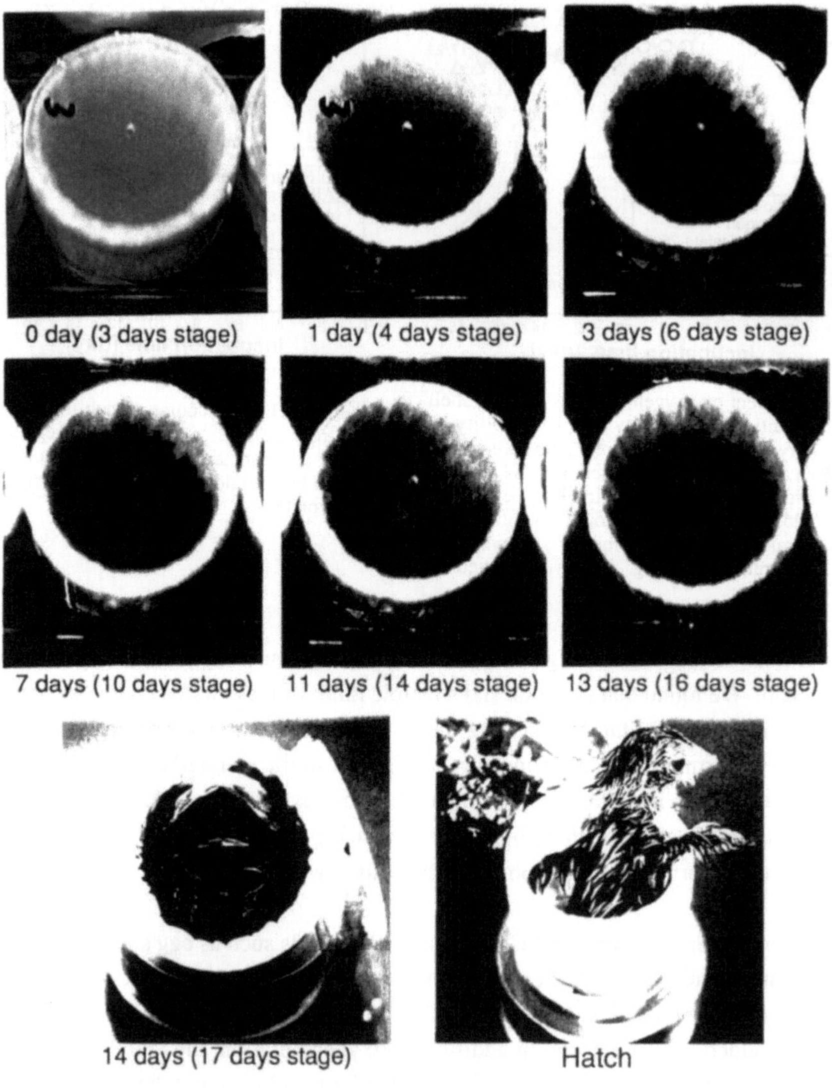

Figure 4. Embryonic development in teflon membrane culture

development. Since the artificial vessel does not have such function, fine egg shell powder as a mineral source was added to the bottom of the culture vessel. By adding the egg shell powder, two out of 18 embryos tested hatched in the teflon membrane culture. For further improvement of the hatchability, Ca lactate which was effective for embryo culture using an chicken egg shell as a culture vessel, was added at the amount of 35 mg per an embryo. By the addition of both egg shell powder and Ca lactate, hatchability increased up to 45%. Figure 4 shows photographs of embryonic development in the teflon membrane culture.

In the case of quail embryo culture using a chicken egg shell, we established culture conditions with 80% hatchability as described above. The hatchability of the teflon membrane culture was lower than that of the egg shell culture. The reason of this difference may be lack of minerals although fine egg shell powder was added to the teflon membrane culture. In fact, the beaks and legs of the embryos of the teflon membrane culture seemed smaller and embryos were also lighter than those of normal incubation and the egg shell culture in average weight (Fig. 5). Therefore, to obtain higher hatchability for the teflon membrane culture, the development of effective mineral supply method is essential.

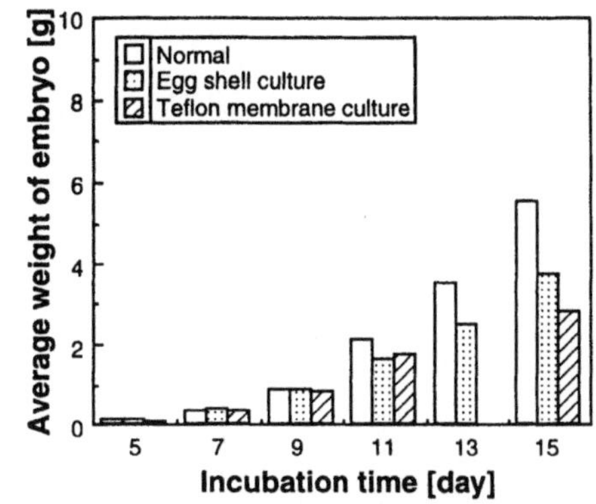

Figure 5. The time course of average weight of embryo in quail embryo cultures. □: normal incubation, ⊡: egg shell culture, ▨: teflon membrane culture.

References

1. Rowlett, K. and Simkiss, K. (1987) Explanted embryo culture: In vitro and in ovo techniques for domestic fowl, *Brit. Poul. Sci.* **28**, 91-101
2. Perry, M.M. (1988) A complete culture system for the chick embryo, *Nature*, **331**, 70-72
3. Naito, M., Nirasawa, K., and Oishi, T. (1990) Development in culture of the chick embryo from fertilized ovum to hatching, *J. of Exp. Zool.*, **254**, 322-326
4. Nirasawa, K., Naito, M., and Oishi, T. (1992) Culture of japanese quail embryos from the blastodermal stage to hatch in chicken egg shells, *Japan. Poul. Sci.* **29**, 139-144
5. Ono, T., Murakami, T., Mochii, M., Agata, K., Kino, K., Otuka, K., Ohta, M., Mizutani, M., Yoshida, M., and Eguchi, G. (1994) A complete culture system for avian transgenesis, supporting quail embryos from the single cell stage to hatching, *Developmental Biology*, **161**, 126-130
6. Kamihira, M., Oguchi, S., and Iijima, S. (1995) The effect of environmental factor in avian embryo culture in egg shell, *Abstract of the 60th annual meeting of SCEJ*, J126
7. Kamihira, M., Oguchi, S., Kitagawa, Y., and Iijima, S. (1996) Development of avian embryonic culture under artificial environment, *Abstract of the 61th annual meeting of SCEJ*, J126
8. Oguchi, S., Kamihira, M., and Iijima, S. (1995) Embryonic development and hatching by quail embryo culture in teflon membrane, *Abstract of the 28th automn meeting of SCEJ*, I204

ENHANCEMENT OF TRANSFECTION EFFICIENCY USING LIGAND-MODIFIED LIPID VESICLES

JUN YOU, MASAMICHI KAMIHIRA AND SHINJI IIJIMA
Department of Biotechnology, Graduate School of Engineering, Nagoya University Chikusa-ku, Nagoya 464-01, Japan

ABSTRACT. We have developed a simple gene transfection procedure mediated by cationic lipid vesicles for mammalian cells, in which a commercially available synthetic cationic surfactant, dimethyldioctadecyl ammonium bromide (DDAB), was used for making lipid vesicles. The lipid vesicles were modified with insulin or galactose residue for receptor mediated gene transfer to enhance transfection efficiency. The insulin-modified lipid vesicle solution mixed with plasmid DNA (pCMVβ) was added to Hela, MDCK, NIH3T3 or HepG2 cells. In all the cell lines tested, the transfection efficiencies were increased 3–4 folds. However, excessive free insulin addition to the medium resulted in decreased transfection efficiency, suggesting that the specific interaction between insulin and its receptor gave increased efficiency. The galactose-modified lipid vesicles and the plasmid DNA solution was added to HepG2 or HuH-6 cells expressing asialoglycoprotein receptors, and the transfection efficiencies were also increased 3–4 folds with these cell lines.

1. Introduction

Gene transfection technique for eukaryotic cells has become important for analysis of gene function, production of recombinant gene products and gene therapy. Over the last 15-20 years, many chemical and physical methods have been developed for the introduction of DNA into mammalian cells. These include calcium phosphate precipitation [1, 2], DEAE-dextran [3], liposomes either alone [4-6] or in combination with reconstituted viral envelopes [7-9], cationic liposomes (lipofection) [10], and electroporation [11-13]. A major drawback with all these methods is that they markedly vary in the transfection efficiency according to cell lines and they also lack cell specificity.

Recently, lipid vesicles are widely used for gene transfer. The method is classified two types; DNA-included liposome and lipid vesicle/DNA complex. In DNA-included liposome method, DNAs are encapsulated into liposomes. Thus, DNA-included liposomes should be prepared prior to transfection. On the other hand, native DNA is just mixed with small lipid vesicles in DNA/lipid vesicle complex method (lipofection), and the DNA uptake is achieved together with incorporation of the lipid vesicles. Although the latter method is very simple compared with DNA-included liposome method, it requires an expensive special reagent.

We have previously shown that a surfactant-mediated transfer method in which a commercially available surfactant form lipid vesicles could introduce a plasmid DNA to cell lines such as CHO and BHK cells with high transfection efficiency. However, the efficiency highly depended on cell lines. In the present study, we report the modification of the lipid vesicles with ligands for cell surface receptors to enhance the transfection efficiency.

2. Materials and Methods

2.1. TISSUE CULTURE

Hela and MDCK cells were cultured in minimum essential medium (MEM) supplemented with 10% fetal bovine serum (FBS). HepG2, HuH-6 and NIH3T3 cells were cultured in Dulbecco's modified Eagle's medium (DMEM) supplemented with 10% FBS. Both cultures were maintained at 37°C in a CO_2 incubator. The media for both cultures routinely included 100 units/ml penicillin and 100 mg/ml streptomycin.

2.2. PLASMID DNA

A recombinant plasmid pCMVβ-gal [14] containing lacZ gene under the control of the cytomegalovirus (CMV) early promoter was used as a reporter plasmid.

2.3. LIGAND-MODIFIED LIPID VESICLES PREPARATION

A cationic surfactant, dimethyl-dioctadecyl ammonium bromide (DDAB) (Sigma Chemical Co.) was used for making lipid vesicles. DDAB was dissolved in double-distilled water by vortexing at 60°C and stable lipid vesicles were formed at room temperature. Then, the DDAB lipid vesicles were modified with ligands such as insulin and galactose as follows (Fig. 1). For preparing insulin-modified lipid vesicles, stearic acid was first activated by *N,N'*-dicyclohexylcarbodiimide and mixed with insulin solution at pH 7.5 to make covalent binding between stearic acid and insulin. After dialysis with double-distilled water, the stearic acid-insulin complex was introduced to the DDAB lipid vesicles by sonication. On the other hand, lactose was converted to the

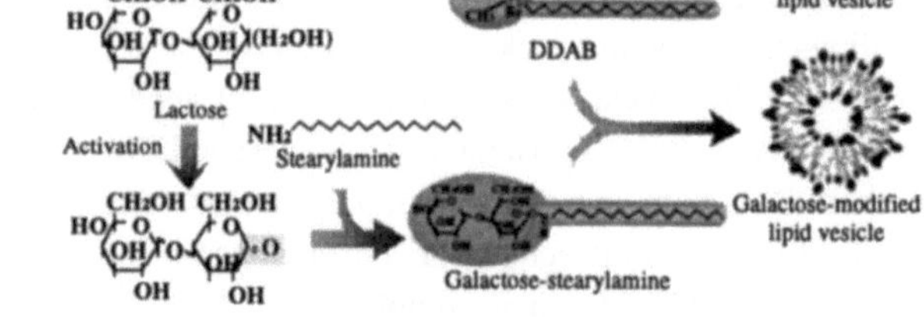

Figure 1. Preparation of ligand-modified lipid vesicles.

lactone and then mixed with stearylamine to make galactose-stearylamine complex. After extraction with chloroform/double-distilled water (1:1, v/v), the galactose-stearylamine complex was introduced to the DDAB lipid vesicles by sonication for making galactose-modified lipid vesicles.

2.4. SURFACTANT-MEDIATED TRANSFECTION

The cells (50 to 70% confluent) were rinsed twice with serum-free Opti-MEM medium (Gibco BRL Co.) to remove serum from the culture. The medium was removed just before the addition of the lipid vesicle/DNA complex. Small aliquots of DNA and lipid vesicle solution were diluted separately in 0.75 ml Opti-MEM medium and then gently mixed and incubated at room temperature for about 30 min to form lipid vesicle/DNA complex. The solution containing the complex was added to the cells in a 35 mm-tissue culture dish. After 6 h, the DNA solution was removed and normal growth medium was added. After 24–48 h cultivation, the culture dishes were washed with phosphate buffered saline solution (PBS) and fixed with 2 ml of an aldehyde solution containing 2% (v/v) formaldehyde and 0.2% glutaraldehyde in PBS for 5 min at room temperature. The cells were rinsed twice with PBS and then stained with 2 ml of X-gal solution (5 mM potassium ferricyanide, 5 mM potassium ferrocyanide, 2 mM $MgCl_2$, and 1 mg/ml 5-bromo-4-chloro-3-indolyl-β-D-galactopyranoside in PBS).

3. Results and Discussion

In our previous study, DDAB was chosen among six commonly available cationic surfactant reagents to make lipid vesicles, since it showed low toxicity and the highest transfection efficiency. In addition, DDAB was easily dissolved in water at 60°C and stable lipid vesicles were formed at room temperature. The transfection efficiency of DDAB lipid vesicle method varied depending on cells and DNAs. When pCMVβ, CHO and BHK cells were used, the efficiency was up to 90% under the suitable conditions. Here, in order to achieve enhancement of transfection efficiency for the cell lines which showed relatively low transfection efficiency and/or have specific receptors, we modified the lipid vesicles with insulin or galactose residue.

The insulin-modified lipid vesicle solution was mixed with plasmid DNA (pCMVβ) to form lipid vesicle/DNA complex and then the complex solution was added to HepG2, NIH3T3, MDCK or Hela cells. As shown in Fig. 2, the transfection efficiency was increased 3–4 folds by using the insulin-modified lipid vesicles in all the cell lines tested. Excessive free insulin addition to the medium resulted in decreased transfection efficiency (Fig. 3). This indicated that the insulin-modified lipid vesicles mediated the plasmid DNA uptake through insulin receptors existing on the cell surface. This procedure may be useful for enhancement of gene transfer, since most of animal cells express insulin receptors on their surface.

An asialoglycoprotein receptor demonstrates many of the features required for a receptor

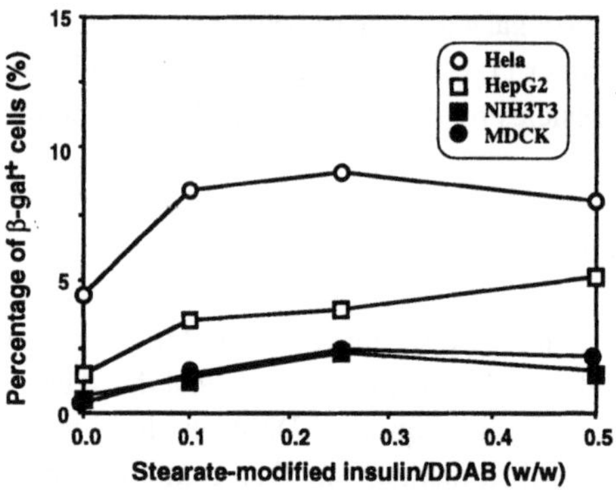

Figure 2. **β-Gal gene transfection using insulin-modified lipid vesicles.** **5 x 10⁴ MDCK, 1 x 10⁵ Hela and NIH3T3, and 2 x 10⁵ HepG2 cells plated the day before the transfection in 35 mm dishes were transfected with 1 µg pCMVβ-gal DNA and 10 µl lipid vesicles (DDAB concentration = 1 mg/ml). β-Galactosidase activity was measured 24 h after the transfection.**

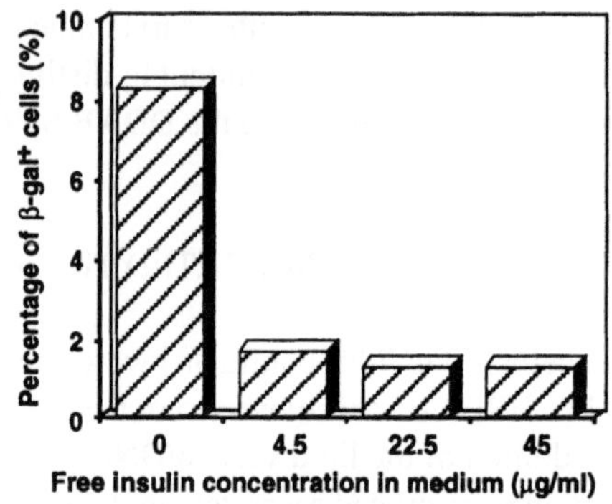

Figure 3. **Effect of insulin as an competitive inhibitor on transfection in Hela cells. 1 x 10⁵ Hela cells plated the day before the transfection in 35 mm dishes were transfected with 1 µg pCMVβ-gal DNA and 10 µl lipid vesicles (Stearate-modified insulin/DDAB = 0.25, w/w; DDAB concentration = 1 mg/ml). β-Galactosidase activity was measured 24 h after the transfection.**

to be used as target for gene transfer into animal tissues: limited tissue distribution, high-affinity binding and a rapid rate of ligand internalization. Asialoglycoprotein receptors are localized on the surface of differentiated hepatocytes and involved in the clearance of desialated proteins from the blood stream. Galactose-terminal glycoproteins can be the ligands for this receptor. Therefore, galactose residue was introduced into DDAB lipid vesicles for cell specific gene transfer. Simply, lipid vesicle/DNA complex prepared by mixing the galactose-modified lipid vesicles and the plasmid DNA (pCMVβ) solution was added to HepG2, HuH-6, Hela and MDCK cells. As shown in Fig. 4, 3–4 folds increased transfection efficiency was observed in HepG2 and HuH-6 cells expressing asialoglycoprotein receptors, and no increase was observed with Hela and MDCK cells

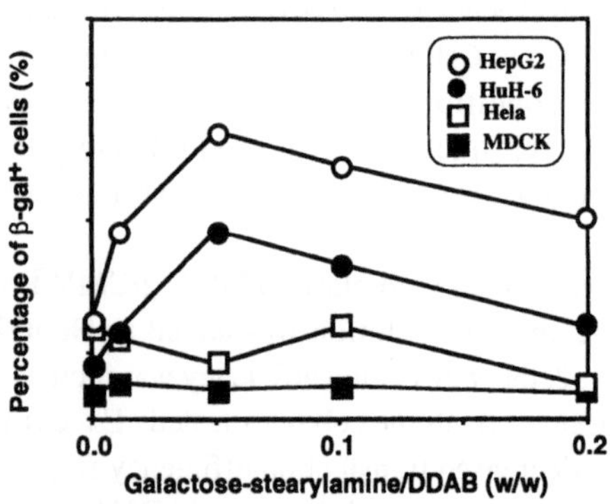

Figure 4. **β-Gal gene transfection using galactose-modified lipid vesicles. 5 x 10⁴ MDCK, 1 x 10⁵ Hela, 2 x 10⁵ HepG2 and HuH-6 cells plated the day before the transfection in 35 mm dishes were transfected with 1 µg pCMVβ-gal DNA and 10 µl lipid vesicles (DDAB concentration = 1 mg/ml). β-Galactosidase activity was measured 24 h after the transfection.**

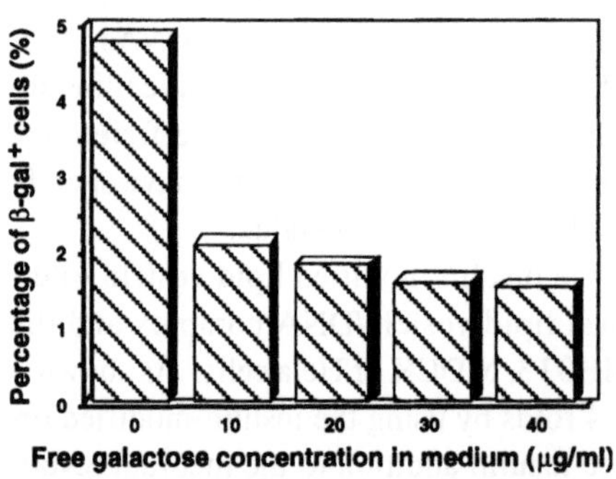

Figure 5. **Effect of galactose as an competitive inhibitor on transfection in HepG2 cells. 2 x 10⁵ HepG2 cells plated the day before the transfec-tion in 35 mm dishes were transfected with 1 µg pCMVβ-gal DNA and 10 µl lipid vesicles (Galactose-stearylamine/DDAB = 0.05, w/w; DDAB concentration = 1 mg/ml). β-Galactosidase activity was measured 24 h after the transfection.**

which do not express asialoglycoprotein receptors. Excessive free galactose addition to the medium resulted in decreased transfection efficiency (Fig. 5), similar to the case of insulin-modified lipid vesicles. This indicated that the asialoglycoprotein receptors were targeted by the galactose-modified lipid vesicles.

As a non-viral DNA delivery system, the ligand-modified lipid vesicles may be useful for enhancement of transfection efficiency and/or cell-type specific gene transfection.

4. References

1. Graham, F.L. and van der Eb, A.J. (1973) A new technique for the assay of infectivity of humanadenovirus 5, *Virology* **52**, 456–467.
2. Chen, C. and Okayama, H. (1987) High-efficiency transformation of mammalian cells by plasmid DNA, *Mol. Cell. Biol.* **7**, 2745–2752.
3. McCutchan, J.H. and Pagano, J.S. (1968) Enhancement of the infectivity of Simian Virus 40 deoxyribonucleic acid with Diethylamino-ethyl-Dextran, *J. Natl. Can. Inst.* **41**, 351–356.
4. Fraley, R., Subramani, S., Berg, P. and Papahadjopoulos, D. (1980) Introduction of liposome-encapsulated SV40 DNA into cells, *J. Biol. Chem.* **255**, 10,431–10,435.
5. Schaefer-Ridder, M., Wang, Y. and Hofschneider, P.H. (1982) Liposomes as gene carriers: efficient transformation of mouse L-cells by thymidine kinase gene, *Science* **215**, 166–168.
6. Itani, T., Ariga, H., Yamaguchi, N., Tadakuma, T. and Yasuda, T. (1987) A simple and efficient liposome method for transfection of DNA into mammalian cells grown in suspension, *Gene* **26**, 267–276.
7. Kaneda, Y., Iwai, K. and Ucheda, T. (1989) Increased expression of DNA cointroduced with nuclear protein in adult rat liver, *Science* **243**, 375–378.
8. Lapidot, M. and Loyter, A. (1990) Fusion-mediated microinjection of liposome-enclosed DNA into cultured cells with the aid of influenza virus glycoproteins, *Exp. Cell. Res.* **189**, 241–246.
9. Nussbaum, O., Rott, R. and Loyter, A. (1992) Fusion of influenza virus particles with liposomes: requirement for cholesterol and virus receptors to allow fusion with and lyis of nautral but not of negatively charged liposomes, *J. Gen. Virol.* **73**, 2831–2837.
10. Felgner, P.L., Gadek, T.R., Holm, M., Roman, R., Chan, H.W., Wenz, M., Northrop, J.P., Ringold, G.M., and Danielsen, M. (1987) Lipofection: a highly efficient, lipid-mediated DNA-transfection procedure, *Proc. Natl. Acad. Sci. USA* **84**, 7413–7417.
11. Kinosita, K. and Tsong, T.Y. (1977) Voltage-induced pore formation and hemolysis of human erythrocytes, *Biochim. Biophys. Acta.* **471**, 227–242.
12. Smithies, O., Gregg, R.G., Boggs, S.S., Koralewski, M.A. and Kucherlapati, R.S. (1985) Insertion of DNA sequences into the human chromosomal β-globin locus by homologous recombination, *Nature* **317**, 230–234.
13. Chu, Hayakawa, H. and Berg, P. (1987) Electroporation for the efficient transfection of mammalian cells with DNA, *Nucleic Acids Res.* **15**, 1311–1326.
14. Macgregor, G.R. and Caskey, C.T. (1989) Construction of plasmids that express *E. coli* β-galactosidase in mammalian cells, *Nucleic Acids Res.* **17**, 2365.

RECOMBINANT PROTEIN PRODUCTION BY CHO CELLS CULTURED IN A CHEMICALLY DEFINED MEDIUM.

STEPHEN F. GORFIEN, JOYCE L. DZIMIAN, MARY LYNN TILKINS, GLENN P. GODWIN AND RICHARD FIKE.
Life Technologies, Inc. 3175 Staley Road., Grand Island, NY 14072. U.S.A.

1. Abstract

Serum-free culture of chinese hamster ovary (CHO) cells has become increasingly common as a way of obtaining high levels of expression of recombinant proteins while simplifying recovery and downstream processing of the product. However, serum-free media may still contain one or more of a variety of animal-derived components including albumin, fetuin, various hormones and other proteins. We have demonstrated that it is possible to eliminate animal-derived proteins from a CHO medium formulation. Plasma protein fractions like albumin and fetuin may be replaced by plant-derived hydrolysates, resulting in medium that is protein-free but still undefined (CHO III PFM). CD CHO Medium is a chemically defined formulation which contains no protein or hydrolysates of either plant or animal origin. Peak cell densities and recombinant protein expression in CD CHO cultures compared favorably to expression in other media, although the maximal cell density and the highest levels of expression were observed at later time points. We were able to successfully supplement the culture with sodium butyrate to increase expression levels at the expense of peak cell density, so for recombinant cell lines showing an inverse relationship between growth and expression of recombinant product, strategies which limit the peak cell density may be useful for increasing expression.

2. Introduction

Reduction or elimination of the serum supplement in media for large scale growth of CHO cells and expression of recombinant product is becoming more common as cost, performance and regulatory concerns associated with the use of serum are raised as issues [1,2]. Since serum-free media may contain various undefined components, including proteins, peptides and protein hydrolysates, these formulations represent an improvement over serum supplemented media, but may not eliminate concerns associated with the use of serum [3]. A further improvement over serum-free formulations are protein-free media. In protein-free formulations, the traditional media supplements like albumin, insulin and transferrin have been replaced by other components. However, unless the formulation is specifically stated to be a protein-free, chemically defined medium, there may be protein hydrolysates of either animal or plant origin which may contribute low molecular weight peptides which stimulate growth and/or expression. The CHO III PFM medium developed in our laboratory is an example of an essentially protein-free medium that is not chemically defined, as it contains a plant-derived hydrolysate. Once the cells were properly adapted to this medium, we were able to demonstrate equivalent growth and superior expression levels when compared to cultures adapted to a serum-free medium, CHO-S-SFM II. However, removal of the plant hydrolysate from the CHO III PFM did not support growth of rCHO cells through repeated subculture, so an entirely new basal formulation was developed that is both protein-free and chemically defined. CD CHO Medium contains no proteins or peptide components of animal, plant or synthetic origin. Our experiences in adapting the cells to various media and regulating growth versus expression are instructive to those wishing to use protein-free and/or defined culture systems.

3. Materials and Methods

3.1. MEDIA, CELLS AND CULTURE CONDITIONS

All media used were formulated without nucleosides and were optimized for suspension culture. *CHO-S-SFM II* is a *low protein* (<100 μg/ml) serum-free medium. *CHO III PFM* is an essentially *protein-free* custom formulation which contains no animal-derived proteins. *CD CHO Medium* is a *chemically defined* formulation which contains no animal or plant derived proteins, peptides or hydrolysates.

CHO cells deficient in dihydrofolate reductase (DHFR–) were obtained from ATCC (CRL-9096, Rockville, MD) and transfected with two plasmids; pSV2dhfr (37146) containing a gene for methotrexate (MTX) resistance (also from ATCC) and pCMVβgal (obtained from Pam Hawley-Nelson, Life Technologies, Inc.) enabling β-Gal expression. The transfection was conducted using LIPOFECTAMINE™ Reagent (Life Technologies, Inc.), and selection was accomplished with 1.2 μM MTX. Cultures were maintained in media supplemented with .3 μM MTX and were routinely subcultured at a density of 3 x 10^5 viable cells/ml every 3 to 4 days in 125 ml plastic shake flasks (20 to 35 ml culture volume) on an orbital shaker platform rotating at 125-135 rpm in a humidified atmosphere of 8% CO_2 in air. The large-scale evaluations of rCHO cells adapted to serum-free suspension culture in CD-CHO were performed in modified 5L Celligen™ Bioreactors (New Brunswick Scientific, Edison, NJ) seeded at densities of 1 to 3 x 10^5 viable cells/ml in a final working volume of 3.8L. Samples were removed daily for the determination of total and viable cell counts, glucose and amino acid utilization, lactate and ammonia accumulation, and recombinant product. The pH, dissolved oxygen, temperature and impeller speed were set to maintain initial settings of 7.40, 50% of air saturation, 37.0°C and 90 rpm, respectively. In one experiment, a concentrated 10x supplement of glucose and glutamine (final concentrations were 3.0 g/L and 1.0 g/L, respectively) was added to the culture on day 9 post-planting.

3.2. RECOMBINANT PROTEIN QUANTITATION AND ENHANCEMENT OF EXPRESSION

rß-Galactosidase was measured in cell lysates using a modification of the method described by Hall, et al [4] and Miller [5]. Briefly, a 2.0 ml sample of cell suspension was collected from each sample and subjected to two freeze-thaw cycles, centrifuged (100 x g for 4 minutes) and the supernatant was saved for rβ-Galactosidase quantitation. 100 µl of each supernatant sample was added to 1.0 ml Z buffer, pH 7.0 (0.06M Na_2HPO_4, 0.04M NaH_2PO_4, 0.01M KCl, 0.01M $MgSO_4 7H_2O$, 0.05M β-Mercaptoethanol), followed by 200 µl of o-nitrophenyl-β-D-galactoside (ONPG, 4.0 mg/ml in dH_20) and incubated at 37°C for 120 minutes. The reaction was halted by addition of 500 µl stop buffer (1M Na_2CO_3) and absorbance at 420 nm was read against a blank of the appropriate medium.

We investigated addition of sodium butyrate as a way to enhance rBeta-Galactosidase expression. Cultures were incubated with either 0.1 mM or 1.0 mM sodium butyrate on days 0 or 2 post planting, or 1.0 mM sodium butyrate on day 4 post planting to assess the effects on recombinant protein expression and growth.

4. Results

rβ-Gal CHO cells adapted easily to CD CHO Medium and once adapted, reached higher total cell densities than cells cultured in CHO-S-SFM II (Figure 1) or cells cultured in CHO III PFM (data not shown). rβ-Gal levels were elevated in CD CHO cultures compared to CHO-S-SFM II cultures (Figure 1). An inverse relationship was observed between cell growth and expression of rβ-Gal in CD CHO cultures which did not express high levels of rβ-Gal until the plateau phase of growth was reached (Figure 2). Addition of sodium butyrate at various concentrations and at different times post-planting was used to enhance rβ-Gal expression (Figure 3). At high levels of sodium butyrate supplementation (1.0 mM), growth as well as expression was limited if the supplement was added on day 0, while addition of 1.0 mM sodium butyrate on day 2 resulted in an enhancement of expression at the expense of growth. A low level of sodium butyrate (0.1mM) resulted in enhanced levels of recombinant protein expression when added at either day 0 or day 2. Figure 4 shows similar growth kinetics and higher rβ-Gal levels in a Celligen bioreactor compared to a shake flask control culture. A glucose and glutamine supplement was added to the bioreactor culture on day 9 post-planting, but failed to increase rβ-gal levels over the peak concentration observed on day 9.

5. Discussion

Industry is now considering alternatives to medium components of animal origin. We have shown that growth of CHO cells is possible in a completely defined medium lacking any animal or plant derived proteins, peptides or hydrolysates. High cell densities were achieved in batch culture in CD CHO Medium compared to a serum-free medium (CHO-S-SFM II) and a protein-free, undefined medium (CHO III PFM). High levels of expression in a DHFR amplified rCHO line cultured in CD CHO Medium occurred after the cells had reached peak cell density. Sodium butyrate has been shown to be a strong inhibitor of cell growth and to induce protein synthesis [6]. Others have shown elevated levels of expression of monoclonal antibodies [7] and recombinant proteins [8,9] following addition of sodium butyrate. In our studies, addition of sodium butyrate was successfully used to limit cell growth and to enhance the level of rβ-Gal expression by rCHO cells. However, concentration and time of addition of the sodium butyrate are important considerations in optimizing protein expression. Successful scale-up to a 5L stirred tank bioreactor was achieved in CD CHO Medium, with higher total levels of expressed product obtained in the bioreactor when compared to a small scale control culture. Our experiences with chemically defined CHO culture present an attractive alternative to serum supplemented, serum-free or protein-free, undefined formulations.

6. References

1. Jayme, D.W., Epstein, D.A., Conrad, D.R. (1988) Fetal bovine serum alternatives, *Nature* 334:547-8.
2. Rosa, M.D. (1989) Serum substitutes: is there a solution? *BioPharm* 2:16-17.
3. Lambert, K.J. and Birch, J.R. (1985) *Animal Cell Biotechnology*, R.E. Spier and J.B. Griffiths, Eds. (Academic Press, London), pp. 85-122.
4. Hall, C.V., Jacob, P.E., Ringlod, G.M., Lee, F. (1983) Expression and regulation of Escherichia coli lacZ gene fusions in mammalian cells. *J Mol Appl Gen* **2(1)**:101-109.
5. Miller, J.H. (1972) *Experiments in Molecular Genetics*. Cold Spring Harbor Laboratory, Cold Spring Harbor, NY.
6. Kruh, J. (1982) Effects of sodium butyrate, a new pharmacological agent, on cells in culture. *Mol Cell Biochem* 42:65-82.
7. Oh, S.K.W., Vig, P., Chua, F., Teo, W.K., Yap, M.G.S. (1993) Substantial overproduction of antibodies by applying osmotic pressure and sodium butyrate. *Biotechnol Bioeng* 42:601-610.

8. Dorner, A.J., Wasley, L.C., Kaufman, R.J. (1989) Increased synthesis of secreted proteins induces expression of glucose-regulated proteins in butyrate-treated chinese hamster ovary cells. *J Biol Chem* 264(34):20602-20607
9. Palermo, D.P., DeGraaf, M.E., Marotti, K.R., Rehberg, E., Post, L.E. (1991) Production of analytical quantities of recombinant proteins in chinese hamster ovary cells using sodium butyrate to elevate gene expression. J *Biotechnol* 19:35-48.

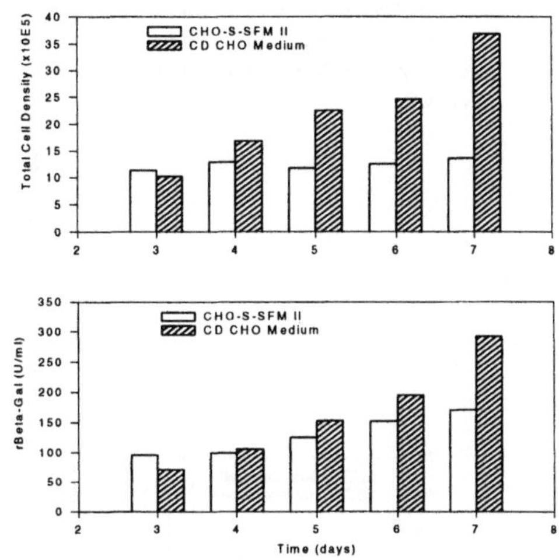

Figure 1. Growth and rbeta-galactosidase expression in CD CHO Medium versus CHO-S-SFM II. Total cell density continued to rise in CD CHO cultures (days 3 through 8, upper panel), while CHO-S-SFM II culture density remained constant; rbeta-galactosidase levels continued to rise in both culture conditions (bottom panel), but CD CHO cultures expressed more total product.

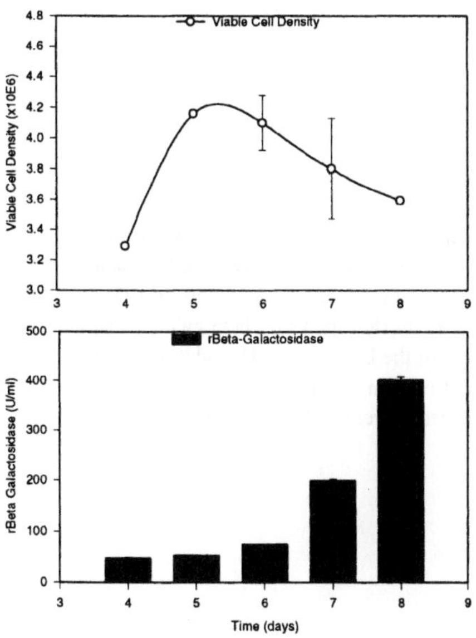

Figure 2. Inverse relationship between growth and expression in a DHFR amplified rCHO line. A rbeta-galactosidase producing CHO line expressed high levels of product (lower panel) only after the plateau phase of the growth curve (upper panel) was reached.

Figure 3. Effect of sodium butyrate concentration and time of addition on growth and expression. Growth of rCHO cells (upper panel) was inhibited by addition of sodium butyrate in a concentration dependent fashion. Expression of rbeta-galactosidase (lower panel) was increased over control levels on day 8 in all conditions except in cultures receiving 1.0 mM sodium butyrate on day 0.

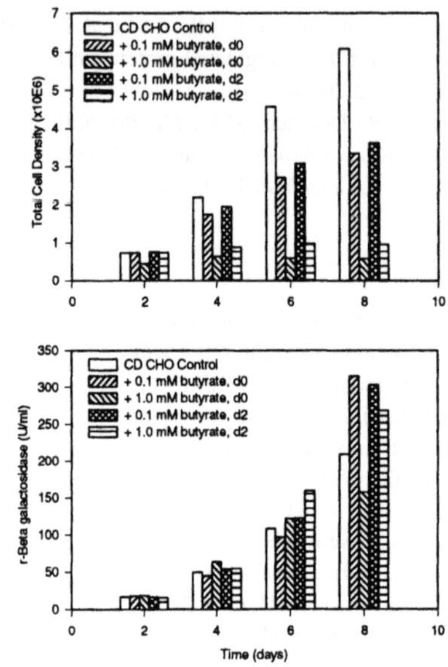

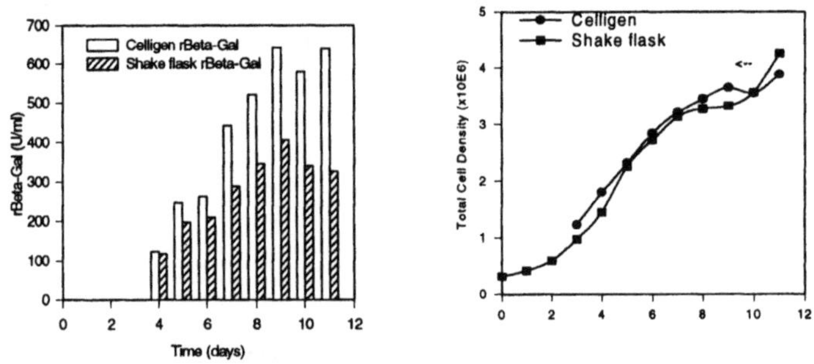

Figure 4. Growth and rbeta-galactosidase expression in a 5L Celligen bioreactor. Growth kinetics of rCHO cells in a 5L stirred tank bioreactor (3.8L working volume) was similar to that observed in a 250 ml (75 ml working volume) shake flask control (left panel). rbeta-galactosidase expression (right panel) was higher in the bioreactor. The addition of the glucose and glutamine supplement (shown by arrow on day 9) was too late to boost productivity over the peak level reached on day 9.

REMOVAL AND INACTIVATION OF PARVOVIRUS FROM PROTEIN SOLUTION BY A BMM MEMBRANE FILTER.

S.Shimokubo[1)], T.Sato[1)], G.Ishikawa[1)], S.Satoh[1)], K.Yamaguchi[2)] and N.Yamamoto[3)]

1)Asahi Chemical Industry Co., LTD. BMM Business Development Department, 5-13, Shibaura 4-Chome, Minato-ku, Tokyo 108, Japan

2)Yamaguchi University School of Medicine, Institute of Laboratory Animals, 1144 Kogushi, Ube-city, Yamaguchi 755, Japan

3)Tokyo Medical & Dental University School of Medicine, Department of Microbiology, 1-5-45, Yushima, Bunkyo-ku, Tokyo 113, Japan

Abstract

Human parvovirus B19 infects humans and causes various diseases. We have developed a BMM membrane which is able to effectively remove parvovirus from protein solution and have demonstrated the removal of porcine parvovirus by BMM15 (membrane mean pore size is 15 nm), using a transmission electron microscope (TEM).

Although there are various methods for virus inactivation, the absence of a lipid envelope in parvoviruses leads to resistance to thermal and solvent detergent treatments. We found that a combination of BMM filtration with ultraviolet (UV) irradiation effectively reduced the infectious virus titers. Thus, virus elimination from protein solution by this combination of virus inactivation and virus removal by BMM filters can be applied in the safe production of drugs.

1. Introduction

Human parvovirus B19 (B19) infects human erythroid cells specifically, and virus replication occurs at high levels in the erythroid progenitor cells. Acute parvovirus infection causes transient aplastic crisis in hemolytic anemia patients and childhood exanthem fifth disease (erythema infectiosum). When pregnant women are infected, there is a risk of hydrops fetalis.

Due to the stability of B19, it can infect through contamination of blood products. As a result, the safety of blood products, especially coagulation factor concentrates (F-VIII, F-IX etc.) and blood transfusion becomes a problem. At present, blood products are produced from pooled blood plasma, with a conceivable risk of parvovirus contamination.

The principle of virus removal by a hollow fiber membrane is based on size exclusion derived from the relative size difference between the virus particle and the membrane pore, and thus the membrane's virus removability does not depend on virus characteristics.

The Bemberg Microporus Membrane (BMM) filter, PLANOVA™ which is composed of regenerated cellulose, was developed for virus removal from bio-drugs and displays high virus removability [1], large filtration capacity and a high recovery rate of proteins. At present, PLANOVA™ are widely utilized in the purification process of bio-drugs and plasma derivatives.

We investigated the removal of porcine parvovirus (PPV), B19 and poliovirus (as a control) from protein solution using BMM filters with mean pore sizes ranging from 8.8 to 35.0 nm. The combination of virus removal by BMM filtration with virus inactivation by UV irradiation was also investigated.

2. Materials & Methods

2.1. MEMBRANE FILTER

BMM filters, PLANOVA™ were provided by Asahi Chem. Ind. Co., Ltd. Japan. Filters used were those with mean pore sizes of 8.8, 14.0, 14.6, 20.0, 21.8, 25.9, 26.7, 29.9, 32.1 and 35.0 nm.

2.2. VIRUS

PPV was prepared as a supernatant of culture medium of ESK cells (porcine kidney) inoculated with PPV stock. B19 was prepared from patients' sera obtained from the Japan Red Cross Blood Center, Fukuoka and suspended in Dullbecco's MEM containing 10% of inactivated FCS. Poliovirus was prepared as a supernatant of cultured medium of FL cells inoculated with the poliovirus stock.

2.3. FILTRATION

Filtration experiments were carried out by the dead-end filtration method using a filtration volume of $3L/m^2$ at a transmembrane pressure of $0.27\ kg/cm^2$ of N_2 gas through pore sizes of 8.8 to 35.0 nm.

2.4. PCR

A PCR technique was utilized for detection of virus DNA so as to compare the amount of virus contamination in the samples [2]. Virus DNA was extracted and serially diluted and subjected to PCR. PCR primer pairs used to amplify PPV virus DNA were determined from the sequence described by Casal et al. [3]. The primer pairs for PCR were 1168 - 1187 and 1489 -1471. In the same way, PCR primer pairs for B19 virus DNA were determined from the sequence described by Astell et al. [4]. The primer pairs for PCR were 799 - 818, 990 - 972. PCR cycles of 1st (94°C-3min, 54°C-1min, 72°C-1min), 2nd-39th (94°C-30sec, 54°C-1min, 72°C-1min), 40th (72°C-5min,4°C) were then performed.

2.5. VIRUS ASSAY

A hemagglutination (HA) assay using avian red blood cells was carried out to determine PPV titers. Agglutination patterns were recorded and end-point titers were thus determined. Poliovirus titer was determined by cytopathic effect (CPE) observation. Virus titer was calculated in accordance with the method of Reed & Muench. Pre- and post filtration values were used to determine logarithmic rejection values (LRV).

2.6. TEM SAMPLE PREPARATION

To determine the location of the virus particles, BMM fibers were fixed for 30 minutes with 2% paraformaldehyde - 2% glutaraldehyde mixture in 0.1M cacodylate buffer (pH7.4), and preserved in tannic acid, then postfixed with osmic acid. Epon 812 resin was used for embedding. Then, ultra-thin sections were prepared and TEM photographs were taken.

2.7. COMBINATION EXPERIMENT OF BMM FILTRATION WITH UV IRRADIATION.

UV irradiation experiments for PPV employed a Spectronics UV lamp (ENF-260 C/J: 250μW/cm^2: wavelength 254nm). The strength of the UV rays was measured by Spectronics digital radiometer (DRC-100X). Virus solution (DMEM+10% FCS) was poured into a 3.5 cm diameter culture dish to achieve a depth of 5 mm and then irradiated with UV rays.

3. Result & Discussion

Figure 1 shows PPV removal by BMM15 using TEM. It was observed that PPV particles were trapped by BMM15 membrane pores.

With regards to membrane structure, the smallest pore size region occurs a little to the outside of the middle portion of the membrane. Most of the PPV particles were trapped in this region, within a 5 μm radius of the outside surface of the BMM membrane.

We investigated PPV removability with various mean pore sizes of BMM filters by HA assay and PCR. With the decrease of filter mean pore size from 35 nm to 9 nm, the PPV removability determined by an HA assay increased from 0 to 7 in LRV value (figure 2). These results indicate that a membrane mean pore size less than 20 nm is required to generate a PPV LRV of more than 4.

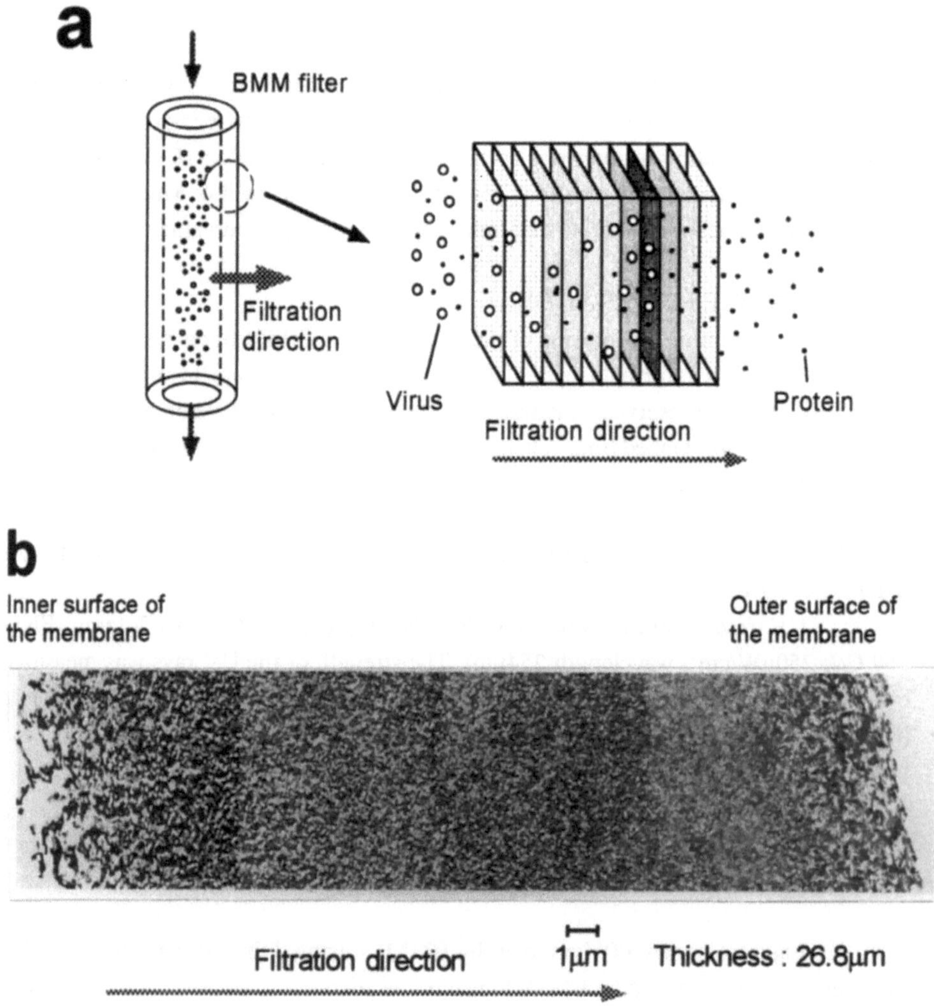

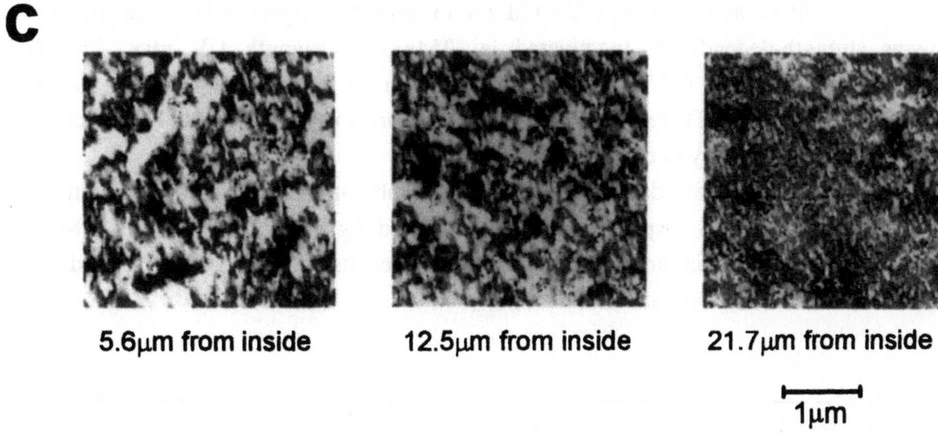

Fig. 1 : The PPV removal by BMM15. a) A schema of virus removal by BMM. b) TEM photograph of the PPV removal by BMM15 (whole). c) TEM photo graph of the PPV removal by BMM15 in each part of the membrane. (A figure shows the distance from the inside of the membrane.)

We also examined the B19 removal by various sizes of BMM filters, and LRV was estimated by the PCR method. B19 was effectively removed by the BMM15 filter, and the LRV comparison of PPV and B19 reveals that BMM filters are more effective in the removal of B19 than PPV. These results may suggest that the B19 virus particle is slightly larger than PPV [5]. In this examination, virus removability increases as filter mean pore size decreases. These results suggest that the virus removal mechanism of BMM filters is based on size exclusion.

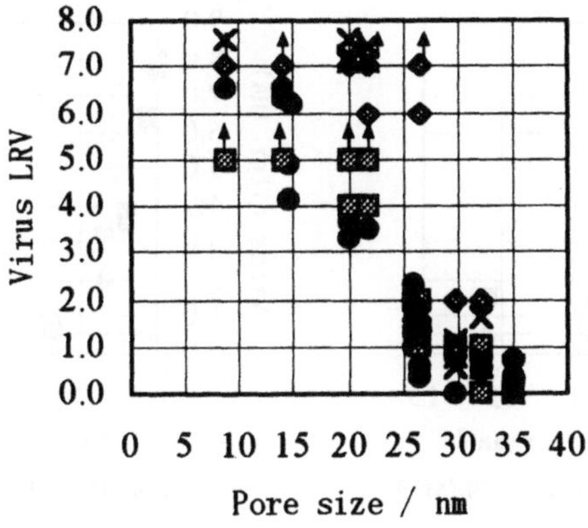

Fig. 2 : Virus removal by BMM filters. ● : PPV (HA), ■ : PPV (PCR), ◆ : B19 (PCR), × : Poliovirus (CPE)

PPV inactivation by UV irradiation is shown in figure 3. It is clear that PPV was strength-dependently inactivated by 254nm wavelength UV rays. The UV irradiation decreases PPV infectious titer, but it must be noted that the UV rays are harmful to proteins. Therefore, when UV irradiation is used in the drug manufacturing process, it is necessary to ensure protein stability.

Then, we considered the effect of the combination of BMM filtration with UV irradiation. Figure 4 displays the result obtained by virus removal with BMM15 followed by UV irradiation. Figure 5 presents the results obtained when UV irradiation preceded virus removal with BMM15.

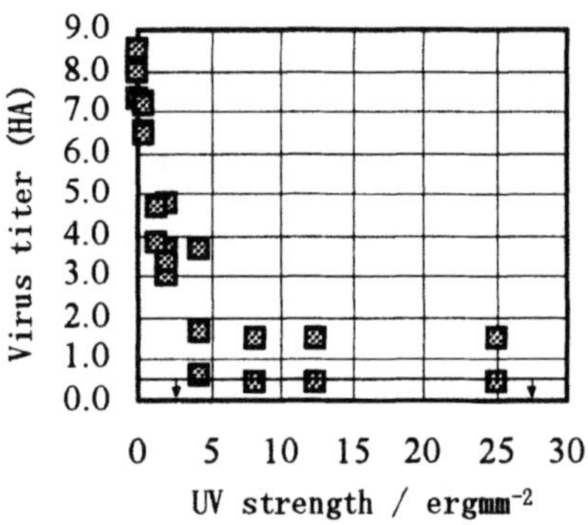

Fig. 3 : UV irradiation on PPV.

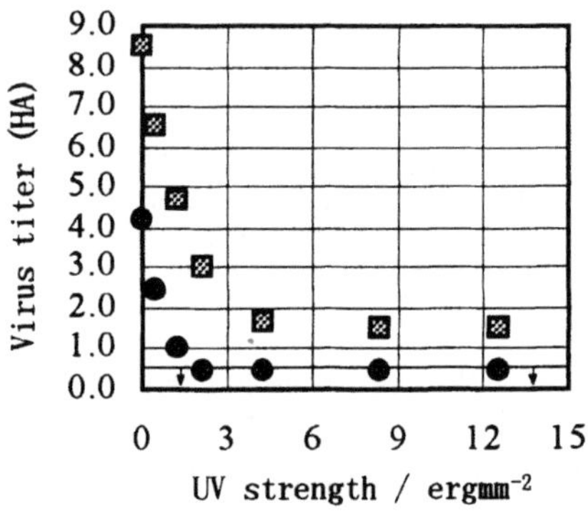

Fig. 4 : BMM15 filtration followed by UV irradiation. ■ : UV irradiation only, ● : BMM15 filtration followed by UV irradiation.

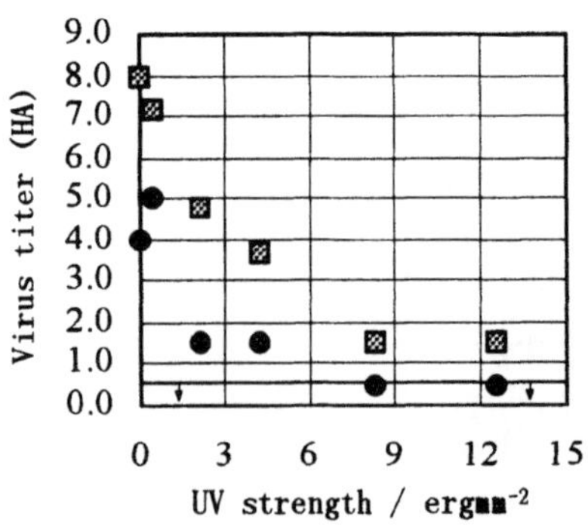

Fig. 5 : UV irradiation followed by BMM15 filtration. ■ : BMM15 filtration only, ● : UV irradiation followed by BMM15 filtration.

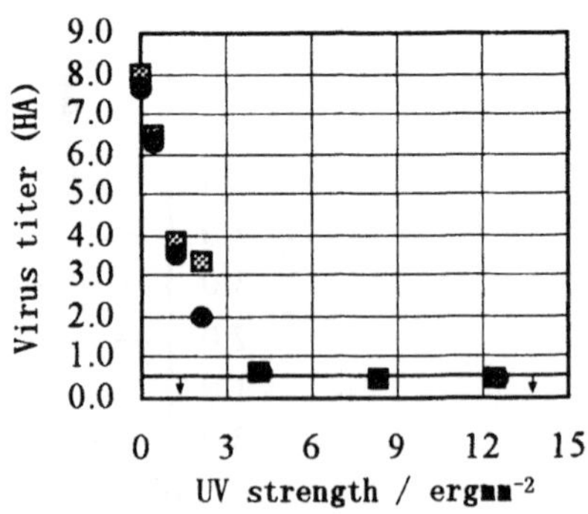

Fig. 6 : BMM35 filtration followed by UV irradiation. ■ : UV irradiation only, ● : BMM35 filtration followed by UV irradiation..

The effect of both virus removal and virus inactivation is regarded as the simple combined effect of the two methods, in the case of "BMM filtration followed by UV irradiation" but not "UV irradiation followed by BMM filtration". In the latter case, even if the virus was inactivated with UV irradiation, virus particles still remained, thus there is no simple combined effect in these two methods. In this situation, BMM filters removed virus below the detection limit even though the virus was not sufficiently inactivated by the UV irradiation step.

Figure 6 shows the result of virus removal with BMM35 followed by UV irradiation. The PPV removal effect of BMM35 was small, but the efficiency of inactivation by UV irradiation was different in each case, regardless of the filtration treatment applied. This may be explained by the removal of aggregated virus particles and impurities by BMM35 filtration.

4. References

1. Hirasaki,T., Ishikawa,G., Nakano,H., Manabe,S. and Yamamoto,N.: Mechanism of virus removal by regenerated cellulose hollow fiber (BMM™), Membrane, **20(2)** (1995), 135 - 142.
2. Durigon,E.L., Erdman,D.D., Gary,G.W., Pallansch,M.A., Torok,T.J. and Anderson,L.J.: Multiple primer pairs for polymerase chain reaction (PCR) amplification of human parvovirus B19 DNA, J. Virol. Meth., **44** (1993), 155 - 165.
3. Ranz,A.I., Manclus,J.J., Diaz-Aroca,E. and Casal,J.I.: Porcine parvovirus : DNA sequence and genome organization, J. Gen. Virol., **70** (1989), 2541 - 2553.
4. Shade,R.O., Blundell,M.C., Cotmore,S.F., Tattersall,P. and Astell,C.R.: Nucleotide sequence and genome organization of human parvovirus B19 isolated from the serum of a child during aplastic crisis, J. Virol., **58** (1986), 921 - 936.
5. Kawase,M., Momoeda,M., Young,N.S. and Kajigaya,S.: Modest truncation of the major capsid protein abrogates B19 parvovirus capsid formation, J. Virol., **69** (1995), 6567 - 6571.

Key Word

BMM filter, Parvovirus, Transmission electron microscope, Virus removal, UV irradiation

MATRICRINE EFFECT OF BASEMENT MEMBRANE COMPONENTS ON *DE NOVO* FORMATION OF MESENCHYMAL TISSUES

NOBUKO KAWAGUCHI[1], ELENI NICODEMOU-LENA[1], KAZUHIRO TORIYAMA[2], SHUHEI TORII[2], YASUO KITAGAWA[1,3]
[1]*Graduate Program of Biochemical Regulation, Graduate School of Agricultural Sciences,* [2]*Deptartment of Plastic Surgery, School of Medicine,* [3]*Nagoya University.*

Subcutaneous injection of EHS extract in combination with 1 μM bFGF induced fat pad formation in the area where the meshwork of basement membrane components was reconstituted *in vivo*. Histological section of the pad showed well developed adipocytes delivered with blood cells. This process was preceded by microvascularization of the area within 2 days. Combination of EHS extract with IGF-1 and growth hormone induced muscle fiber in addition to fat tissue. Injection of erythropoietin together with EHS extract resulted in angiogenesis at periphery of the reconstituted basement membrane.

1. Introduction

Typical basement membrane (BM) is a continuous sheet-like structure mainly composed of type IV collagen, laminin, perlecan and underlay various epithelia and endothelium (1-4). By affecting proliferation, differentiation and migration of the cells, BM has a crucial role in the organization of epithelium and endothelium. In the case of vascular system, the lining of endothelial cells is separated from surrounding smooth muscle cells by BM. BM is important for the organization of mesenchymal tissues as well as the muscle membrane surrounding muscle fibers and the immunochemically stainable BM-like structure in fat tissue. BM is mportant for capillary formation where thin BM-like structure surrounds the endothelial cells forming the walls of capillaries.

Engelbreth-Holm-Swarm (EHS) tumor of mouse has been a good source of BM components since this tumor extracellularly accumulates large amount of BM components. By destroying self association of BM components with 2M urea solution, mixture of major BM components such as type IV collagen, laminin and perlecan can be extracted from EHS tumor (5). Meshwork structure of BM can be reconstituted by re-naturing the extracted proteins by dialyzing out urea at 4 °C and raising the temperature to above 25 °C (Matrigel). Taking advantage of this nature of the EHS tumor extract, we can reconstitute BM-like structure *in vivo* by subcutaneous injection of the extract which polymerize in the mesenchym of mice body. When various growth factors were mixed with EHS tumor extract, we found that angiogenesis occurs at the site of injection which results in the inducible formation of adipose tissue and muscle fibers.

By defining this phenomena to be "matricrine effect" of BM components, we introduce our preliminary experiments of injecting bFGF in combination with BM components. We also tested the angiogenic effect of erythropoietin (Epo).

2. Materials and Methods

2.1 EHS Extract Preparation and Injection EHS extract (Matrigel) was prepared from the transformed mouse EHS tumor as described by Kleinman (5). The tumor was excised from the mice, washed with high salt to remove blood components and then extracted with 2 M urea. Urea was dialyzed out at 4°C. The extract was mixed with various growth factors at indicated concentrations and was injected subcutaneously into BALB/c nude mice in the midabdominal region of head. BM components polymerized and reconstituted the network *in vivo* within 30 min after the injection.

2.2 Assay of Fat Pad Formation and Angiogenesis At indicated day, fat pad or plugs formed by reconstituted BM components were removed and weighed. For the preparation of histological sections, the tissues were fixed in 4% paraformaldehyde in phosphate buffered saline and embedded in parafin. The section were stained with Mayer's Hematoxylin-Eosin.

3. Results

In this study, we first tried to reproduce the result of Green and Kehind (6) who showed that subcutaneous injection of cultured 3T3-F442A cells (pre-adipocyte) into BALB/c nude mice resulted in fat pad formation. When the cells were injected over the sternum, newly formed fat pad became visible from outside of the body within 1 week and it grew until 5 weeks. Histological section of the pad showed well differentiated adipocytes with abundant supply of capillaries. In order to know the effect of BM components on this fat pad formation, F442A cells were injected after suspensing in EHS tumor extract. As shown in Fig.1, the weight of newly formed pad was two fold of injecting the extract alone. Addition of bFGF to the cell suspension at final concentration of 1μg/ml markedly increased the weight, showing that bFGF (directly or indirectly) stimulated proliferation and differentiation of F442A cells. To our surprise, injection of the extract in combination with bFGF alone (without F442A cells) gave rise to fat pad formation.Histological section of such pad showed well developed adipocytes within reconstituted BM network.This result showed that BM components in combination with bFGF created new local condition in mesenchyme and the endogenous precursor of adipocytes could migrate, proliferate and differentiate to form fat tissue. We confirmed that injection of bFGF alone did not show this effect.

We tested the injection of matrigel in combination with other growth factors such as PDGF, growth hormone, insulin and IGF-1. Among these growth factors, bFGF formed the largest pad. To our surprise, however, growth hormone, insulin and IGF-1 induced muscle fiber formation in addition to adipose tissue.

Histochemical study of the earlier stage of this fat pad formation suggested that the microvasculization preceds the migration and differentiation of adipose cells. Delivery of blood cells into the area of reconstituted BM was observed as early as 2 days after the injection. Since bFGF has a strong angiogenic effect, the induced formation of fat pad suggested to be the concequence of this microvascularization.

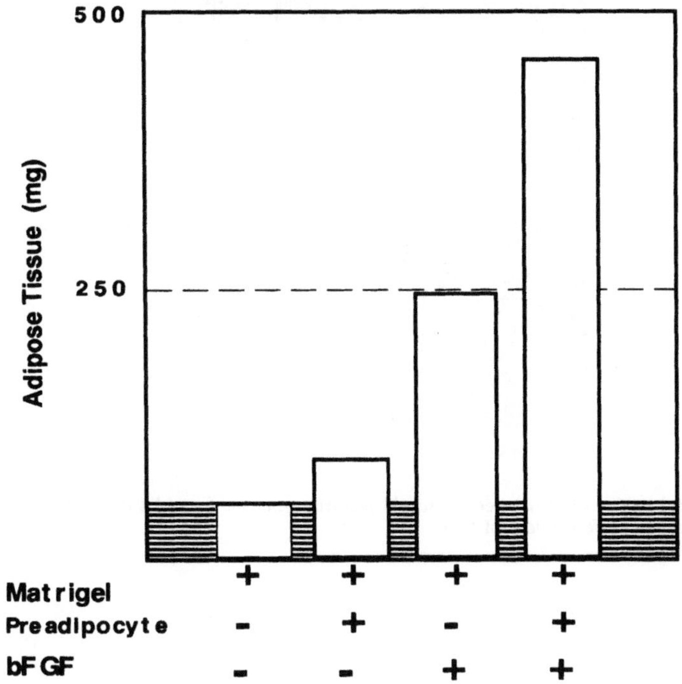

Figure 1. Weight of fat pad fromed 5 weeks after injecting EHS tumor extract in combination with 3T3-F442A cells and bFGF

Since erythropoietin (Epo) has been shown to stimulate the growth of cultured endothelial cells, we tested injection of Epo in combination with EHS tumor extract. Histological section of the plugs in Fig.2 shows that EHS extract mixed with 0.01-0.03 U/ml Epo can induce migrating cells from dermal side in dose dependent manner (Fig.2A and B). In contrast, no migration of cells can be observed at higher concentration of Epo (Fig.2C and D). This is probably due to down regulation of Epo-R which may desensitize the responding cells to Epo. When the effect of Epo at lower concentrations was compared to the injection of bFGF in combination with the EHS extract, the cell migration was less evident than 100 ng/ml bFGF but more evident than 1 or 10 ng/ml bFGF. Effect of Epo was distinct from that of bFGF in that the cell migration was observed only in periphery of the reconstituted BM.

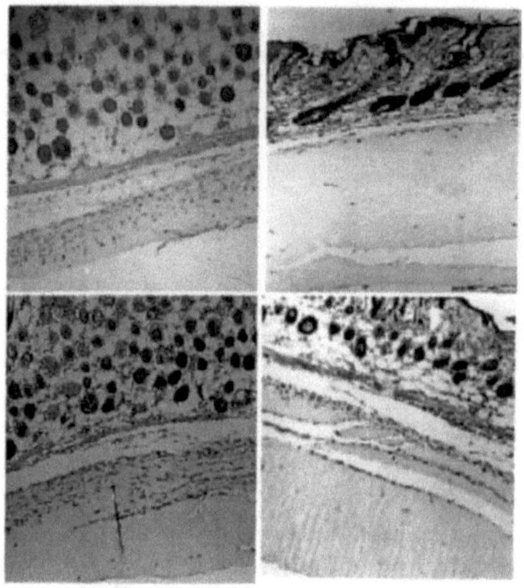

Figure 2. Histological section of *in vivo* reconstituted basement membane formed 4 days after the injection of EHS extract containing different dose of Epo.

4. Discussion

Our results showed that BM-like structure reconstituted in the mesenchyme created a region for endothelial cells to invaded and the consequent blood supply induced migration of the mesenchymal stem cells which differentiated into adipocytes and muscle cells. Since injection of either EHS extract or growth factor alone did not show this effect, we can assume that the reconstituted BM stably accumulated growth factors which could affect the cells in a long duration of time. By mimicking the words like "endocrine", "paracrine" and "autocrine", we would like to define this effect of BM components as "matricrine". The major heparansulfate proteoglycan of EHS tumor extract, perlecan, may be the responsible molecule. However, our preliminary identification of such molecule by injecting purified components showed that separation of the extract destroyed the "matricrine" effect.

References

1) Aratani, Y. and Kitagawa, Y. *FEBS Lett.*, **235**, 129 (1988)
2) Aratani, Y.,and Kitagawa, Y. *J. Biol. Chem.*, **263**, 16163 (1988)
3) Ono, M., Aratani, Y., Kitagawa, I.and Kitagawa, Y. *Exp. Cell Res.*, **187**, 309 (1990)
4) Aratani, Y., Sugimoto, E. and Kitagawa, Y. *FEBS Lett.* **218**, 47 (1987)
5) Kleinman, H.K., McGarvay, M.L., Hassell, J.R. et al. *Biochemistry*, **25**, 312 (1986)
6) Green. H. and Kehinde, O. *J. Cell Physiol.*, **101**, 169 (1979)

HIGH-CELL-DENSITY SUSPENSION PERFUSION CULTURE OF PORCINE HEPATOCYTE SPHEROIDS FOR A BIOARTIFICIAL LIVER

Y. Sakai[1*], K. Naruse[2], I. Nagashima[2],
T. Muto[2] and M. Suzuki[1]
[1] *Institute of Industrial Science, University of Tokyo*
7-22-1 Roppongi, Minato-ku Tokyo 106, Japan
[2] *Fourth Department of Surgery, University of Tokyo*
7-3-1 Hongo, Bunkyo-ku, Tokyo113, Japan
**e-mail: yasuyuki@yari.iis.u-tokyo.ac.jp*

1. Introduction

A bioartificial liver system (BAL) is expected to become an effective treatment for patients with acute liver failure or for patients waiting for liver transplantation. Four groups have already performed actual human clinical trials. However, except for the trial by Rozga *et al.* [1], the life-saving effect was not so significant. Thus, along with increase in immobilized cell numbers (up to $2X10^{10}$ cells, 10% of whole human liver cells), enhancement in cellular functional capacity is still required. The present requirement from the clinical sides of the BAL system is very simple, that is, it should function only up to 6 hour's plasma perfusion and should be disposable in order to avoid acute immuno-rejection reaction [1]. Therefore, maximal metabolic performance even in the short term is a crucial subject. In this study, we have developed a prototype of a new BAL module immobilizing 2.5% of human liver cells, based on the suspension perfusion culture of porcine hepatocyte spheroids.

2. Material and Methods

Figure 1 shows a newly developed bioartificial liver module, which is a 150-mL volume spin filter-type perfusion culture bioreactor. It is in one fifth scale of that for human clinical application. Porcine hepatocyte spheroids ($4X10^9$ cells) formed by 40 hours of 1-L-scale suspension culture (mean diameter is over 200 μm) [2] were inoculated into the module. Oxygenated culture medium at a DO level of 20-24ppm was supplied from the bottom, mixed with hepatocyte spheroids and filtered with rotating stainless cages (pore size is 50 μm). Immediately after the inoculation, we started perfusion with 1-L volume of culture medium at a flow rate of 60 mL/min for up to 12 hours. The medium was William's medium E supplemented with 10%-BS, hormones, antibiotics, 1 mM-NH_4Cl and 200 μM-7-ethoxycoumarin. As a control experiment, we performed small-scale rotational culture of porcine spheroids.

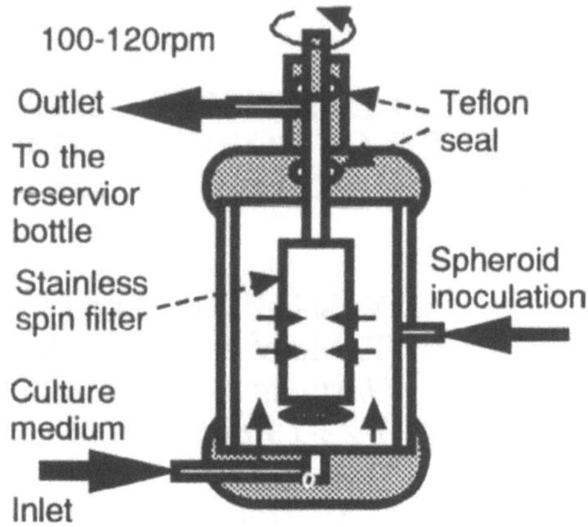

Figure 1 Spheroid/Suspension BAL

3. Results and Discussion

Approximately 80% of the porcine hepatocyte spheroids inoculated was retained within the module after 12 hours of perfusion culture. Cells leaked from the module were dead single cells and small spheroids which were heavily damaged by perfusion. **Figure 2** shows the changes in ammonium and urea concentrations in the reservoir bottle of module culture and in the medium of dish-level culture as a function of perfusion time. We can directly estimate the functional deterioration when the cell number was increased to the clinical scale module, because the ratio of medium volume to cell number was kept constant between the two cultures of 250 times the difference in cell number. Compared with dish-level cultures, the cells in a bioreactor have almost the same ammonium removal capacity and slightly-reduced urea synthesis capacity. **Figure 3** shows cytochrome p450IA1 activity demonstrated by 7-ethoxycoumarin transformation to 7-hydroxycoumarin. They were measured and estimated by fluorescences at two different emission wavelengths [3]. There was no significant deterioration in the function when scaled-up to module-level cultures.

We had been worried about excess spheroid aggregation or spheroid destruction due to very-high-cell-density and vigorous stirring in this study. Upon morphological observation, however, no such phenomenon occurred at least up to 12 hours. Neither spheroid aggregation nor functional deterioration was observed. Although suspension perfusion cell culture has very good mass transfer rates between the cells and perfused medium and was previously examined in actual human clinical trials [4], bioartificial liver modules based on

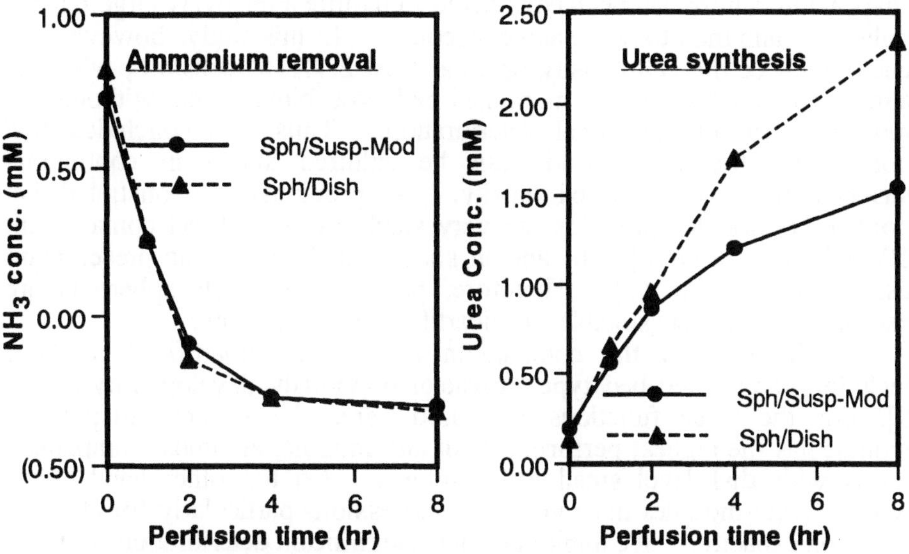

Figure 2 Ammonium removal and urea synthesis of the module- and dish-level cultures of spheroids

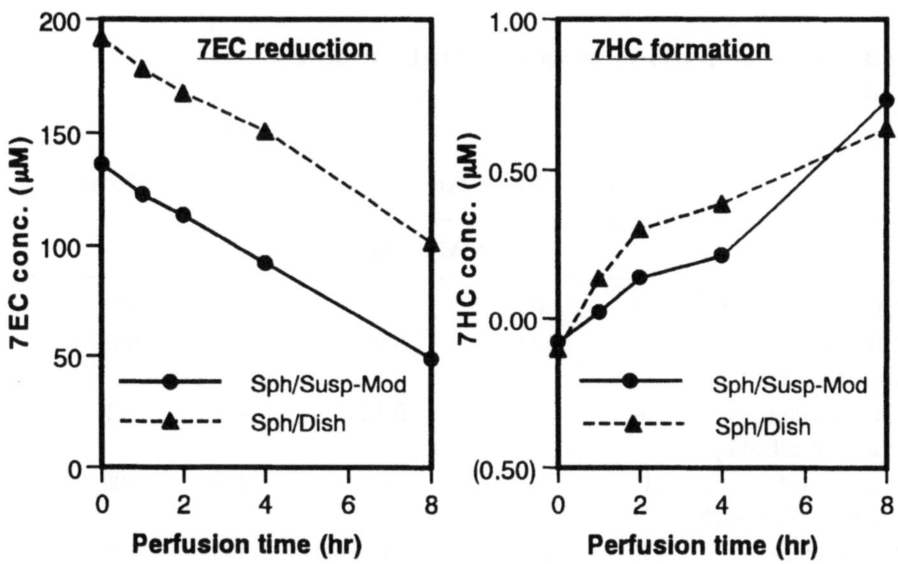

Figure 3 7-Ethoxycoumarin metabolism of the module- and dish-level cultutes of spheroids

suspension culture had been considered to require a relatively large plasma volume, and thus to be disadvantageous. In this study, however, we could enhance the cell density up to at least 2.7X10^7 cells/mL, which is almost comparable to that in packed-bed type bioreactors, without any morphological or functional deterioration. Thus, we concluded that porcine hepatocyte spheroids can be cultured stably in continuous suspension at very high cell density. We previously demonstrated that porcine hepatocyte spheroids are very stable even in direct contact with 100% human plasma [5, 6] and in short-term hypothermic preservation using UW solution [7]. Therefore, porcine hepatocyte spheroids are excellent biocatalysts suitable for bioartificial liver systems.

Table 1 is the comparison of various bioartificial modules including the packed-bed type bioreactor previously developed by us [8, 9]. For the three functions measured here, the spheroid/suspension module has the highest performance of any module, at almost comparable levels with dish-level small-scale cultures. On the other hand, non-woven fabric modules have very low expressions particularly for albumin secretion capacity. We think that such a significant deterioration is due to the difficulty in large-scale cell inoculation at very high-cell density, which is needed to set up a compact bioreactor using normal hepatocytes of low growth activity.

Table 1 Comparison of various BAL modules

	Module			Dish	
	Spheroid/ Susp.	Packed-bed		Sph.	Mono.
		Sph./NWF	Sing./NWF		
Cell No. (10^{10} cells)	0.4	0.6	1.0	10^{-4}	10^{-4}
Cell density (10^7-cells/cm^3)	2.7	3.0	5.0	0.3	0.05
Alb. secretion (μg/10^6-cells·8h)	2.4	0.22	0.12	4.5	1.2
NH_3 removal (nmol/10^6-cells·2h)	210	98	63	247	108
Urea synthesis (nmol/10^6-cells·2h)	170	139	84	193	181

In summary, the spheroid/suspension bioartificial liver module developed showed significantly improved in vitro cellular capacity, when compared with previous packed-bed-type bioartificial liver modules. It adequately meets the demand of the clinical side, that is, maximal metabolic activity even in the short term. We are now preparing ex-vivo perfusion experiments using this spheroid/ suspension bioartificial liver module. The results may be promising because 2-times enhancement in animal survival time has already been obtained by our previous non-woven fabric bioartificial livers [10].

References

[1] Kamlot A, Rozga J, Watanabe FD, Demetriou AA (1996) Review: Artificial liver support systems. *Biotechnol Bioeng* **50**, 382.

[2] Sakai Y, Naruse K, Nagashima I, Muto T, Suzuki M (1996) Large-scale preparation and functions of porcine hepatocyte spheroids. *Int J Artif Organs* **19**, 294.

[3] Sakai Y, Ichikawa K, Sakoda A, Suzuki M (1996) Quantitative comparison of rat hepatocyte functions in two improved culture systems with or without rat liver epithelial cell line. *Cytotechnol*, in press.

[4] Margulis MS, Erukhimov EA, Andreiman LA, Viksna LM (1989) Temporary organ substitution by hemoperfusion through suspension of active donor hepatocytes in a total complex of intensive therapy in patients with acute hepatic failure. *Resuscitation* **18**, 85.

[5] Sakai Y, Naruse K, Nagashima I, Muto T, Suzuki M (1996) Functional stability of porcine hepatocyte spheroids in various culture systems under 100% porcine and human plasma conditions. *Artif Organs* **20**, 56.

[6] Sakai Y, Naruse K, Nagashima I, Muto T, Suzuki M (1996) In vitro function of porcine hepatocyte spheroids in 100% human plasma. *Cell Transplant* **5**, S41.

[7] Sakai Y, Naruse K, Nagashima I, Muto T, Suzuki M (1996) Short-term hypothermic preservation of porcine hepatocyte spheroids using UW solution. *Cell Transplant* **5**, 501.

[8] Naruse K, Sakai Y, Nagashima I, Jiang GX, Suzuki M, Muto T (1996) Development of a new bioartificial liver module filled with porcine hepatocytes immobilized on non-woven fabrics. *Int J Artif Organs* **6**, 347, 1996.

[9] Naruse K, Sakai Y, Nagashima I, Jiang GX, Suzuki M, Muto T (1996) Comparison of porcine hepatocyte spheroids and single hepatocytes in the non-woven fabric bioartificial liver module. *Int J Artif Organs*, in press.

[10] Naruse K, Nagashima I, Sakai Y, Jiang GX, Suzuki M, Muto T (1995) Investigation of homogenous perfusion experiment by the bioartificial liver module filled with porcine hepatocytes. *Xth World Congr Intern Soc Artif Organs, Taipei*, 161, 1995.

COLD-INDUCIBLE GENE EXPRESSION IN FISH CELLS

Molecular Cloning and Gene Expression of Rainbow Trout Ferritin H

M. YAMASHITA, N. OJIMA, AND T. SAKAMOTO
National Research Institute of Fisheries Science
Fukuura, Kanazawa, Yokohama 236, Japan

1. Introduction

Fishes vary widely in their responses to cold temperature [1]. Many fish species from temperate regions generally acclimate to cold temperatures and survive cold stress [1]. In contrast, tropical fishes are very cold sensitive, and are injured or killed by exposure to a cold temperature [1, 2]. One of the most drastic examples of cold acclimation is increased frost tolerance. Low molecular mass hydrophilic proteins were found at high concentrations in the serum of polar-dwelling marine fishes that winter in the subfreezing climate [3]. Other changes that occur during cold acclimation were found to be alterations in lipid composition and the appearance of new isozymes [1]. However, the molecular mechanisms and exact roles of such cold acclimation processes are almost completely unknown.

Elucidation of the functions of cold-regulated genes and the mechanism responsible for their expression are now the focus of intense research in the field of cold acclimation of fish. Since a limited number of changes in gene expression occurs during the cold acclimation of rainbow trout cultured cells [4], the cold-regulated cDNA clones were isolated and characterized from a subtracted cDNA library of the cold-treated trout cells in the present study [5]. These clones were identified to encode the ferritin H isoforms and their expression was responsive to cold-acclimation of the trout cells.

2. Materials and Methods

Rainbow trout gonadal fibroblast cell line RTG-2 cells were maintained in Leibovitz L-15 medium containing 8 % fetal bovine serum and an antibiotic mixture at 22 °C. Cold-treatment of the culture was a shift of incubation temperature to 4 °C.

Poly(A) RNA was extracted from control RTG-2 cells cultured at 22 °C and the cold-treated cells at 4 °C for 1 day. The first-strand cDNA of cold-treated cells was synthesized with oligo(dT) primer adapter: GACTCGAGTCGACATCGATTTTTTTTT-TTTTTTTT. For subtraction hybridization, a double stranded cDNA for control cells was biotinylated with biotinyl-dCTP by random primer labeling method. The mixture of the biotinylated control DNA and the oligo(dT) primer adapter-primed first-strand cDNA from the cold-treated cells was heated at 100 °C for 1 min and then hybridized at 68 °C for 20 h. After the addition of streptavidin, the mixture was extracted with phenol-chloroform to remove the mixture of streptavidin and the biotinylated nucleic acids into the phenol layer and the interface. After the second subtraction step, the subtracted cDNA was converted to double-stranded cDNA. The double-stranded subtracted cDNA was ligated to adapter (GCGGTGACCCGGGAGATCTGAATTC and GAATTCAGATC) with a T4 DNA ligase. Following the phenol-chloroform extraction, fragments of DNA greater than approximately 300 bp were separated from excess adaptors and small DNA fragments by a spun column. The eluate was used as a template for PCR amplification. The amplified products were subcloned into *Eco*RI-cut pBluescript II vector.

To determine the importance of ferritin H for cell growth, a set of antisense (5'-TGGCGGATCTGAGACTCCAT-3') and sense (5'-ATGGAGTCTCAGATCCGCCA-3') 20mer-deoxyribonucleotides corresponding to the translation initiation site of rainbow trout ferritin H mRNA (amino acid residue No. 1-7) was prepared. Aliquots of the oligonucleotides dissolved in water were then added to a medium to a final concentration of 1.6, 8.0, 16 or 30 μM. After the cells were incubated at 22 °C for 2 days in the antisense or sense oligonucleotide, they were transferred to 4 °C for 1 day or cultured at 22 °C as a control.

3. Results

The PCR-based subtractive hybridization method was employed to identify cold-inducible genes in the rainbow trout cells. A subtracted cDNA library from the cold-treated cells was constructed, and differentially screened for cold-induced cDNAs. cDNA clones whose inserts produced a stronger signal with the cold-induced probe than with control probe were selected. Induction of positive clones was confirmed by Northern blotting. So far, 48 clones of the subtracted library have been screened; 8 clones showing strong signals with the cold-induced probe and cross-hybridizing among them was used for further studies. Complete nucleotide sequences were determined for these cDNA clones.

Sequence analysis classified the 8 clones obtained into three independent sets of cDNAs with homology to the mammalian ferritin H subunit. These cDNA clones are about 900 bp long and each of the three sequences, tentatively named as ferritin H-1, H-2 and H-3, contains an open reading frame of 176 amino acids flanked by approximately 200 nucleotides on the 5' as well as 3' sides.

Northern hybridization was used to analyze the mRNA levels of ferritin H isoforms in rainbow trout cells during cold acclimation. The mRNAs of the ferritin H-1, H-2 and H-3 started to increase during the first 1 day of the cold-treatment and remained high during the whole seven-day long acclimation period. Nuclear run-on transcription assays showed that the transcription of the three ferritin H genes was up-regulated 2-3 fold by cold-treatment for 24 h, normalized by the gene transcription rate of 28S-rRNA. This finding apparently indicates that the ferritin H genes are regulated by the cold-treatment at the transcriptional level.

To characterize the cold-induction of ferritin H, antiserum was raised by immunizing a rabbit with a synthetic peptide corresponding to the first 20 residues of the N-terminal amino acid sequence. The ferritin levels started to increase during the first 5 days of cold treatment, and then remained high during the whole 14-day-long acclimation period. The expression of ferritin H is highly responsive to changes in temperature.

For antisense-oligo experiments, both antisense and sense strands were designed to the region of the translation initiation site of ferritin H mRNA. The 20mer-oligonucleotides, antisense and sense strands, were added to cell cultures, and the level of ferritin H was assayed by immunoblotting, normalized to HSC70 levels. The addition of the antisense oligonucleotide resulted in a decrease in the ferritin H level and 8-30 μM antisense oligonucleotide repressed the ferritin H synthesis almost completely, whereas cells treated with the sense oligonucleotide showed the same level of ferritin H as the control.

When the trout cells were subjected to 4 °C under the condition of a decreased ferritin H level, obtained by the addition of the antisense oligonucleotide, cell growth was apparently inhibited; 41% inhibition of cell growth was observed on the addition of the antisense oligonucleotide at 8.0 μM to a culture, 46% inhibition was observed with 16 μM, and 51% inhibition was observed with 30 μM. In comparison, the cells treated with the antisense oligonucleotide at 1.6 to16 μM showed normal cell growth at 22 °C, while 15% inhibition was observed with 30 μM. The sense oligonucleotide at 1.6 to 16 μM did not show any obvious effect on cell growth, supporting the specific effect of the antisense oligonucleotide on cells.

4. Discussion

Cold acclimation in rainbow trout cells is considered to be mediated by alterations in the mRNAs and proteins present in cold-treated cells. Several mRNAs and proteins were found that change in abundance during cold-treatment. To elucidate which genes are induced, a subtracted cDNA library from the cold-treated rainbow trout cells was constructed and screened to isolate the cDNA induced in the cold-treated cells. cDNA clones containing the entire open reading frame of protein produced in large quantity during cold acclimation were identified to encode ferritin H isoforms. Northern blot analysis and nuclear run-on transcription assay showed that the transcription and accumulation of the mRNAs of trout ferritin H isoforms were regulated by cold temperature. The mRNAs of the ferritin H isoforms significantly accumulated in the cold-treated cells. This finding indicates that the expression of three members of ferritin H in the trout cells is altered with cold temperature.

The antisense oligonucleotide experiment showed that suppression of the elevated ferritin H expression during cold acclimation significantly reduced cell growth. This finding indicates an association between the induction of ferritin H and cellular mechanisms during cold acclimation of the trout cells. Since ferritin stores and detoxifies intracellular iron, and its biosynthesis is regulated by the extracellular iron level [6], ferritin H in cold acclimated cells might participate in the metabolic compensation for the delay of iron uptake and/or iron oxidation upon a shift to a cold temperature. Figure 1 shows the hypothetical mechanism of cold-induction of ferritin genes. The pathway of this response is different from the more general heat shock response. Further work, now in progress in our laboratory, may elucidate an upstream regulatory element and a nuclear factor that mediate inducibility of the ferritin H gene in the cold-acclimated rainbow trout cells.

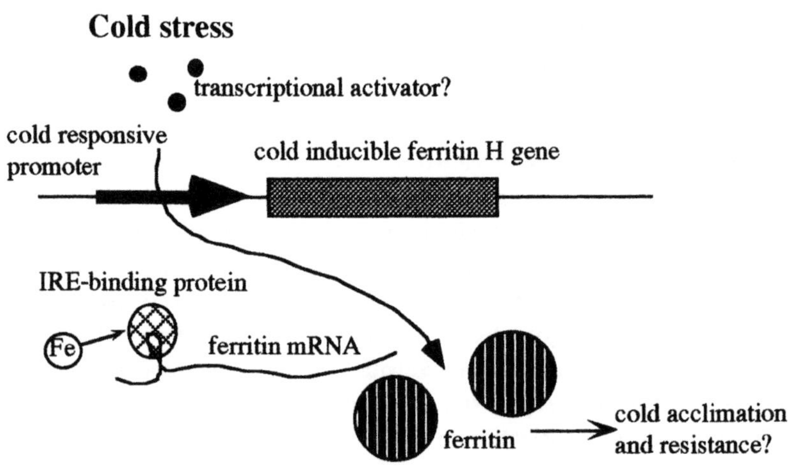

Figure 1. Schematic model of the cold-inducible expression of the rainbow trout ferritin H gene.

5. References

1. Johnston, I.A.: Cold adaptation in marine organisms, *Philos. Trans. R. Soc. Lond. Biol.* **326** (1990), 655-666, discussion 666-667.
2. Mosser, D. D., Heikkila, J. J., and Bols, N. C.: Temperature ranges over which rainbow trout fibroblasts survive and synthesize heat-shock protein, *J. Cell. Physiol.* **128** (1986), 432-440.
3. Davies, P. L. and Hew, C. L.: Biochemistry of fish antifreeze proteins, *FASEB J.* **4** (1990), 2460-2468.
4. Yamashita, M., Ojima, N., and Sakamoto, T.: Induction of proteins in response to cold acclimation of rainbow trout cells, *FEBS Lett.* **382** (1996), 261-264.
5. Yamashita, M., Ojima, N., and Sakamoto, T.: Molecular Cloning and Cold-Inducible Gene Expression of Ferritin H Subunit Isoforms in Rainbow Trout Cells, *J. Biol. Chem.* **271** (1996), 26908-26913.
6. Theil, E. C.: Regulation of ferritin and transferrin receptor mRNAs, *J. Biol. Chem.* **265** (1990), 4771-4774.

DROSOPHILA CELL LINES AS THE SOURCE OF BASEMENT MEMBRANE COMPONENTS

CHINO KUMAGAI[1], MOJGAN AZIMI[1], TATSUHIKO KADOWAKI[2] and YASUO KITAGAWA[1,2]
[1]*Graduate Program of Biochemical Regulation, Graduate School of Agricultural Sciences, Nagoya University and* [2]*Nagoya University BioScience Center, Chikusa, Nagoya, 464-01 Japan*

Drosophila cell lines Kc 167, Schneider (both originated from hemocyte precursor) and Clone 8 (originated from wing disc) secrete large amount of laminin and type IV collagen. In addition to these mammalian equivalents of basement membrane components, they secrete *Drosophila* specific extracellular matrix proteins such as peroxidasin, glutactin, tiggrin and papilin and unknown polypeptide of 300 kDa. Kc 167 was the best in terms of the productivity and active growth in a serum-free medium (HyQ CCM3). Kc 167 secreted large amount of basement membrane components (1.2 mg from 1 g cells) comparable to that of mouse EHS tumor (6 mg from 1 g tumor). Secreted proteins could be effectively separated by sucrose density sedimentation. *Drosophila* basement membrane components showed an inhibitory effect to the attachment of bovine aortic endothelial cells to culture dishes.

1. Introduction

Basement membrane (BM) is a special extracellular matrix which forms the continuous sheet underlying various epithelia and endothelia. It has a crucial role in organization of tissues by affecting proliferation, differentiation and migration of the cells (1-3). It also surrounds muscle fibers, fat cells and peripheral nerves and functions as the filter of macromolecules in kidney glomerulus. Among major components of BMs, type IV collagen forms framework of the sheet structure of BM by head-to-head and tail-to-tail association, laminins affects cellular function by interacting with integrins at cell surface and perlecan concentrates growth factors such as TGFβ in the area of BM by its ability of binding to various growth factors. Reflecting crucial role of BMs *in vivo*, many *in vitro* experiments have shown dramatic effects of BM components on various cell lines. These finding opened many ways of application of BM components for effective cultivation of differentiated animal cell lines. For this purpose, Engelbreth-Holm-Swarm (EHS) tumor of mouse has been a only source of BM components since this tumor accumulates large amount of BM components extracellularly.

We report here that a *Drosophila* cell line (Kc167) is another good source of BM components since they secrete BM components into the medium in amount comparable to EHS tumor.

2. Materials and Methods

2.1 CELL CULTURES

Kc 167 cells were cultured with serum-free HyQ CCM3 medium (Hyclone Lab. Inc.), Schneider 2 cells with Schneider medium (Gibco BRL) containing 10% fetal bovine serum (FCS) and Clone 8 cells with M3 medium (Sigma) containing 2% FCS, 2.5% fly extract and 0.125 units/ml insulin. All cell lines were maintained at 25°C.

2.2 LABELING CELLS WITH [^{35}S] METHIONINE

Confluent Kc 167 cells were dispersed and transferred to 60 mm dishes. After attachment of the cells to dishes, the medium was replaced with methionine-free M3 medium (Sigma) containing 0.25 mCi/ml [^{35}S]Pro-mix (Amersham) and incubated for 4 hours at 25 °C. Labeled medium was concentrated by 10 % trichloroacetic acid precipitation and analyzed by SDS-polyacrylamide gel electrophoresis (PAGE) as described(1-3).

2.3 CELL ATTACHMENT ASSAY

Kc167 medium was concentrated by PM30 or PM100 (Amicon) and dialyzed against phosphate-buffered saline (PBS). Bovine aortic endothelial cells (BAEC) were cultured on the dishes (6-well, Nunc) coated with concentrated Kc 167 medium, mouse type I collagen solution or Matrigel (dialyzed urea extract from EHS tumor) and in Dalbecco's modified Eagles medium containing 10% FCS.

3. Results

3.1 EXTRACELLULAR PROTEINS PRODUCED BY *DROSOPHILA* CELL LINES

Drosophila cell lines Kc 167 cells, Schneider 2 cells (both originated from hemocyte precursor) and Clone 8 (originated from wing disc) secrete large amount of extracellular proteins. When the proteins secreted into cultured medium from [^{35}S]- methionine labeled-cells were concentrated by trichloroacetic acid-precipitation and separated by two-dimensional SDS-PAGE in which non-reducing and reducing electrophoresis are combined, limited numbers of polypeptides were detected within the migration area of more than 100 kDa. Figure 1 depicts the result for Kc 167 cells as the representative. Based on the molecular mass and the disulfide-bonded complex formation, the detected polypeptides were identified as laminin chains, type IV collagen chain, glutactin, tiggrin, papilin and peroxidasin (4). Type IV collagen secreted from Kc 167 appeared to be linked by non-reducible bonding to form a trimer which further dimerizes into a hexamer (Fig.1), while type IV collagen produced by Clone 8 cells appeared to form only a disulfide-bonded trimer (not shown). Schneider cells produced no type IV collagen and peroxidasin (not shown). Kc 167 specifically produced an unknown 300 kDa polypeptide which can form homopolymer by disulfide-bonding (Fig.1).

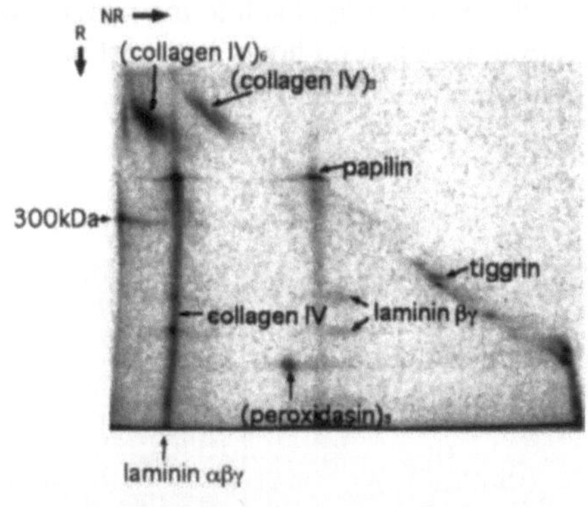

Fig.1 Non-reducing and reducing two dimentional SDS-PAGE of proteins secreted from Kc 167. Radiolabeled medium from Kc 167 culture was electrophoresed from *left* to *right* under non-reducing condition and from *top* to *bottom* under reducing condition.

3.2 EXTRACELLULAR PROTEIN PRODUCTIVITY OF KC 167 CELLS

The extracellular protein production by Kc 167 cells was further evaluated. On usual tissue culture dishes, Kc 167 cells attach to the dish surface during initial growth but become detached after reaching the confluence. SDS-PAGE of normalized volume of conditioned medium showed that amount of secreted proteins increased during the attached growth but it reached a plateau after the confluence (not shown). Kc 167 cells can grow in suspension as well. When the cells were seeded in a 100 ml suspension culture at density of 1×10^6 cells/ml, the maximum density of 1×10^7 cells/ml was attained within a week to give a packed cell volume of 1 ml. Concentration of the medium by a ultra-filtration apparatus which cut off the molecules less than 30 or 100 kDa gave the protein amount of 22 or 1.2 mg, respectively. This amount is comparable to 6 mg of BM proteins extracted from 1g of EHS tumor. Due to self-association of mouse type IV collagen and laminin, the concentrated extract from EHS tumor spontaneously forms a gel (Matrigel) at the temperature higher than 4 °C. In contrast, *Drosophila* BM proteins have poor ability of self-association and the macromolecules in conditioned medium of Kc 167 medium could be easily separated by sucrose density gradient centrifugation (not shown).

3.3 ANTI-ADHESIVE EFFECT OF *DROSOPHILA* EXTRACELLULAR PROTEINS TO MAMMALIAN CELLS

To examine the effect of *Drosophila* extracellular proteins to mammalian cells, tissue culture dishes were preliminary treated either with phosphate-buffered saline (PBS), concentrated medium from Kc 167, mouse type I collagen solution or extract from EHS tumor (Matrigel). Within 4 hours after seeding bovine aortic endothelial cells (BAEC) on these dishes, they strongly attached to the surface treated with type I collagen and Matrigel but did not spread as on the control surface (Fig.2). During 1 day of culture, the cells showed active migration on type I collagen and Matrigel and tended to form microvascular structure on Matrigel

(Fig.2). On the other hand, the cells could attach to the surface treated with *Drosophila* extracellular during first several hours but most of them detached from the dishes after 1 day of culture.

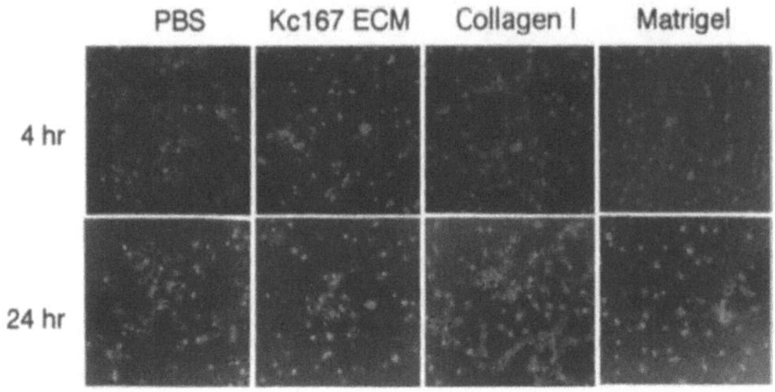

Fig.2 Anti-adhesive effect of *Drosophila* extracellular proteins to BAEC.
Tissue culture dishes were preliminary treated either with phosphate-buffered saline (PBS), concentrated medium from Kc 167, mouse type I collagen solution or extract from EHS tumor (Matrigel) and BAEC were cultured on them for 4 hours or 1 day.

4. Disccusion

In mammalian laminin family, five α, three β and two γ chain variants have been identified and they are assembled into, at least, twelve different $\alpha\beta\gamma$ trimers. Despite such variety of mammalian laminins, only laminin-1 with the composition of $\alpha1\beta1\gamma1$ has been exclusively studied because of EHS tumor as the abundant source of this molecule. Shortage of the sample supply has limited the study of other laminin molecules. In this study, we have shown that Kc 167 cells can supply *Drosophila* laminin in an amount comparable to EHS tumor. In addition, *Drosophila* BM proteins can be easily purified due to their low ability of self-association. Thus, evaluation of Kc 167 cells as the appropriate source of *Drosophila* laminin may facilitate the study of this molecule.

Concentrated medium from Kc 167 cells showed anti-adhesive effect to a mammalian cell line; BAEC. This may suggest disadvantage of *Drosophila* extracellular proteins for the cultivation of mammalian cells. However, this anti-adhesive effect suggests that they have affinity to some type of cell surface receptor which is essential for the attachment to tissue culture dishes. Since such anti-adhesive effect of extracellular matrix proteins is important to understand the cell migration during tissue remodeling, further analysis of this effect may open a novel application *Drosophila* extracellular proteins to animal cell cultivation.

5. References

1) Morita, A., Sugimoto, E.,and Kitagawa, Y. (1985) *Biochem. J.*, **229**, 259-264
2) Aratani, Y., and Kitagawa, Y.(1988) *J.Biol.Chem.*, **263**, 16163-16169
3) Tokida, Y., Aratani, Y., Morita, A., and Kitagawa, Y. (1990) *J. Biol. Chem.*, **265**, 18123-18129
4) Fessler,J.H., Nelson, R.E., and Fessler, L.I. (1994) *Methods in Cell Biology*, **44**, 303-328

THE BINDING EFFECT OF IMMUNE COMPLEX BY RAT RENAL GLOMERULAR EPITHELIAL CELL EXPRESSED WITH HUMAN CR1 GENE

P.C.Wang, S.Kimura, Y. Gao and M. Matsumura
Institute of Applied Biochemistry, University of Tsukuba
1-1-1 Tennodai, Tsukuba, Ibaraki 305 Japan

Abstract

This study discloses a novel concept different from both the traditional artificial kidney and the cell therapy, to cure the chronic renal disease, particularly the disease of SLE (systemic lupus erythematosus), which is regarded as an autoimmune disease caused by the existance of excessive amount of immune complex in the patients' serum. In this study, a full length of human tonsil CR1 cDNA was transfected into nomal rat glomerular epithelial cells to enable the CR1 expression in a long term (more than 1 year). As compared with the non-CR1 expressed glomerular epithelial cells, the CR1 expressed cells showed a higher binding effect to immune complex in a long period (more than 60 minutes), and the binding effect was demonstrated to be mediated by complement of serum. Anti-CR1 antibody blocking the binding effect indicates that CR1 plays an important role on immune-complex binding, and the long period binding suggests that CR1 on the renal epithelial cells may possess different function from that on erythrocyte which transports and processes immune complex in the circulating blood in a short period (less than 10 minutes).

key words: immune complex, glomerular cell, CR1 expression, binding

Introducation

Patients with systemic lupus erythematosus (SLE) demonstrated a loss of complement receptor type 1 (CR1, a receptor of complements C3b and C4b) on glomerular epithelial cells (1) and showed a deposition of DNA/anti DNA antibody immune complexes on glomeruli (2). However the biological function of CR1 and the clearance of immune complexes are poorly understood because the normal human or rat glomerular epithelial cells lose the expression of CR1 in the cell culture.

CR1 is a polymorphic membrane glycoprotein expressed on human erythrocytes, peripheral leukocytes, plasma and renal glomerular podocytes (3). CR1 binds complement C3b and C4b that have covalently attached to immune complexes and the consequences of these interactions depend on the cell type bearing the receptor. Erythrocytes CR1 binds immune complexes for transport to liver. CR1 on neutrophils and monocytes internalizes bound complexes either by adsorptive endocytosis or by phagocytosis and CR1 in plasma may regulate the activation of complement. Recent studies proving that the injured glomerular podocytes releasing membrane-bound CR1 into the urine suggested that CR1 may serve the trapping of immune complex. The polymorphism of CR1 was reported containing 3 allotypes, A/F, B/S and C/F' corresponding to 190, 220 and 160kD, respectively, which relates to the specificity of CR1 binding site to the ligand. CR1 has been detected from the podocytes of normal human kidney in vivo. But its expression was lost in the cell in culture, thus the investigation of its biological function remained difficult in the in vitro experimental system.

In order to clarify the biological function of CR1 on the glomerular podocytes and for the further use on cell therapy, establised cell line is considered a useful experimental system. However, the human glomerular cell line has not yet been established. The only one normal cell line SGE1 was establihsed from rat renal glomeruli (4), but CR1 could not be detected in cell culture and therefore cannot be used for evaluating the correlation between CR1 and immune complex. To overcome the deficiency of the loss of CR1 on the culture cells, a full length of human tonsillar CR1 cDNA (5) was transfected into SGE1 cells and the expresssion of CR1 was compared with that of the non-cDNA transfected cells. Furthermore, the CR1 expressed cells were used to investigate the binding effect of pathogenic immune complex of DNA/anti-DNA antibody, and the factor mediated the binding effect between the immune complex and CR1 on the renal epithelial cells was clarified.

Material and Methods

Cells, plasmids and media

Rat glomerular epithelial cell line SGE1 was gifted by Dr. T. Okigaki (11). Cells were cultured in type I collagen coated culture dishes with DHSF medium (per liter: D-MEM, 5g; HamF12, 5.3g; $NaHCO_3$, 1.9g; ITS, 6.25mg; EGF, 1μg) as well as 10% FBS. Plasmid pCR1 containing a full length of CR1 cDNA from human tonsillar library was cultivated in L-broth added with ampicillin and tetracyclin. Plasmid pSV2neo possessing a neomycin marker is used for the cotransfection of pCR1 into SGE1 cells. The selective medium for the transfected SGE1 was DHSF with neomycin G418 (Sigma), and the cells were grown in the incubator at 37°C with 5% CO_2 humidity.

DNA manupulation and electroporation

All of the DNA manupulations and the purification of plasmid were conducted using standard methods described by Sambrook et al. (6). The cotransfection of pCR1 and pSV2neo into SGE1 was carried out by electroporation at 960µF, 270volt by using Gene Pulser(Bio Rad). The electroporated cells were cultured in the selective medium and positive clones were selected for the analysis of CR1 expression.

Analysis of CR1 expression

Flow cytometry was used to analyze the CR1 expression. Cells reacted with FITC-labeled mouse anti-human CR1 monoclonal antibody were subject to FACS Vantage (Becton Dickinson), and cells reacted with IgG were used as control. Propidium Iodide (PI) was reacted with cells so as to distinguish dead cells from viable cells. SDS-PAGE and Western blotting were conducted to identify the molecular weight of the expressed CR1. A non-isotope ECL kit (Amersham) was used to visualize CR1 band on the blotted nitrocellulose membrane.

Immune complex (DNA/anti-DNA-antibody)

Plasmid pUC18 DNA was used as antigen, and was digested by restriction enzymes EcoR1 and Apal1 to form various fragments of DNA (219bp, 497bp, 724bp and 1246bp). Anti-DNA-antibody was raised from NZB/W mouse with autoimmune disease, and the construction of hybridoma which produces anti-DNA-antibody is described in other report of this proceedings. The formation of immune complex was conducted by reacting the prepared anti-DNA-antibody with ^{32}P isotope labeled DNA at $37^{o}C$ for 1hr.

Binding assay of immune complex by CR1 expressed cells

The immune complexes of various sizes of isotope labeled DNA (70 pM)/anti-DNA-antibody ($1x10^{4}$ pM) were reacted with SGE1 cells ($1x10^{6}$) in gelatin veronal buffer at $37^{o}C$ at various time intervals (3-60 min), and then centrifuged to remove the supernanat. The supernatant and pellet were subject to scintillation counter respectively, to estimate the radioactivity. The binding effect is shown as the following equation.

$$\text{binding effect(\%)} = \frac{\text{radioactivity of pellet}}{\text{radioactivity of pellet and supernatant}}$$

Results and discussion

The electroporation caused one half of the 10^{7} SGE1 cells inviable, and the subsequent selective medium excluded the non-transfected cells. Finally, 170 positive clones were screened by limiting dilution and some of them were randomly picked up for the analysis of CR1 expression. Figure1 shows the fluorescence intensity analysis on the positive clone by flow cytometry. A significant increasing fluorescence intensity was observed on the positive clone (black peak) as compared with the control (white peak). Figure 2 shows the western blotting of two positive clones as well as the control. The expressed CR1 on SGE1 cells revealing a significant band above 200kD coincided with one of the three known allotypes of CR1 (220KD), which is the normal allotype without losing any C3b binding sites. Thus the positive transfected cell is considered a suitable experimental system for the investigation on the correlation between CR1 and immune complex.

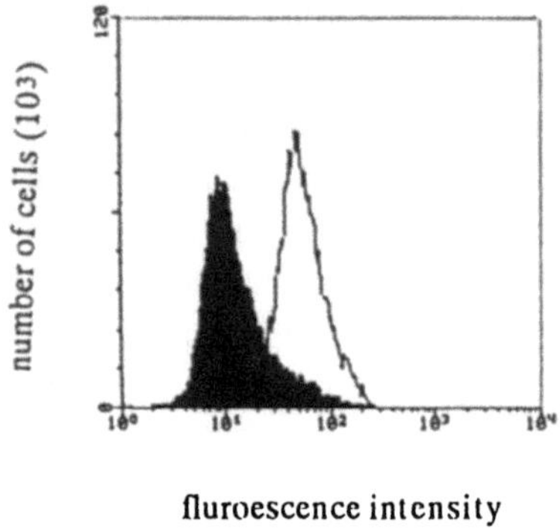

Fig. 1 Flow cytometry analysis of positve clone after 1 year culture

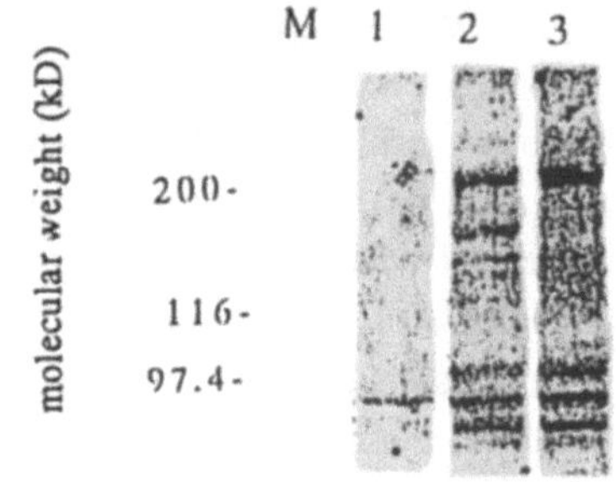

Fig. 2 Western blotting analysis of CR1 expression on SGE1 cells.
Lane M, molecular weight marker; Lane 1, control; Lanes 2 and 3, positive clones

Time course of immune complex bound to SGE1 cells is shown in Fig.3. The binding effect of the CR1-expressed cells reached to the peak in 3-5 minutes after the cells reacted with immune complex, and then declined gradually. However, the binding effect did not vanish even after 1 hr, which is different from the short binding (10 min) of CR1 on erythrocyte. This fact may imply that CR1 on the glomerular epithelial cells plays a different role from that on erythrocyte. The control cells also showed the same binding tendency as the positive CR1-expressed cells, but with only 1/2 binding effect, which might be due to the physical attachment between cells and immune complex. Figure 4 shows that preculture of anti-CR1-antibody with cells blocked the binding of immune complex. This fact suggests that CR1 is essentail to the immune complex bound.

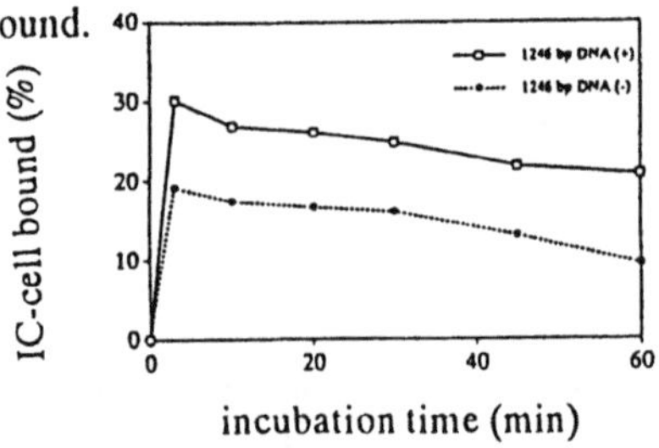

Fig. 3 Time course of the bidning of immune complex to SGE1
□ SGE1 bearing CR1, ● control

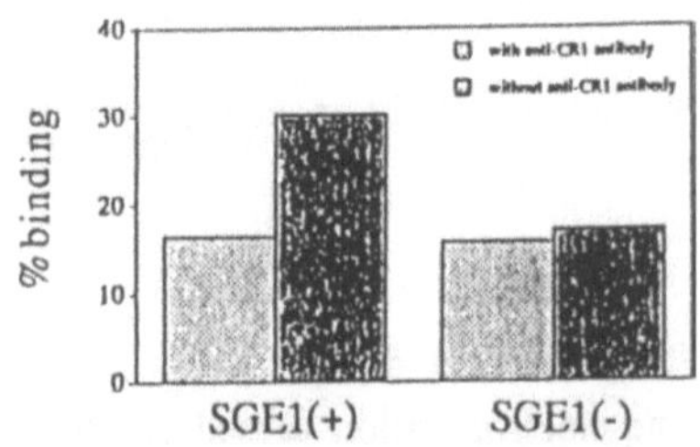

Fig. 4 The binding of immune complex to SGE1 preincubated in the presence and absence of anti-CR1 antibody.
SGE(+): cells bearing CR1, SGE1(-): control

Since complement in the serum is known to opsonize the immune complex to the CR1 on other types of cells, the binding effect in the presence of the inactivated serum was investigated. The result showed that the inactivated serum abolished the immune complex binding (Fig. 5), which may suggest that the immune complex bound is mediated by the complement. In conclusion, the CR1-expressed cells are capable of trapping immune complex through complement. The possibility of using these cells for in vivo therapy of SLE will be investigated in the future.

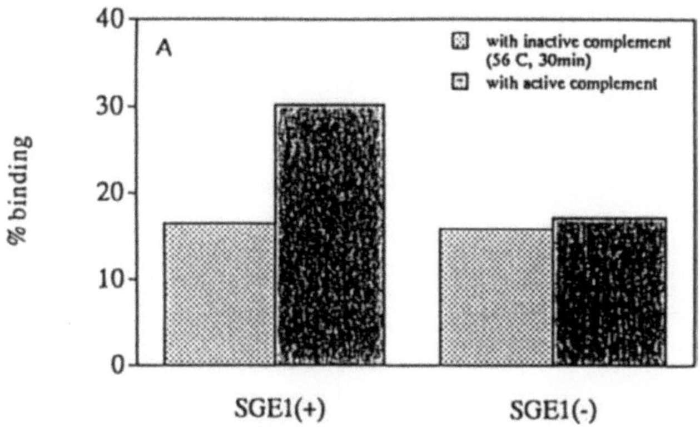

Fig. 5 The binding of immune complex to SGE1 in the presence of inactive and active complement.
SGE1(+): cells bearing CR1, SGE1(-): control

Reference

(1) Kazatchkine, M.D., Fearon, D.T., Apply, M.D., Mandate, C. and Bariety, J. (1982) Immunohistochemical study of the human glomerular C3b receptor in normal kidney and in seventy-five cases of renal diseases*J.Clin.Invest.*, 69, 900-912.

(2) Miyakawa, Y., Yamada, A.,. Kosaka, K., Wsuda, F., Kosugi, E., and Mayumi, M. (1981) Defective immune-adherence (C3b) receptor on erythrocytes from patients with systemic lupus erythematosus. *Lancet.*, 2, 493-497.

(3) Fearon, D.T. and Wong,W.W. (1983) Complement ligand-receptor interactions that mediate biological responses*Annu. Rev. Immunol.*, 1, 243-171.

(4) Yamada, M., Kawaguchi, M., Takamiya, H.,Wada H., and Okigaki, T.(1988) Establishment and characterization of an epithelial cell line, SGE1, from isolated rat renal glomeruli.*Cell Struct. Funct.*, 13, 495-513.

(5) Wong, W.W. (1990) Structural and functional correlation of the human complement recptor type 1. *J. Invest. Dermatology*, 94, 64S-67S.

(6) Sambrook. J., Fritsch, E.F. and Maniatis, T. (1989) Molecular Cloning. *Cold Spring Harbor Laboratory Press.* New York

PREPARATION AND CHARACTERIZATION OF DNA/ANTI DNA ANTIBODY IMMUNE COMPLEX USING MONOCLONAL ANTIBODIES

Y. GAO[1,2], P.C. WANG[1], A. TAJIMA[1], K. YASUDA[2]
M. MATSUMURA[1]
1. Institute of Applied Biochemistry, University of Tsukuba, 1-1-1 Tennodai, Tsukuba city, Ibaraki 305, Japan.
2. Biomaterial, Co., Ltd., 1-21 Junka, Fukui 910, Japan.

ABSTRACT

DNA/anti-DNA immune complex (IC) was formed using monoclonal antibody and low molecular homogeneous DNA, and the pathogenicity was studied both *in vitro* and *in vivo*. IC was shown having high ability of activation and binding to the complement, whereas the RBC binding capacity was shown relatively low. The results of sucrose density gradient ultracentrifugation indicated that the sedimentation coefficient of the IC was between 10S and 23S. Organ distribution and uptake studies indicated that IC has high affinity to kidney. We concluded that IC prepared *in vitro* has same pathogenicity as that found in patient of active autoimmune disease, and therefore could be applied to many studies *in vivo* and *in vitro*.

Keywords anti-DNA SLE immune complex NZB/W mice complement receptor

INTRODUCTION

Circulating immune complex is believed to be responsible for many different symptoms and tissue damage in the autoimmune diseases such as systemic lupus erythematosus (SLE). It is known that IC consists of DNA and anti-DNA antibody and can activate complement through the classical complement activation pathway. In general, complement conjugated IC could be bound to erythrocyte through the complement receptor and thereafter be transported to the reticuloendothelial system and finally be degraded (Cornacoff et al., 1983). IC that does not be removed from blood circulation will deposit to the kidney and other tissues, and cause inflammation.

Recently, many attempts have been done to remove the IC from circulation, by using an immuno-absorbent device. For evaluating the immuno-absorbent, serum from active SLE patients or experimental animals (Terman et al., 1976), such as rabbits immunized with DNA in advance, were used as the source of IC. IgG purified form the serum above was also served as an *in vitro* preparation of IC. However, the above mentioned materials might cause the following disadvantages: the materials from patients and animals are donor dependent, and have different characteristics of IC that will make the evaluation complicated. In addition, these materials may be dangerous due to the possibility of the infection by HIV and hepatitis virus.

In this study, A simple *in vitro* preparation of IC using monoclonal anti-DNA antibodies was introduced, and the characteristic was investigated both *in vitro* and *in vivo*.

MATERIALS AND METHODS

Preparation of immune complexes. The anti-DNA antibody (IgG2a, κ) producing hybridoma was produced by fusion of the BALB/c P3U1 myeloma cells with spleen cells of unimmunized 7-month-old female NZB/W F1 mice (SLC), using the PEG method. The antibody was purified from culture medium with affinity chromatography on protein G-Sepharose (Pharmacia).

To generate low molecular DNA fragments, plasmid pUC18 DNA was digested with restriction nucleases EcoR1 and ApaL1. DNA fragment of 219 base pairs (bp) was electroeluted from the agarose gel and purified with phenol and chloroform. DNA was radioactively labeled with 32P-dCTP (ICN Pharmaceuticals, Inc.) using DNA labeling kit (Wako Pure Chemical) to a specific activity of 0.77×10^8 cpm/μg. All the *in vitro* and *in vivo* experiments with 32P-DNA were performed within 5 days after the labeling reaction.

Immune complexes were formed at various Ab/DNA ratios. 32P-DNA was incubated with Ab in PBS(-) buffer at 37°C for 1-4 hrs. Siliconized microcentrifuge tubes were used in all assays.

Binding assays. DNA binding by Ab was measured by the Farr assay described elsewhere (Taylor et al., 1980) with the following modifications. The initial reaction volume of DNA and Ab at various ratios was 96 μl totally. Then 96 μl of a 50 % ammonium sulfate solution was added, and the mixture was incubated at 4°C for 1 hour. Subsequently, the sample tubes were spun at 10,000 *g* for 10 min and the radioactivity (32P-DNA) was measured by scintillation counter. The total radioactivity of supernatant and pellet was regarded as a total cpm. Finally, 96 μl (1/2 total volume) of the supernatant was transferred to a new tube and counted and compared to the total cpm per assay. The binding activity by Ab was expressed as follows: percentage DNA bound = 100 x (1/2 total cpm - cpm of 1/2 volume supernatant)/total cpm.

The RBC binding assay was performed as follows. 96 μl of IC was mixed with 2 μl of 0.1M $CaCl_2$, 2 μl of 0.1M $MgCl_2$ and 50 μl guinea-pig serum that was 4 times diluted in gvb (gelatin-veronal buffer: 0.1 % gelatin, 5 mM sodium barbitol), and then added with 100 μl of washed normal human erythrocyte at a concentration of 1.25×10^9/ml in gvb^{2+} (gvb containing 0.5 $MgCl_2$ and 0.15 mM $CaCl_2$). The mixture was incubated at 37°C for 5 min and spun at 8,000 *g* for 1 min. After the total cpm was counted the supernatant was removed. Cell pellet was washed twice with gvb^{2+} and the radioactivity was counted. The radioactivity of 125 μl of the supernatant (1/2 total volume) was also counted and the IC binding results were expressed as the percentage of radioactivity bound to RBC as follows: percentage IC bound to RBC = 100 x (cpm in cell pellet/total cpm).

Complement activation capacity of IC was investigated by the C3b incorporation test. Complement treated IC was incubated with rabbit anti-C3b antibody (Nordic Immunological Laboratories) at 37°C for 2 hrs. Then goat anti-rabbit immunoglobulin was added and incubated at 37°C for one more hour and then at 4°C for 2 hrs. The samples were spun at 8,000 *g* for 15 min. The total cpm and the cpm of 1/2 vol of supernatant were counted. Complement activation capacity was expressed as the percentage radioactivity of the precipitation as follows: percentage IC that incorporated C3b = 100 x (1/2 total cpm - cpm in 1/2 vol supernatant)/total cpm.

Sucrose density gradient ultracentrifugation. The sizes of IC were measured by sucrose density gradient ultracentrifugation (SDGC). Samples were run at 40,000 rpm for 16 hrs. Human serum albumin, IgG and IgM were used as standard markers of 4.6S, 7S, 19S, respectively. Fractions were harvested and the radioactivities were counted. The data were expressed as cpm per milliliter of sucrose gradient media that were collected from the bottom of the tubes.

Organ uptake of complexes. Three to five months old female BALB/c mice (SLC), with no autoimmune disease and no demonstrable anti-DNA antibody activity in blood circulation, were used on the study of organ distribution. Animals were injected with IC at a dose of 40 to 60 μg 32p-DNA in 0.2-0.3 ml of PBS(-) per body, via the lateral tail vein, and were sacrificed by cervical dislocation at 40 min after injection. The liver, spleen, kidney and lung were dissected, weighted and the radioactivity was counted to estimated the IC uptake by soft tissue. Total body content of radioactivity was measured by summing the content of the radioactivity in all parts of body. The radioactivity measurements were normalized to give cpm counts/mg tissue.

RESULTS AND DISCUSSION

In the preliminary studies, the DNA and 32P-DNA were tested for the antibody binding capacity. There was no significant change in the binding capacity after DNA being labeled with isotope.

32P-DNA/anti-DNA IC formed with a wide range of Ab/DNA ratios were prepared and the percentage of DNA bound to antibody was investigated. Figure 1 shows the percentage of DNA bound at different ratios of Ab/DNA. When IC was formed at high Ab/DNA ratios (≥10 mg IgG/μg DNA), the percentage of DNA bound to antibody was above 80 %. Control experiments using non-anti-DNA antibody showed no DNA bound (data not shown), which indicated that monoclonal antibody prepared in this work was high specificity to the DNA. The following experiments were performed at an Ab/DNA ratio between 10 and 20 mg IgG/μg DNA unless indicated.

It is known that IC could be divided into 3 types, the large, small and very small. There are many evidences suggesting that large IC is cleared rapidly from the circulation by Fc mediated (Finbloom and Plotz, 1979) and complement mediated (Cornacoff et al., 1883; Takahashi et al., 1985) pathways in the body. The very small IC does not activate the complement and may be considered of non-pathogenic (Horgan and Taylor 1984). However the small IC in the circulation can only be cleared much more slowly than the large IC, and therefore may have greater chances of deposition in tissues (Emlen and Burdick, 1985). Furthermore, investigators have

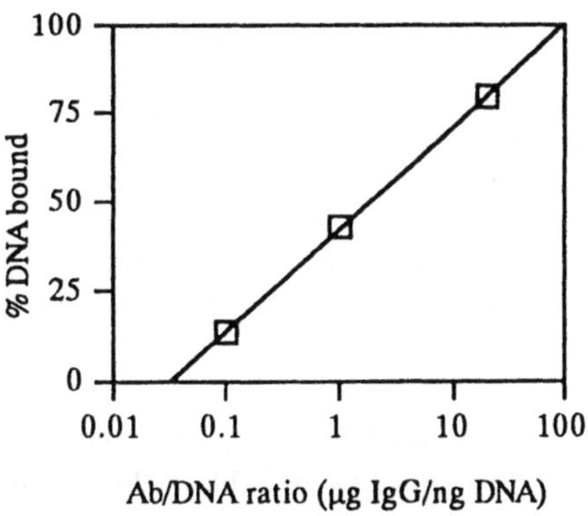

Figure 1. Percentage of 32P-DNA bound to anti-DNA antibody at various Ab/DNA ratios.

demonstrated that small IC in SLE consisted of low molecular DNA between 30 and 240 bp (Ikebe et al., 1983). It is considered that the IC prepared *in vitro* could be small IC formed with low molecular of DNA.

Figure 2 shows the percentage of IC bound by RBC and the percentage of IC that activated and incorporated to the complement. The experiments were performed at an Ab/DNA ratio of 20 mg IgG/μg DNA, and the DNA size was 219 bp. The result reveals the percentage of IC incorporated to C3b was about 60 %. However, the IC bound by RBC was only 20 to 25 % as compared with that bound by C3b. These data suggest that the IC prepared in this study was able to activate the complement and incorporate to the complement component of C3b, but the ability to bind C3b receptor of RBC was not high. Therefore IC prepared in this study were considered pathogenic.

In order to reveal the sedimentation coefficients (S value) of the IC, sucrose density gradient ultercentifugation was performed at 274,000 *g* for 16 hours, at 5-40 % w/v linear sucrose density gradient in PBS(-). IC was fractionated and the S value was calculated corresponding to the standard marker. Figure 3 shows the distribution pattern and the S value of IC in the sucrose density gradient after ultracentrifugation. The S value was between 10S and 23S. These sizes extremely match the sizes that were reported to be the most pathogenic IC in SLE patients (Bruneau et al., 1977), and may suggest that the IC was small. Analysis of the SDGC confirmed that the *in vitro* results of the IC incorporation to complement and binding to RBC.

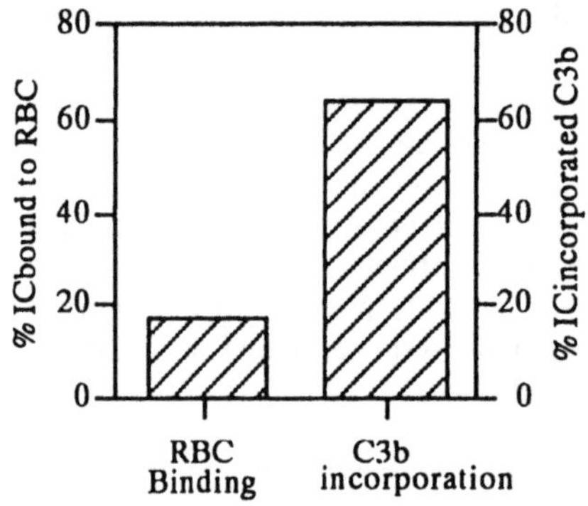

Figure 2. Complement component C3b activation and incorporation of IC and RBC binding of IC. Results are expressed as percentage of IC bound or incorporation. Data are represented as the mean ± S.D. of three to four separate determinations.

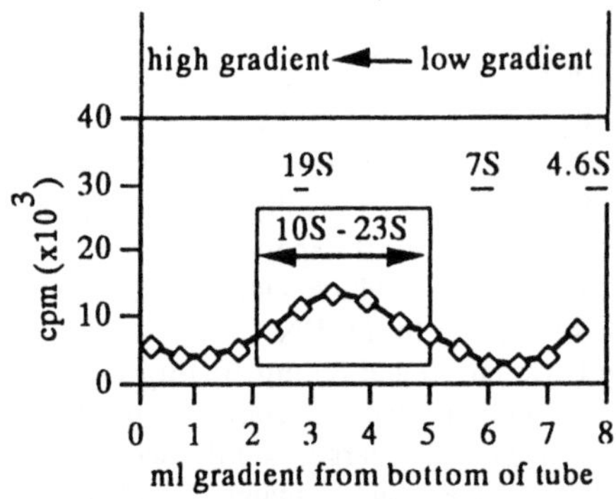

Figure 3. Sucrose density gradient centrifugation of IC formed with DNA of 219 base pairs. The numbers inside the gragh indicate the marker of human serum albumin, IgG and IgM which were 4.6S, 7S and 19S respectively.

It was observed that the radioactivity of IC remained in blood circulation dropped rapidly (data not shown). Table 1 shows the organ distribution and the uptake of IC at the time of 40 min after administration. The liver uptaken and retained about 60 % radioactivity of total body content, suggested that IC was transported and degraded mainly in liver. Though kidney took small portion of IC, the cpm contained per mg tissue showed relatively high than other tissues other than the liver. These results may suggest that the IC prepared in this study had high affinity to the kidney. Hence it may be pathogenic *in vivo*.

Table 1. Organ distribution and body uptake of IC

organs	% total body content	cpm/g (x10^3)
liver	36.9	60.0
spleen	18.3	3.0
kidneys	23.1	10.0
lung	12.6	3.0

Taken together, we have prepared the immune complexes *in vitro* using monoclonal antibody and low molecular DNA. The IC was of high complement activation ability and low RBC binding capacity. The S value was between 10S and 23S. Furthermore, the IC was of high affinity to the kidney. We concluded that the IC prepared *in vitro* in this study are likely to be pathogenic similar to those from SLE patients, and therefore could be expected to be applied in many *in vitro* and *in vivo* studies.

Acknowledgment. We thank the Dr. Tohru Okigaki of Shigei Medical Research Institute, Japan for his helpful discussions and Dr Kazuhiro Yoshihara of National Institute of Animal Health, Japan for his helpful suggestion to the preparation of the anti-DNA hybridoma.

REFERENCES

Bruneau C. D., Edmonds J. P., Hughes G. R. V. and Aarden L. (1977) Detection and characterization of DNA/anti-DNA complexes in a patient with systemic lupus erythematosus. *Clin. exp. Immun. 28, 433-436.*

Cornacoff J. B., Hebert L. A., Smead W. L., Van Aman M. E., Birminghan D. J. and Waxman F. J. (1883) Primate erythrocyte-immune complex-clearing mechanism. *J. clim. Invest. 71, 236-247.*

Emlen W. and Burdick F. (1985) Clearance of Model DNA -anti-DNA immune complexes in mice. *Arthritis Rheum 28. S40.*

Finbloom D. S. and Plotz P. H. (1979) Studies of reticuloendothelial function in the mouse with model immune complexes. II. Serum clearance, tissue uptake, and reticuloendothelial saturation in NZB/W mice. *J. Immun 123, 1600-1603.*

Horgan C. and Taylor R. P. (1984) Quantitative analyses of complement fixation in three immune complex systems. *Immunology 52, 752-759.*

Ikebe K., Gupta R. C. and Tan E. M. (1983) Characterization of DNA in polyethylene glycol precipitated immune complexes from serum of patients with systemic lupus erythematosus. *Clin. exp. Immun. 53, 169-177.*

Takahashi N., Fujita T., Takata Y. and Tamura N. (1985) Interaction of complement solubilized complexes with mouse peritoneal mcrophages and their clearance and tissue uptake. *Clin. exp. Immun. 61, 176-182.*

Taylor R. P., Waller S. J., Haden C. and Addis D. J. (1980) Stability of DNA/anti-DNA complexes. III. Kinetic control of immune complex size. *J. Immun 124, 2571-2577.*

Terman D. S., Stewart I., Robinette J., Carr R. and Harbeck R. (1976) Specific removal of DNA antibodies *in vivo* with an extracorporeal immuno-absorbent. Clin. exp. Immun. *24, 231-237.*

MORPHOLOGY AND FUNCTION OF RAT HEPATOCYTES CULTURED ON EXPANDED POLYTETRAFLUOROETHYLENE SURFACE.

T.TOKIWA[1], J. KANO[1], X-Y. MENG[1], M. KODAMA[1], T. MATSUMURA[2]
*[1]Bionic design group, National Institute for Advanced Interdisciplinary Research, Tsukuba, Ibraki 305, [2]Research Institute of Innovative Technology for the Earth and Meiji Milk Production Co., Cell Technology Center, Odawara 250, Japan**

Abstract

For the development of a bioreactor using hepatocytes, particularly for a hybrid-type artificial liver, we study culture substrates in this paper with particular attention to a porous material, expanded polytetrafluoroethylene (ePTFE). Hepatocytes obtained from adult rats were attached to a substrate and grown in a chemically defined medium either with fetal calf serum (FCS, 10%), or with bovine serum albumin (BSA, 0.003- 0.3%). When cultured in a FCS-supplemented medium, hepatocytes spread poorly, retaining their cuboidal shape, and later reorganized themselves into multilayer aggregates on the ePTFE surface. The characteristics of hepatocytes were influenced by the modification of the ePTFE surface as well as the culture medium. Hepatocytes grown on a polyvinylalcohol (PVA)-treated ePTFE surface formed many more multilayer aggregates than those grown on a PVA-untreated ePTFE surface. Such highly multilayered aggregates were also noted when the cells were cultivated in a BSA-supplemented medium. The hepatocytes in all these aggregate cultures exhibited high albumin expression capability as compared with those in monolayer cultures. On the other hand, when cultured in a FCS-free or a BSA-free medium, hepatocytes were attached, spread well and then formed monolayers on both a PVA-treated and -untreated ePTFE surface as did the cells on a collagen-coated polystyrene surface. We concluded from these results that functional hepatocyte cultures can be obtained in multilayer aggregates immobilized on an ePTFE surface in a serum-supplemented medium.

Introduction

The growth and function of cultured hepatocytes are known to depend on the mode of cell to cell interaction and cell to substrate interaction. A porous material, expanded polytetrafluoroethylene (ePTFE) has an extensively large surface area and surface complexity, but is chemically inert. ePTFE has been studied as a possible material for vascular substitute[1]. However, few studies on the feasibility of using ePTFE for culturing isolated hepatocytes on its surface have been done so far. The present work examines ePTFE in its suitability as a culture substrate in terms of the morphology and albumin expression of rat hepatocytes cultured on its surface in vitro. The results will provide information on its application in tissue engineering such as for use in a hybrid-type artificial liver device.

*This work was supported in part by the New Energy Industrial Technology Development Organization (NEDO)

Materials and Methods

Parenchymal hepatocytes were isolated from male Wistar rats by perfusion of the liver in situ with collagenase following the methods of Seglen[2]. Hepatocytes were dispersed in a chemically defined medium [Williams E with dexamethasone (10^{-6} M), insulin (10^{-7} M), epidermal growth factor (25 ng/ml) and 100 U/ml penicillin/100 ug/ml streptomycin] supplemented with or without 10% fetal calf serum (FCS) or bovine albumin fraction V (BSA) and were plated onto 30-mm Millicell culture plate inserts attached by ePTFE membranes. The inserts with these membranes were placed in 6 well polystyrene plates containing 3 ml of culture medium. Control plates were coated with 0.03% type I collagen. Cells were cultured in a humidified atmosphere of 5% CO_2 and 95% air at 37 ^{o}C. Polyvinylalcohol (PVA)-treated (hydrophilic) and -untreated ePTFE (hydrophobic) membranes (approximately 30 μm of fibril length) were provided by Sumitomo Denko, Co., Osaka, Japan. PVA in the membranes was cross-linked by electronic beam irradiation. For immunocytochemical staining, samples were fixed in 10% neutral buffered formalin and then processed to paraffin wax for sectioning and immunocytochemistry was performed by anti-rat albumin rabbit antibody (Cappel, Cochranville, PA). Primary antibody was visualized by using labeled streptavidin biotin DAKO LSAB kit (DAKO, Carpinteria, CA).

Results and Discussion

Hepatocytes obtained from adult rats were attached to substrates and grown in a chemically defined medium either with 10% FCS, or with 0.003 to 0.3% BSA. When cultured in a FCS-supplemented medium on the surface of ePTFE, hepatocytes spread poorly, retaining their cuboidal shape, and later reorganized themselves into multilayer aggregates. The characteristics of hepatocytes were influenced by the modification of the ePTFE surface as well as the culture medium. Hepatocytes grown on a PVA-treated ePTFE surface formed many more multilayer aggregates than those grown on PVA-untreated ePTFE surface. Such highly multilayered aggregates were also noted when the cells were cultivated in a BSA-supplemented medium. Scanning electron microscopy revealed that multilayer aggregates formed on a PVA-treated ePTFE surface had three dimensional structures. On the other hand, when cultured in a FCS-free or a BSA-free medium, hepatocytes were attached, spread well and then formed monolayers

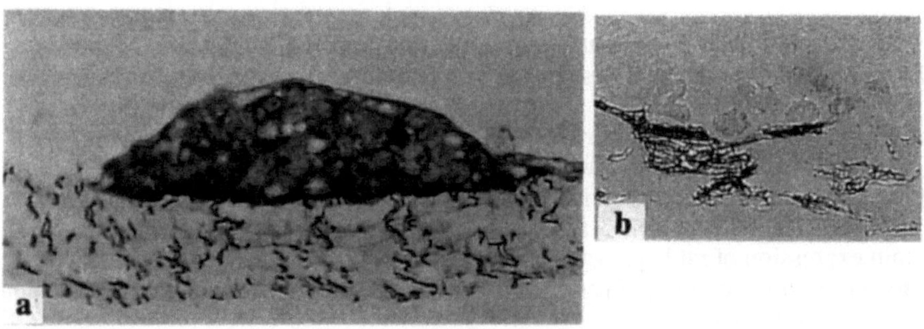

Fig. 1. Immunocytochemical staining of multilayer hepatocyte aggregates in section. Positive staining for albumin (a) in Day 9 multilayer aggregates can be seen compared to primary antibody minus control (b). x 200.

on both PVA-treated and -untreated ePTFE surfaces as on a collagen-coated polystyrene surface. The precise mechanism of multilayer aggregate formation on a PVA-treated ePTFE surface remains to be determined. It is well known that texture, porosity and other physical properties and surface charge, hydrophobicity and other chemical properties of the polymer substrates can influence cell behaviors such as cell adhesion and growth[3]. It is also noted that limiting cell spreading on substates may be important to maintain hepatic functions[4]. It is likey that proteins such as serum albumin are adsorbed to PVA-treated ePTFE surfaces and thereby limit cell spreading, and affect subsequent cell differentiation or formation of multilayer aggreagates. One of the primary goals of tissue engineering is to sustain the cell specific functions of cultured cells. Immunocytochemical studies using sections show that multilayer aggregates are positively stained as compared with primary antibody-minus control (Fig. 1) and with the cells in monolayer culture, suggesting that hepatocytes by forming multilayer aggregates adapt themselves to a three dimensional environment which is reminiscent of that normally seen in vivo.

Conclusion

Functional hepatocyte cultures can be obtained in multilayer aggregates immobilized on an ePTFE surface in a serum-supplemented medium. The present results suggest that μμμa porous PVA-treated ePTFE membrane is useful as a substrate for applications in tissue engineering .

References

1. Hirabayashi, K., Saitoh, H., Ijima, H., Takenawa, T., Kodama, M. and Hori, M. (1992) Influence of fibril length upon ePTFE graft healing and host modification of the implant. J. of Biomed. Mat.Res. 26, 1433-1447

2. Seglen, P.O. (1976) Preparation of isolated rat liver cells. Methods Cell Biol. 13, 29-83

3. Baldwin, S.P. and Saltzman, W.M. (1996) Polymers for tissue engineering. TRIP4, 177-182.

4. Tong, J.Z., Sarrazin, S., Cassio, D., Gauthier, F. and Alvarez, F. (1994) Application of spheroid culture to human hepatocytes and maintenance of their differentiation. Biol. Cell 81, 77-81.

EVALUATION OF ECTOPIC BONE FORMATION BY DEMINERALIZED ALLOGRAFTS IN 0.6HCL USING A CONTROLLED-HEAT ULTRASONIC CLEANER

H. I. SHIN*, C. J. KIM, H. J. JANG, S. H. LEE
*Department of Oral Pathology * and Oral & Maxillofacial Surgery, School of Dentistry, Kyungpook National University*
101 Dongin 2-Ga, Joong-Gu, Taegu, Korea

Introduction

Bone grafting techniques have been used for centuries, with varying degrees of success, for reconstructing lost or defective osseous tissues. While autogenous bone grafts are regarded as the most reliable bone regenerative material, they may require a second site surgical procedure and may not always produce an adequate quantity of bone[1]. Therefore, there is considerable interest in the development of alternatives to fresh autogenous bone grafts.

The search for a suitable alternative to autogenous grafts has led to the use of allogeneic bone of various preparations[2]. Since the first report of Urist[3] in 1965, many studies have focused on the use of demineralized bone as an osteoinductive bone substitute, because it is known that demineralization exposes the bone morphogenetic protein within the organic bone matrix so that this can stimulate the differentiation of osteoprogenitor cells into osteoblasts. By intense efforts, nowadays, several types of demineralized bone preparation are used clinically with successful osteoinduction[4]. However, these preparation protocols adopting conventional Urist's[5] or Reddi and Huggins'[6] protocol with slight variations are very complicated. They require demineralization with dilute hydrochloric acid and use organic solvents, as well as wash in physiologic phosphate buffers with enzyme inhibitors and lyophilization procedures.

In this study, we evaluated a more simple and rapid protocol for an allogeneic, low-antigenic bone graft that has osteoinductive as well as osteoconductive characteristics in a clinical situation.

Materials and Methods

Demineralized bone matrix was prepared by sterile technique from the femur and tibia of male Sprague-Dawley rats weighing about 250g. Dissected diaphyses were cut 5mm in length and the marrow was removed. The cortical tubes were demineralized in heated 0.6N HCl for at 50°C, 60°C, 70°C and 80°C for 10, 20, and 30 mins, respectively using a controlled-heat ultrasonic cleaner. The demineralized allografts at each temperature, except for 70°C and 80°C due to lysis of demineralized collagen bone matrix, were implanted in dorsal subcutaneous pouches of 54 Sprague-Dawley rats weighing about 200g. The undemineralized bone allografts treated in distilled water in a similar demineralizing process served as controls. The implants were harvested at 1, 2, and 3 weeks and then prepared for routine light microscopic observation.

Results

The demineralization using a controlled-heat ultrasonic cleaner in 0.6N HCl over 70°C caused lysis of the collagenous matrix, thus the allografts treated at 70°C and 80°C were excluded in this evaluation.

The histologic analysis revealed active ectopic bone formation in the demineralized allografts at 60°C for 20 and 30 mins(Fig. 1), whereas the demineralized allografts at 50°C for 30mins(Fig. 2) and at 60°C for 10 mins(Fig. 1) induced favorable bone formation. However, the demineralized allografts at 50°C for 10 and 20 mins(Fig. 2) showed relatively poor osteoinductive activity. The ectopic bone formation induced by the demineralized allografts proceeded through the endochondral bone formation mode confined within a fibrous capsule, while the undemineralized allografts of each group induced no ectopic bone formation(Fig. 3, 4). All implants of each group revealed no remarkable inflammatory or immunologic reaction through all experimental periods.

Discussion

It is generally accepted that demineralization of allogeneic bone implants and elimination of antigens are essential to ensure optimal osteoinduction[7]. The experimental studies on demineralized bone matrix as a bone substitute are also very promising, but methods of storage and sterilization of these grafts are still unclear[8]. In this investigation, we used a controlled-heat ultrasonic cleaner to get more rapid and sufficient demineralization with the additional effect of sterilization. Although BMP may have significant clinical potential as a bone graft substitute, the success of this application will depend on the production of sufficient recombinant human BMP and the development of a suitable delivery system for this osteoinductive factor[9]. In this point of view, the investigation of a new simple and rapid treatment which maximizes the BMP activity in bone itself may be another useful tool for bone grafting procedures.

The convention preparation protocols require more than 3 hours at 4°C and need additional time for dehydration and washing as well as lyophilization. For use in tissue banking, the Reddi and Huggins'[6] preparation is completed within one day and Urist's[5] protocol requires five to six days for completion of various steps, including lyophilization, whereas our protocol can be completed within 30 minutes without additional steps. This protocol consists of only demineralization using a controlled-heat ultrasonic cleaner and washing procedure. In histologic analysis, the allografts treated in 60°C 0.6N HCl for 20 and 30 minutes revealed active ectopic bone formation through endochondral mode similar to as in BMP-IBM implants, and there were no remarkable inflammatory or immunologic reaction. These findings suggest that complete demineralization is necessary to get optimal osteoinduction and that complete demineralization using a controlled-heat ultrasonic cleaner in 60°C 0.6N HCl is feasible within 30 minutes. Therefore, in cases, when the patient's own bone is tried as an implant for the reconstruction of function and esthetics after partial or total resective tumor surgery, or when a similar shape to that which is being replaced is necessary, our protocol may be applicable due to rapid processing and efficient osteoinductive potentiality.

In addition, the low-heat treatment may reduce antigenicity by destroying the cell viability and subcellular particles. The most significant source of antigens in bone allografts appears to be the cells bearing cell surface antigens and when a temperature more than 47°C causes devitalization of bone cells[10]. When banked bone is used, freezing and freeze-drying procedures also destroy subcellular particles as well as cell viability, thereby decreasing the tissue immunizing capacity of the preserved bone and leaving the matrix intact for transplantation[11,12]. The low-heat treatment and the acid used for demineralization also produce a sterilizing effect, because it is known that simply freezing the bone allograft reduces the risk of transinfection to 1 in 8 million and that the chemical agents used in processing produce a virucidal effect[13].

In conclusion, the findings of our investigation suggest that the simply and rapidly demineralized allografts in 60°C 0.6N HCl for more than 20 minutes using a controlled-heat ultrasonic cleaner may enhance the healing capacity with a minimizing of antigenicity as well as exhibiting a sterilizing effect.

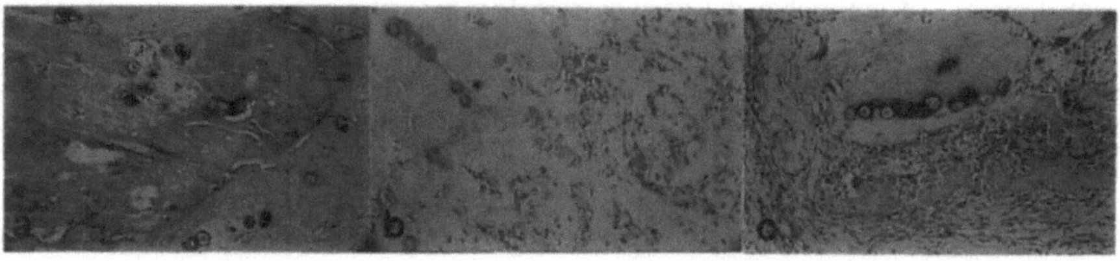

Fig. 1 Photomicrographs of 3 weeks after implantation of allografts demineralized in 0.6N HCl at 50°C for 10 mins(a), 20 mins(b), and 30 mins(c) using a controlled-heat ultrasonic cleaner. The endochondral bone induction is more favorable in c) than a) & b). (H&E, x 200)

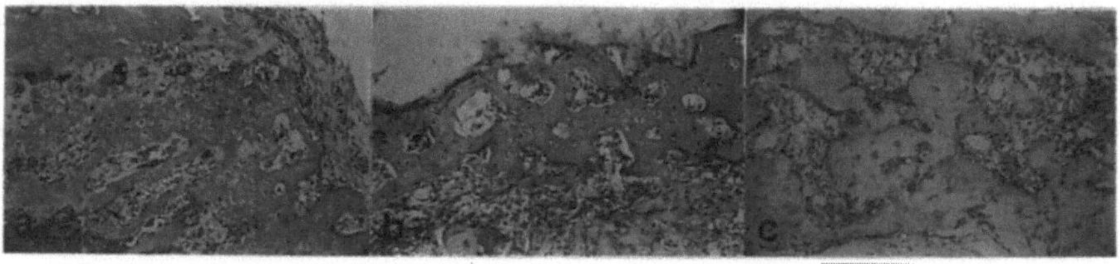

Fig. 2 Photomicrographs of 3 weeks after implantation of allografts demineralized in 0.6N HCl at 60°C for 10 mins(a), 20 mins(b), and 30 mins(c) using a controlled-heat ultrasonic cleaner. The ossicle formation containing marrow is more intensive in b) & c) than a). (H&E, x 200)

Fig. 3 Photomicrographs of 3 weeks after implantation of heat-treated allografts in distilled water at 50°C for 10 mins(a), 20 mins(b), and 30 mins(c) using a controlled-heat ultrasonic cleaner. They show only focal osteoclastic resorption without osteoinductive evidence. (H&E, x 200)

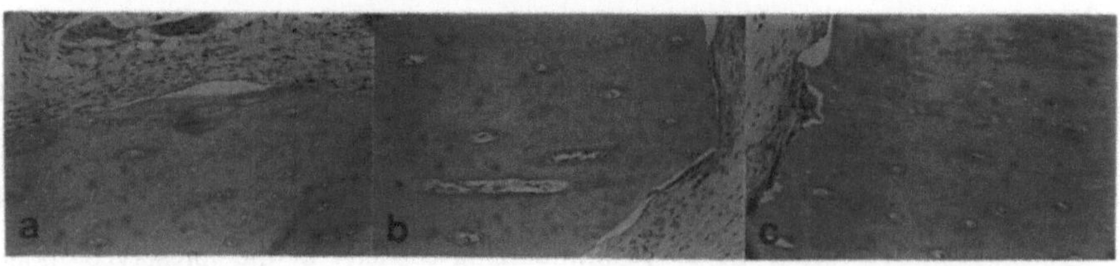

Fig. 4 Photomicrographs of 3 weeks after implantation of heat-treated allografts in distilled water at 60°C for 10 mins(a), 20 mins(b), and 30 mins(c) using a controlled-heat ultrasonic cleaner. They reveal no remarkable findings except for osteoclastic activity. (H&E, x 200)

References

1. Ellis, E.(1988) Surgical reconstruction of defects of the jaws, in Peterson, L.J. et al (eds), *Contemporary Oral & Maxillofacial Surgery*. St. Louis, C.V. Mosby Co., pp. 651-662.
2. Ellis, E. and Dinn, D. : Homologous bone in maxillofacial surgery, *J. Oral Maxillo -fac Surg.* **51** (1993), 1181-1193.
3. Urist, M.R.: Bone : Formation by autoinduction, *Science* **150** (1965), 893-899.
4. Kubler, N., Reuther, J., Kirchner, J., Priessnitz, B., and Sebald, W.: Osteoinductive, morphologic and biomechanical properties of autolyzed antigen-extracted allogeneic human bone, *J. Oral Maxillofac Surg.* **51** (1993), 1340-1357.
5. Urist, M.R. , Mikulski, A., and Boyd, s.D.: A chemosterilized antigen-extracted auto -digested alloimplant for bone banks, *Arch Surg.* **110** (1975), 416-428.
6. Reddi, A.H. and Huggins,C.: Biochemical sequences in the transformation of normal fibroblasts in a adolescent rats, *Proc Natl Acad Sci USA* **69** (1972), 1601-1605.
7. Oberg, S. and Rosenquist, J. B.: Bone healing after implantation of hydroxyapatite granules and blocks (interpore 200) combined with autolyzed antigen - extracted allogeneic bone and fibrin glue, *Int. J. Oral Maxillofac Surg.* **23** (1994), 110-114.
8. Dupoivieux, L, Costes, V., Jammet, P., and Souyris, F.: Experimental study on demineralized bone matrix(DBM) and coral as bone graft substitutes in maxillofacial surgery, *Int. J. Oral Maxillofac Surg.* **23** (1994), 395-398.
9. Miyamoto, S., Takaoka, K., Okada, T., et al.: Polylactic acid -polyethylene glucol block copolymer : A new biodegradable synthetic carrier for bone morphogenetic protein, *Clin. Orthop.* **294** (1993), 333-343.
10. Erksson, R. A. and Alberktsson, T .: The effect of heat on bone regeneration : An experimental study in the rabbit using the bone growth chamber, *J. Oral Maxillofac Surg.* **42** (1984), 705-711.
11. Hyatt, G. W.: The bone homograft, *Am Acad Orthop Surg Instructional Course Lect.* **17** (1968), 133-148.
12. Chalmers, J.: Transplantation immunity in bone homografting, *J. Bone Joint Surg.* **41B** (1959), 160.
13. Mellonig, J. T., Prewett, A. B., and Moyer, M. P.: HIV inactivation in a bone allograft, *J. Periodontol.* **63** (1992), 979-983.

Reorganization of vascular wall cells *in vitro*

--- effect of VEGF gradient ---

Katsuko FURUKAWA[1)], Takashi USHIDA[1, 2)] and Tetsuya TATEISHI[1, 2)]

Graduate School of Medicine, University of Tsukuba[1)],

National Institute for Advanced Interdisciplinary Research[2)]

Tennoudai 1-1-1[1)], and Higashi 1-1-4[2)], Tsukuba, Ibaraki, 305, Japan

[Abstract]

To reconstitute the hierarchic structure similar to native vessels, it is an essential step to move endothelial cells (EC) selectively toward the material surface and form a monolayer of EC. For this purpose, we studied the effect of vascular endothelial growth factor (VEGF) as a chemoattractant on selective chemotaxis of EC in a co-culture system. VEGF stimulated EC migration approximately 3 times as much as that without VEGF, while it did not stimulate smooth muscle cell(SM) migration. Therefore, VEGF was suggested as a useful tool for reconstituting the hierarchic structure. However, in the co-culture system EC migration was dramatically decreased and totally suppressed despite the presence of VEGF. The results suggest that it is necessary to inhibit this suppression effect for reconstituting blood vessels *in vitro*. [Keywords: endothelial cells, smooth muscle cells, VEGF, artificial vessels and tissue engineering]

[Introduction]

The native vessel has a characteristic hierarchic structure. EC exist on the luminal surface and play significant roles in antithrombogenicity. Although many researchers have investigated artificial materials with EC to implove their antithrombogenicity (Sbarbati, 1991), these attempts have not been successful thus far, because the seeded cells were washed off the surface by the blood stream after implantation (Prendiville, 1991). Since it was reported that seeding of smooth muscle cells enhanced endotheliarization of vascular grafts (Yue, 1988), artificial materials with a base of feeder cells including smooth muscle cells and fibroblasts have attracted great interest.

A model of a blood vessel constituted from EC, SM and fibroblasts (FB) *in vitro* was first reported by Weinberg and Bell in 1986 (Weinberg, 1986). However, this model still has many troublesome problems in respect to complicated and time consuming procedures, involving parallel isolation and culture of the 3 types of cell, and manual sequential deposition of each type of cell to reconstitute the hierarchic structure (Furukawa, 1996).

In order to solve these problems, we investigate whether the 3 types of cell constituting vessels can recognize themselves spontaneously *in vitro* into the structure similar to native vessels. Because VEGF is a specific endothelial cell mitogen and angiogenic factor (Koch, 1994, Sakamoto, 1995), we examined whether VEGF can induce the hierarchic structure of native vessels efficiently *in vitro* in the co-culture system of EC and SM.

[Materials and Methods]

1. Cell culture

Bovine thoracic aorta endothelial cells (EC) were incubated with alpha modified Eagle medium (MEM) containing 10% fetal bovine serum (FBS) in a humidified atmosphere of 5% CO_2 in air at 37 ℃. When they became confluent, EC were subcultured with 0.05% trypsin-0.01% EDTA solution. EC were identified by positive fluorescence by uptaking of

1,1'-dioctadecyl-3,3,3'3'-tetramethylindocarbocyanine perchlorate labeled acetylated low density lipoprotein (Dil-Ac-LDL). Rabbit aortic smooth muscle cells (SM), the kind gift of Dr. Sasaki et al. (Life Science Center, Biochemical Research Lab., Asahi Chemical Industry, Nobeoka, Miyazaki, Japan) were incubated with MEM containing 10% FBS in a humidified atmosphere of 5% CO_2 in air at 37 ℃ as described in (Sasaki, 1989). Before they became confluent, SM were subcultured with 0.05% trypsin-0.01%EDTA solution to proliferate the cells (synthetic phenotype). In our experiments, only SM of contractile phenotype prepared using the following methods described by Sasaki et al. were used. SM of proliferative phenotype were incubated more than 5 days after they became confluent and then tripsinized and seeded on a 75cm^2-flask at a density of 3.3x10^6 cells per flask in the MEM supplemented with 0.5% FBS. After 2 days, the SM represented a typical contractile phenotype and were used for the to following chemotactic assays.

2. Chemotactic assay

Freshly trypsinized EC and/or SM of contractile phenotype (2.0x10^5 cells/0.33 cm^2) were seeded to the upper compartment of Transwell (3 μm pore size) coated with collagen and incubated for 2 hours at 37 ℃ to attach to the surface of the filter. VEGF (16 ng/mL) were added to the lower compartment and incubated for 4 hours. After removal of the cells from the upper surface, the filters were fixed and stained with Hoechst 33342 to count the cell number. The number of cells on the lower surface of the filter was determined by counting the number of nuclei per 20 high-power field (x400).

All experimental data were represented as mean±standard error. Student's t-test was used to determine significance ($p<0.05$) of differences.

[Results and Discussion]

If we can make EC migrate to the surface faster than SM and FB, and form a monolayer of EC, reconstitution of a hierarchic structure similar to native vessel would be facilitated. Therefore, we examined whether EC can reach the surface through filters faster than SM. Because VEGF is a chemoattractant for EC in the iso-culture system of EC, we examined whether VEGF can attract only EC to the surface in the co-culture system of EC and SM.

The number of EC migrating to the lower surface of the filter in response to VEGF (28.5±3.82) is approximately 3 times as many as that without VEGF (10.0±1.66 cells/field) in the iso-culture system of EC. VEGF stimulated significantly EC migration through the filter (Figure 1, $p<0.01$, $n=5$). The number of SM migrating in response to medium supplemented with VEGF (1.72±0.536 cells/field) is the same as that without VEGF (1.43±0.308 cells/field) in the iso-culture system of SM. VEGF did not stimulate SM migration through the filter (Figure 2, $p>0.05$, $n=5$).

In the co-culture system, the number of both cells migrating decreased (Figure 3B, 1.09±0.159 cells/field, $p>0.05$, $n=5$) as compared with the iso-culture system (Figure 3A). The effect of VEGF on the number of EC migrating was totally suppressed in the co-culture system (Figure 4B, 1.46±0.577 cells/field), although in the iso-culture system of only EC the number of cells migrating significantly was increased by VEGF (Figure 4A). It was suggested that inhibition of the suppression effect was necessary to reconstitute the hierarchic structure similar to native blood vessels *in vitro*.

[References]

Furukawa K, Ushida T and Tateishi T (1996) *In vitro* reconstitution of blood vessel wall with cultured endothelial cells, smooth muscle cells and fibroblasts. Jpn. J. Artif. Organs **25**: 714-719.

Koch AE, Harlow LA, Haines GK, Amento EP, Unemori EN, Wong WL, Pope RM and Ferrara N (1994) Vascular endothelial growth factor. J. Immunol.: **152**, 4149-4156.

Prendiville EJ, Coleman JE, Callow AD, Gould KE, Laliberte-Verdon S, Ramberg K and Connolly J (1991) Increased *in vitro* incubation time of endothelial cells on fibronectin-treated ePTFE increases cell retention in blood flow. Eur. J. Surg. **5**: 311-319.

Sakamoto T, Ishibashi T, Kimura H, Yoshikawa H, Spee C, Harris MS, Hinton DR and Ryan SJ (1995) Effect of tecogalan sodium on angiogenesis in vitro by choroidal endothelial cells. Invest. Ophthalmol. Vis. Sci. **36:** 1076-1083.

Sasaki Y, Uchida T and Sasaki Y (1989) A variant derived from rabbit aortic smooth muscle: Phenotype modulation and restoration of smooth muscle characteristics in cells in culture. J Biochem. **106:** 1009-1018.

Sbarbati R, Ginnessi D, Cenni MC, Lazzerini G, Verni F.and De Caterina R (1991) Pyrolytic carbon coating enhance teflon and dacron fabric compatibility with endothelial cell growth. Biomaterials **14**: 491-498.

Yue X, van der Lei B, Schakenraad JM, van Oene GH, Kuit JH, Feijen J and Wildevuur RH (1988) Smooth muscle cells seeding in biodegradable grafts in rats. Surgery **103**: 206-212.

Weinberg CB and Bell E (1986) A blood vessel model constructed from collagen and cultured vascular cells. Science **231:** 397-400.

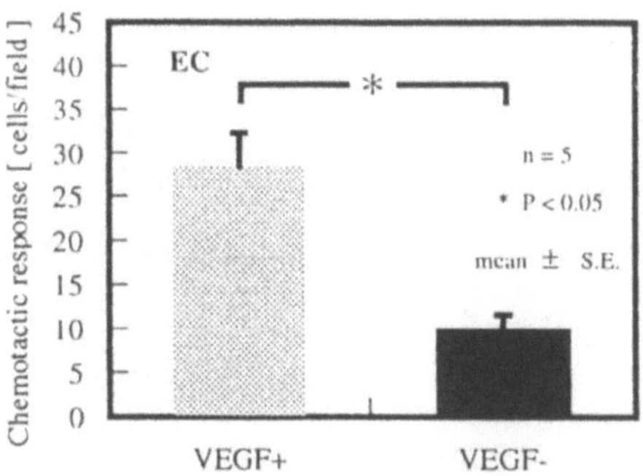

Figure 1 Effect of VEGF on EC migration.

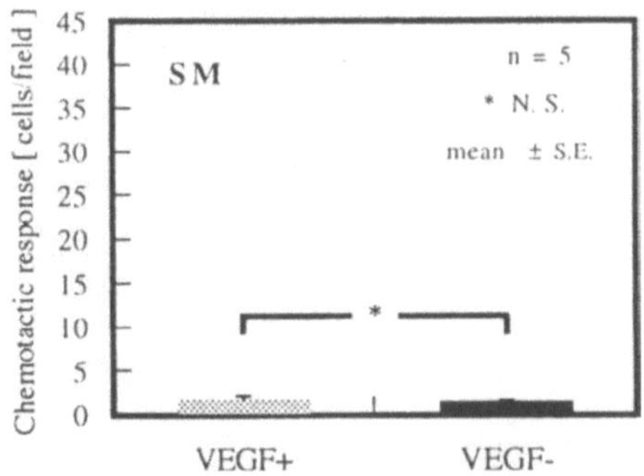

Fig. 2 Effect of VEGF on SM migration.

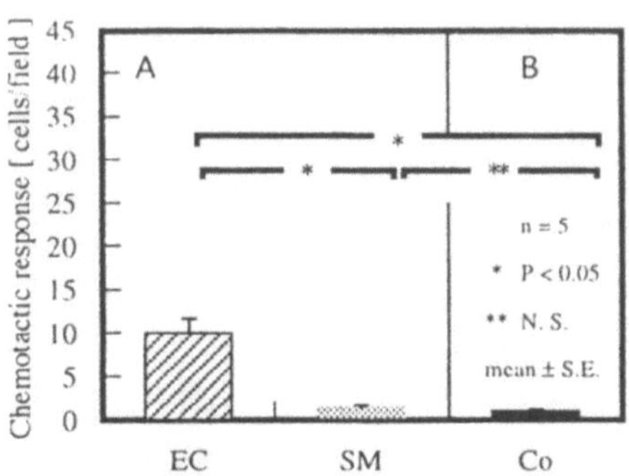

Fig. 3 Effect of co-cultivation on cell migration in the medium without VEGF. A: separated culture of EC or SM. B: co-culture of EC and SM.

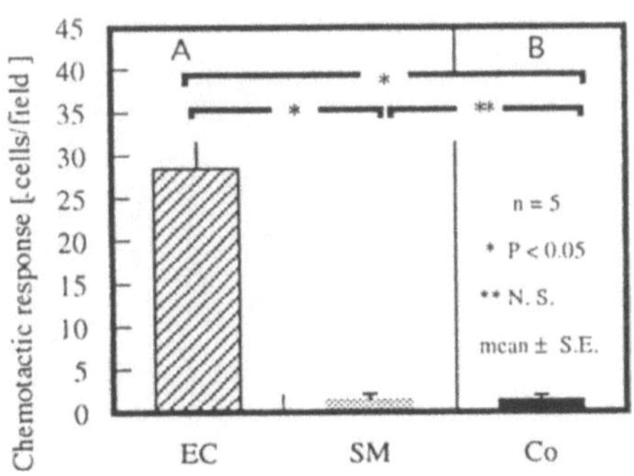

Fig. 4 Effect of co-cultivation on cell migration in the medium supplemented with VEGF. A: separated culture of EC or SM. B: co-culture of EC and SM.

EFFECTS OF CELL MORPHOLOGY ON EXPRESSION OF LIVER SPECIFIC FUNCTION AND HORMONE RESPONSE IN PRIMARY HEPATOCYTE CULTURE

RYUJI HAMAMOTO, KEISUKE YAMADA,
MASAMICHI KAMIHIRA AND SHINJI IIJIMA
Department of Biotechnology, Graduate School of Engineering, Nagoya University,
Chikusa-ku, Nagoya, 464-01, Japan

ABSTRACT. We compared the spheroid formation (multi-cellular aggregation) of hepatocytes and the expression of liver specific functions such as albumin secretion when hepatocytes were cultured with various extracellular materials. We found that poly-D-lysine and Eudragit enhanced spheroid formation and hepatocytes in spheroids exhibited higher liver specific functions comparing to monolayer culture. The results indicated that cell-cell interaction caused by spheroid formation was a key factor to promote the expression of liver specific function. On the other hand, HGF response ability was declined when hepatocytes formed spheroids. We then examined that the relations between the expression of transcription factors and liver specific function during the spheroid formation. The expression of a CCAAT enhancer binding protein (C/EBP-α) increased and the high expression level was maintained when hepatocytes formed spheroids. The effects of cAMP and calcium ionophore on albumin secretion were also examined, and it was found that the increase of calcium ion concentration in the cells was important for the expression of the liver function.

1. Introduction

The cell-cell and cell-matrix interactions have been recognized as an important factor to regulate growth and differentiation of cells [1]. In primary hepatocyte culture, it has been observed that enhanced liver-specific functions could be maintained for long term when the cells formed multi-cellular aggregates called spheroid [2,3]. However, the molecular mechanisms of this phenomenon and the correlation with signal transduction during the spheroid formation have not been clarified yet.

In this study, to characterize this phenomenon, we first observed the expression of liver specific function and the change in HGF response ability when hepatocytes were cultured with various extracellular substrata. Thereafter, we studied on two transcription factors, a CCAAT enhancer-binding protein, C/EBP-α, which has been proposed to have a role in maintaining differentiation of cells [4], and LAP (C/EBP-β), which is a major albumin promoter binding protein that confers liver-specific gene expression [5]. The expression of these transcription factors and liver specific function during the spheroid formation was measured.

2. Materials and Methods

Materials

Collagen type IV (Becton Dickinson), fibronectin (Becton Dickinson), laminin (Upstate Biotechnology), Eudragit S100 (Röhm Pharma) and poly-D-lysine (SIGMA) were used as

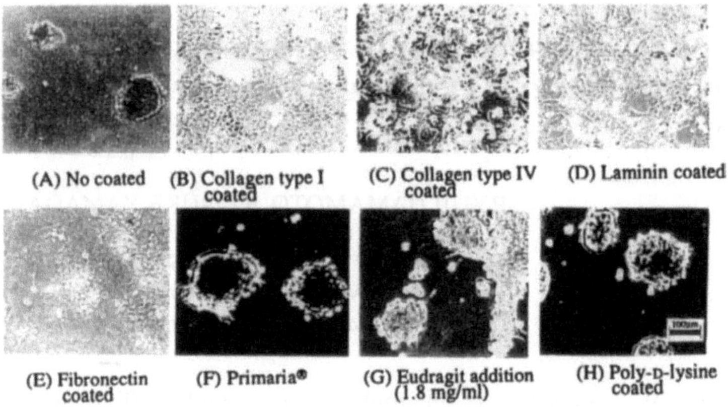

Figure 1. Phase-contrast microscopic observation of hepatocytes at 4 days.

an extracellular matrix.

Tissue culture plastic dishes (IWAKI, 3000-035) were used for coating the extracellular matrices (collagen type IV, laminin, fibronectin and poly-D-lysine). The dishes were also used for the cultures in the presence of Eudragit and no coated control. Collagen type I coated dishes were obtained from IWAKI (4000-010), and Primaria® dishes were purchased from Becton Deckinson (3801).

Isolation and culture of hepatocytes.
Rat hepatocytes were obtained from male Sprague-Dawley rats (6–7 weeks old) using collagenase perfusion method [6]. Williams' medium E, supplemented with 0.1 μM $CuSO_4 \cdot 5H_2O$, 25 nM Na_2SeO_3, 1.0 μM dexamethasone, 20 ng/ml EGF (SIGMA), 20 μg/ml egg yolk lipoprotein, 48 μg/ml gentamicin sulfate and 100 μg/ml chloramphenicol, was used as culture medium. In some experiments, 10 ng/ml hepatocyte growth factor (HGF; R&D systems) was added to the medium. The cells were seeded at a density of approximately 5.0×10^5 cells per a 35 mm-diameter plastic dish and cultured in a CO_2 incubator.

Measurements of DNA synthesis and albumin secretion.
DNA synthesis was evaluated from the incorporation of ^{3}H-thymidine by a standard procedure [7]. Albumin concentration in the medium was measured by enzyme-linked immunosorbent assay (ELISA) using peroxidase conjugated anti-rat albumin antibody (Cappel; 55776) [8].

Detection of transcription factors.
Nuclear proteins were isolated from hepatocytes according to the procedure of Greenberg and Ziff with slight modification [9]. Proteins were fractionated by SDS-PAGE on a 12% gel and electrophoretically transferred to PVDF membranes (Hybond™-PVDF; Amersham). After blocking the membrane with 5% nonfat dry milk, transcription factors were detected using anti-C/EBP-α antibody (Santa Cruz Biotechnology; #sc-61) and anti-LAP (C/EBP-β) antibody (Santa Cruz Biotechnology; #sc-150). The blots were developed by using ECL detection system (Amersham) and exposed to X-ray film for 1 to 30 min for visualization.

Detection of c-MET.
The detection of c-MET was performed by a procedure of Naldini *et al.* [10] Hepatocytes were lysed with 1 ml of lysis buffer (10 mM Tris-HCl, pH 7.8, 1% NP-40, 0.15 M NaCl, 1

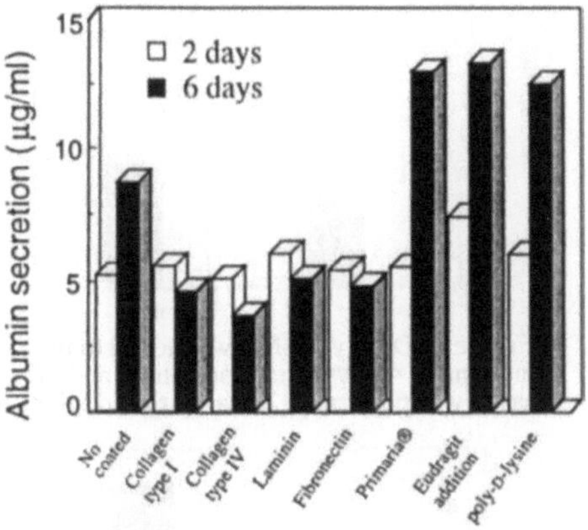

Figure 2. Albumin secretion of hepatocytes cultured on various dishes at 2 and 6 days.

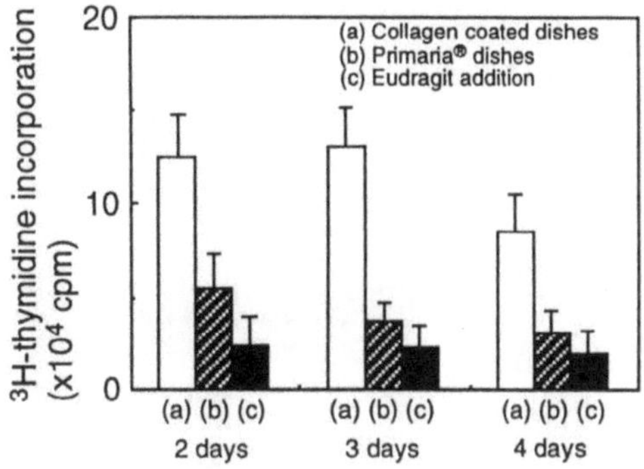

Figure 3. ^{3}H-thymidine incorporation of hepatocytes after HGF addition.

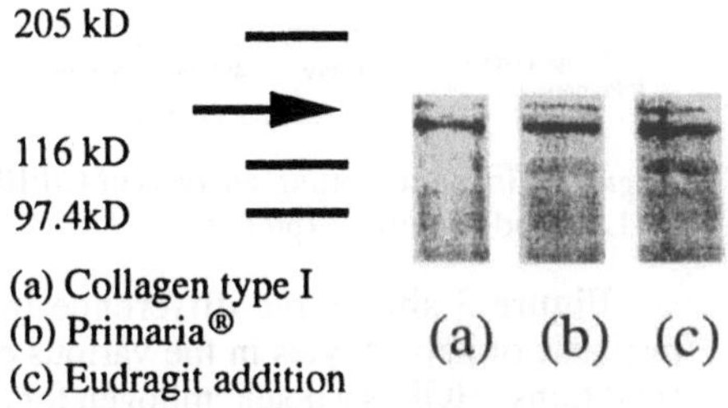

Figure 4. Immunobloting analysis of c-MET (HGF receptor) on hepatocytes cultured at 3 days.

mM EDTA, 10 µg/ml aprotinin, 1 mM PMSF, 10 µg/ml pepstatin A, 50 µg/ml leupeptin) and the solution was immunoprecipitated with anti-MET antibody (Santa Cruz Biotechnology; #sc-162). The samples were subjected to immunoblotting mentioned above using the anti-MET antibody.

3. Results and Discussion

Figure 1 shows the photographs of adult hepatocytes at 4 days. When hepatocytes were cultured on collagen type I, type IV, laminin and fibronectin coated dishes, the cells formed monolayer (Fig. 1B, C, D and E). On the other hand, the cells formed multi-cellular aggregation, spheroids, on Primaria® dish, Eudragit addition and poly-D-lysine coated dish. In the case that hepatocytes were cultured on no coated dish, they formed cell aggregation but the size was smaller comparing with the other spheroid cultures.

Figure 2 presents the albumin secretion of hepatocytes cultured with the various conditions. When hepatocytes formed spheroids (Primaria® dish, Eudragit addition and poly-D-lysine coated dish), enhanced albumin secretion was observed on day 6 compared with that on day 2. On the other hand, when hepatocytes were cultured as monolayer (collagen type I, type IV, laminin and fibronectin coated dishes), albumin secretion level decreased gradually. Even small cell aggregation formed on no coated dish exhibited higher secretion level than that with the monolayer cultures at 6 days. It is well known that hepatocytes express highly differentiated function on Englebreth-Horm-Swarm mouse sarcoma tumor gel (EHS gel) which contains laminin and collagen type IV [2]. The cells formed spheroids on EHS gel. In this study, when hepatocytes were cultured on laminin and collagen type IV coated dishes, the cells were cultured as monolayer and high albumin secretion have not been observed. These results may indicate that the induction of the high liver specific function observed on EHS gel is not due to the signals caused by attachment to the extracellular matrix, but is attributable to the cell-cell interaction by aggregate formation.

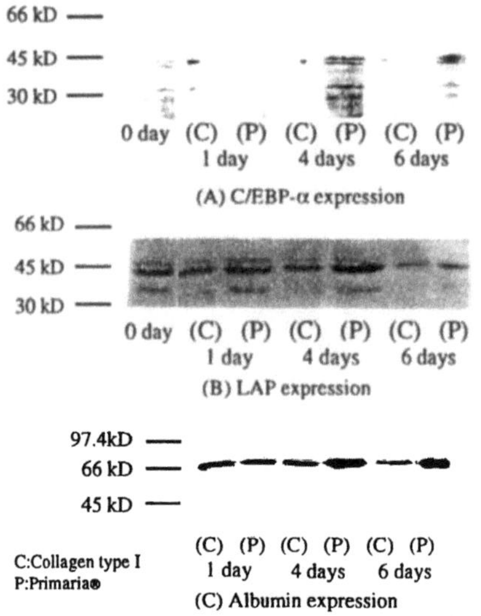

Figure 5. Immunoblotting analysis of C/EBP-α, LAP and Albumin expression.

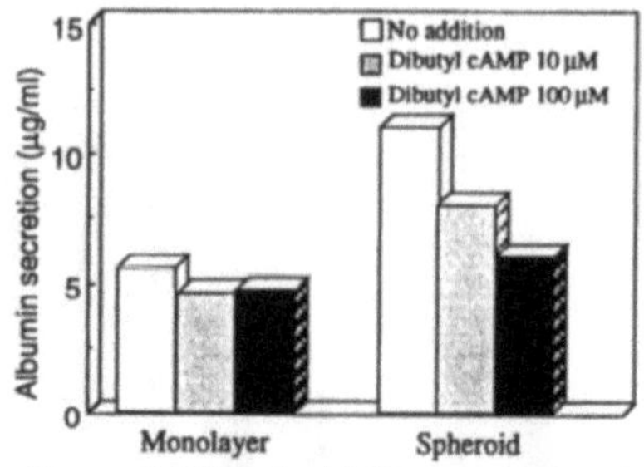

Figure 6. Dibutyl cAMP was added to the medium at 4 days after inoculation, and the assay was performed 24h after the addition.

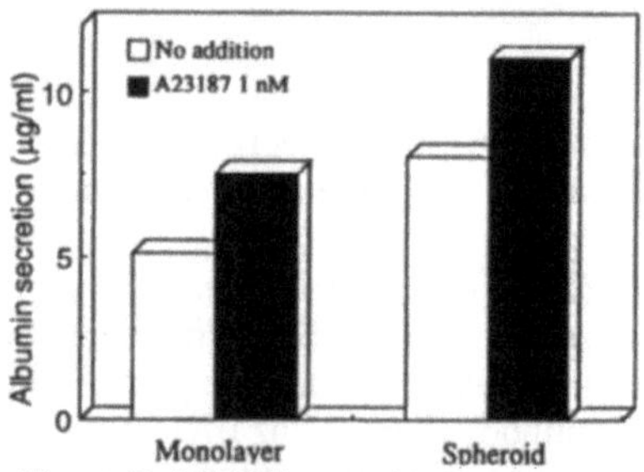

Figure 7. A23187 was added to the medium at 4 days after inoculation, and the assay was performed 24h after the addition.

Figure 3 shows the difference in HGF response of hepatocytes in the various culture conditions. HGF is a potent mitogen for mature hepatocyte and may act as an important regulator of cell growth in response to injury, such as partial hepatectomy and hepatitis [11,12]. In this study, hepatocytes were cultured on collagen type I coated and Primaria® dishes, and in the presence of Eudragit, and the difference in HGF response was observed from a view point of ^{3}H-thymidine incorporation of hepatocytes. When hepatocytes formed spheroids on Primaria® dish and in the presence of Eudragit, the uptake of ^{3}H-thymidine after HGF addition was repressed drastically compared with that of the monolayer culture. In order to clarify the molecular mechanism of this phenomenon, we examined the expression and phosphorylation of c-MET (HGF receptor). Figure 4 presents the expression of c-MET detected by immunoblotting using anti-MET antibody. No remarkable difference in the c-MET expression was observed depending on spheroid formation. Tyrosine phosphorylation and down regulation of c-MET after HGF addition were observed for both monolayer and spheroid cells (data not shown). These results show that the growth stimulation signals induced by HGF addition was prevented at some point in the signal pathway from the receptor to nucleus when hepatocytes formed spheroids.

In order to obtain some informations about the mechanisms for enhancement of the liver specific functions in the spheroid cultures, the expression of transcription factors, C/EBP-α and LAP, was examined (Figure 5). C/EBP-α is a CCAAT enhancer binding protein related to the expression of liver specific functions and differentiation. When hepatocytes formed spheroids, higher expression of C/EBP-α was observed compared to the monolayer culture (Fig. 5A). On the other hand, no difference in the expression of LAP was observed.

For the purpose of identifying second messenger concerning with signal transduction induced by spheroid formation, we also studied the effects of dibutyl cAMP and calcium ionophore (A23187) on albumin secretion (Fig. 6, 7). Both cAMP and calcium ion play an important role for controlling of growth and differentiation in various cell phenotype [13, 14]. When dibutyl cAMP was added to the medium in the spheroid cultures, the albumin secretion decreased as the concentration increased, but no such decrease was observed in the monolayer culture. However, both in the monolayer and spheroid cultures, the albumin secretion increased when the calcium ionophore was added to the medium. In particular, albumin secretion increased by the ionophore even if hepatocytes were cultured as monolayer. This suggests that calcium ion may have a potent function to express the liver specific function.

As mentioned above, spheroid formation is the most important factor to induce high liver specific functions of hepatocytes. Although the molecular mechanism of the enhancement of liver specific functions in spheroid cultures is not still clear, it is one of the possible mechanism that spheroid formation may lead to increase of calcium ion concentration in the cells, that activates transcription factors for liver specific functions. Analysis of the molecular mechanism of the expression of liver specific functions induced by spheroid formation is now under study.

References

1. Ben-Ze'ev A., Robinson, G. S., Bucher, N. L. R. and Farmer S. R. (1988) Cell-cell and cell matrix interactions differentially regulate the expression of hepatic and cytoskeletal genes in primary cultures of rat hepatocytes, *Proc. Natl. Acad. Sci. USA* **85**, 2161-2165
2. Landry, J., Bernier, D., Ouellet, C., Goyette, R. and Marceau, N. (1985) Spherodical aggregate culture of rat liver cells: Histotypic reorganization, biomatrix deposition, and maintenance of functional activities, *J. Cell Biol.* **101**, 914-923
3. Koide, N., Shinji, T., tanabe, T., Asano, K., Kawaguchi, M., Sakaguchi, K., Koide, Y., Mori, M. and Tsuji, T. (1989) Continued high albumin production by multicellular spheroids of adult rat hepatocytes formed in the presence of liver-derived proteoglycans, *Biochem. Biophys. Res. Comm.* **161**, 385-391
4. Umelk, R. M., Friedman, A. D. and McKnight, S. L. (1991) CCAAT/enhancer binding protein: a component of a differentiation switch, *Science* **251**, 288-292
5. Buck, M., Turler, H. and Chojkier. (1994) LAP (NF-IL-6), a tissue-specific transcriptional activator, is inhibitor of hepatoma cell proliferation, *EMBO J.* **13**, 851-860
6. Seglen, P. O. (1976) Preparation of isolated rat liver cells, *Methods Cell Biol.* **13**, 29-83
7. Bhora, F. M., Kothary, P. C., Imanishi, H., Eckhauser, F. E. and Raper, S. E. (1994) Vasopression stimulation DNA synthesis in cultured rat hepatocytes, *J. surgi. Res.* **57**, 706-710
8. Lee, J., Morgan, J. R. and Tompkins, R. G. (1992) The importance of proline on long-term hepatocyte in a collagen gel sandwich configuration: Regulation of protein secretion, *Biotechnol. Bioeng.* **40**, 298-305
9. Greenberg, M. E. and Ziff, E. B. (1984) Stimulation of 3T3 cells induces transcription factor of the c-Fos proto-oncogene, *Nature (London)* **311**, 433-437
10. Naldini, L. E., Vigna, R., Narsimhan, G., Gaudino, G, Zarnegar, R., Michalopoulos, G. K. and Comoglio, P. M. (1991) Hepatocyte growth factor (HGF) stimulates the tyrosine kinase activity of the receptor encoded by the proto-oncogene c-MET, *Oncogene.* **125**, 313-320
11. Tashiro, K., Hagiya, M., Nishizawa, T., Seki, T., Shimonishi, M., Shimizu, S. and Nakamura, T. (1990) Deduced primary structure of rat hepatocyte growth factor and expression of the mRNA in rat tissues, *Proc. Natl. Acad. Sci. USA* **87**, 3200-3204
12. Hamamoto, R., Kamihira. M. and Iijima, S. (1995) Effects of HGF and TGF-β1 addition in rat primary hepatocyte culture, *Symposium Series of Chem. Eng.*, **44**, 138-142
13. Mine, T., Kojima, I., Ogata, E. and Nakamura, T. (1991) Comparison of effects of HGF and EGF on cellular calcium in rat hepatocytes, *Biochem. Biophys. Res. Comm.* **181**, 1173-1180
14. Taylor, S. S. (1989) cAMP-dependent protein kinase: model for an enzyme family, *J. Biol. Chem.* **264**, 8443-8446

HIV VECTOR MEDIATED GENE TRANSFER INTO CD4 POSITIVE CELLS

T. SHIMADA AND K. MIYAKE
Department of Biochemistry and Molecular Biology, Nippon Medical School, Tokyo 113, Japan

1. Introduction

Retroviral vectors derived from Moloney murine leukemic virus (MoMLV) with the amphotropic envelope are the most commonly used vector in clinical trials of human gene therapy (Miller, 1993). The advantages of this vector system include broad host range, stable integration into the chromosomes, and availability of packaging cell lines. The host range of retrovirus is determined by the interaction between the virus envelope and the cell surface receptor. The receptor for amphotropic retrovirus has been recently identified as a sodium-dependent phosphate transporter which is widely distributed on various tissues and species (Miller et al. 1994; Van Zeijl et al. 1994). However, the level of expression of the protein appears to vary and might limit viral entry into particular target cells such as T cells and hematopoietic stem cells. Another disadvantage of retrovirus mediated gene transfer is inability to infect growth arrested cells. It has been reported that breakdown of the nuclear envelope at mitosis is necessary for nuclear transport of the oncoretroviral preintegration complex (Humphries and Temin, 1974; Miller et al. 1990). Because most of the target cells are quiescent, this requirement for cell division has limited the applicability of retroviral vectors for therapeutic gene transfer.

Human immunodeficiency virus (HIV) is an etiological agent of acquired immunodeficiency syndrome (AIDS). HIV is a human retrovirus and belongs to the lentivirus subfamily. CD4 molecules on human helper T lymphocytes have been identified as a major receptor for HIV infection (Maddon et al. 1986). Therefore, HIV infect specifically CD4 positive cells and destroy the immune system of the patients. An important feature of HIV is that HIV and probably other lentiviruses are able to infect non-dividing cells (Brice Weinberg et al. 1991). This property seems to be due to the existence of a nuclear localization signal (NLS) in the viral matrix protein which enable the viral complex actively transport through the nucleopore (Gallay et al. 1995). It has been recently reported that HIV based vectors pseudotyped with the VSV envelope or the MoMLV envelope can transduce non-proliferating cells (Naldini et al. 1996).

These properties of HIV suggest that HIV based vectors may be useful for gene therapy applications. We have developed a packaging system for production of safe HIV vectors (Shimada et al. 1991). In the present paper, we show that HIV vectors are capable of targeted gene transfer into CD4+cells and gene transfer into non-dividing cells. In addition, we describe a new strategy to use the HIV vector for gene transfer into non-lymphoid cells.

1. Targeted gene transfer into CD4 positive cells

We designed packaging and vector plasmids for producing HIV vectors (Shimada et al. 1991). These plasmids were co-introduced into monkey kidney Cos cells and the culture medium containing recombinant HIV vectors were collected after 48 hours. We first tested for replication competent HIV using the HTLV-1 transformed human T cell line, MT-2, that is highly sensitive to cytopathic effects of HIV. However, syncytium formation was not observed in MT-2 cells cultured with the conditioned medium of the transfected Cos cells, indicating that no detectable infectious viruses were contaminated in this packaging system.

We studied tissue specificity of HIV vectors. CD4+HeLa cells and CD4-parental HeLa cells were incubated with the HXN vector and selected in the presence of G418 for two weeks. Resistant colonies appeared only in CD4 positive cells. No colonies were detected in CD4-cells. When 10 ug/ml soluble CD4 molecules were co-incubated with virus vectors, colony formation was completely inhibited. Soluble CD4 at this concentration did not inhibit cell growth. These results clearly showed that gene transfer by HIV vectors is mediated by CD4 on the target cell surfaces. Therefore, HIV vectors can be used for targeted gene transfer into CD4 positive T lymphocytes.

T-lymphocytes are an important target for gene therapy of genetic diseases (Blaese et al. 1995), cancer (Rosenberg et al. 1990), and AIDS (Baltimore, 1988). However, human lymphocytes are relatively resistant to retroviral mediated gene transfer. HIV vectors may be useful for these therapeutic applications.

2. Gene transfer into non-dividing cells

CD4+HeLa cells (CD4H) were arrested at the G2 phase by gamma irradiation and at the G1/S phase by aphidicolin treatment. The arrested state of the cells was verified by flow cytometry analysis. These non-dividing cells were incubated with the HIV-based vector containing the β-gal gene and stained with X-gal after two days (Fig. 1). Approximately the same numbers of blue cells were detected in G2- and G1/S- arrested CD4H cells. The transduction efficiency of HIV vectors in these non-dividing cells was comparable to that in dividing cells. In a control experiment, the non-dividing cells were incubated with the MoMLV based retrovirus vector carrying the β-gal gene. The number of blue cells in the non-dividing cells was less than 1 % of that in the dividing cells, indicating that cell proliferation is necessary for transduction with MoMLV vectors as reported previously (Lewis and Emerman, 1994).

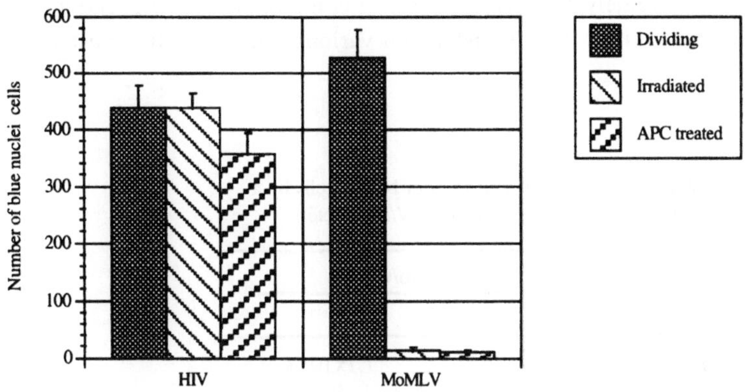

Figure 1. Transduction efficiencies of HIV and MoMLV retroviral vectors in non-dividing cells Proliferating, irradiated, and APC treated CD4H cells were incubated with HIV/β-gal or MoMLV/β–gal vectors.

We next examined whether HIV vectors are stably integrated into the chromosomal DNA or exist as unintegrated episomal DNAs in non-dividing cells after gene transfer. The junction sequences including the HIV LTR and the flanking sequences were amplified by the linker mediated PCR (LMPCR) technique (Wattel et al. 1995). The sequences of these amplified DNA demonstrated that the LTRs lost 2 bp from the ends and linked to heterogenous sequences. These structural features are exactly the same as those reported for wild-type HIV (Muesing et al. 1985). Therefore, wild-type HIV and recombinant HIV vectors appear to integrate into the host genome through a common mechanism. These results indicate that HIV vectors are able to integrate stably into the chromosome of non-dividing cells.

3. Gene transfer into non-lymphoid cells by HIV vectors

The stable gene transfer into non-dividing cells suggests that HIV vectors may be useful for gene therapy of various target diseases in addition to AIDS. To use the HIV vector for gene transfer into CD4-cells, we developed a method to expand the host range of HIV vectors (Miyake et al. 1996). In this strategy, target cells are first incubated with the adenovirus vector containing the CD4 gene (Adex1CACD4) followed by incubation with the HIV vector (HXN). Transient but high level expression of CD4 could be achieved in both adherent and floating cells. HIV vectors could be stably integrated into these transiently CD4 positive cells (Table 1). Because adenovirus vectors can transduce many types of cells at extremely high efficiencies, the combination of the adenovirus and HIV vectors could be used for gene transfer into various types of cells arrested in the cell cycle.

Table 1. Transduction efficiency of Non-T cells by a combination of Adeno/CD4 and HIV/neo vectors

Cells	Adeno/CD4	
	-	+
HeLa	0	1.1x10*4
3T3	0	0
K562	0	51/96
Raji	0	28/96

Cells were first transduced with Adeno/CD4 followed by transduction with HXN. For HeLa and 3T3 cells, after G418 selection for two weeks, the number of resistant colonies was counted. For K562 and Raji cells, the transduced cells were dispensed into 96-well culture plates in the G418 containing medium. Wells containing G418 resistant cells were counted two weeks after transduction.

4. Conclusion

We developed a packaging system for producing replication incompetent helper free HIV vectors which are capable of targeted and efficient gene transfer into CD4+cells. HIV vectors could stably integrate into the chromosome of non-dividing cells. Transient expression of CD4 by the adenovirus vector rendered non-lymphoid cells susceptible to HIV vector mediated gene transfer. The combination of the adenovirus vector with the CD4 gene and the HIV vector might be used for stable gene transfer into non-dividing non-T cells.

References

Baltimore, D.(1988). Intracellular immunization. Nature 335, 395-396.

Blaese, R. M., Culver, K. W., Miller, A. D., Carter, C. S., Fleisher, T., Clerici, M., Shearer, G., Chang, L., Chiang, Y., Tolstoshev, P., Greenblatt, J. J., Rosenberg, S. A., Klein, H., Berger, M., Mullen, C. A., Ramsey, W. J., Muul, L., Morgan, R. A., and Anderson, W. F.(1995). T lymphocyte-directed gene therapy for ADA-SCID: Initial trial results after 4 years. Science 270, 475-480.

Brice Weinberg, J., Matthews, T. J., Cullen, B. R., and Malim, M. H.(1991). Productive human immunodeficiency virus type 1 (HIV-1) infection of nonproliferating human monocytes. J. Exp. Med. 174, 1477-1482.

Gallay, P., Swingler, S., Aiken, C., and Trono, D.(1995). HIV-1 infection of nondividing cells: C-terminal tyrosine phosphorylation of the viral protein is a key regulator. Cell 80, 379-388.

Humphries, E. H. and Temin, H. M.(1974). Requirment for cell division for initiation of transcription of Rous sarcoma virus RNA. J. Virol. 14, 531-546.

Lewis, P. F. and Emerman, M.(1994). Passage through mitosis is required for oncoretroviruses but not for the human immunodeficiency virus. J. Virol. 68, 510-516.

Maddon, P. J., Dalgleish, A. G., McDougal, J. S., Clapham, P. R., Weiss, R. A., and Axel, R.(1986). The T4 gene encodes the AIDS virus receptor and is expressed in the immune system and the brain. Cell 47, 333-348.

Miller, A. D.(1993). Retrovirus packaging cells. Hum. Gene Ther. 1, 5-14.

Miller, D. G., Adam, M. A., and Miller, A. D.(1990). Gene transfer by retrovirus vectors occurs only in cells that are actively replicating at the time of infection. Mol. Cell. Biol. 10, 4239-4242.

Miller, D. G., Edwards, R. H., and Miller, A. D.(1994). Cloning of the cellular receptor for the amphotropic murine retrovirus reveals homology to that for gibbon ape leukemia virus. Proc. Natl. Acad. Sci. U. S. A. 91, 78-82.

Miyake, K., Tohyama, T., and Shimada, T.(1996). Two-step gene transfer using an adenoviral vector carrying the CD4 gene and human immunodeficiency viral vectors. Hum. Gene Ther. 7, 2281-2289.

Muesing, M. A., Smith, D. H., Cabradilla, C. D., Benton, C. V., Lasky, L. A., and Capon, D. J.(1985). Nucleic acid structure and expression of the human AIDS/lymphadenopathy retrovirus. Nature 313, 450-458.

Naldini, L., Blomer, U., Gallay, P., Ory, D., Mulligan, R., Gage, F. H., Verma, I. M., and Trono, D.(1996). In vivo gene delevery and stable transduction of nondividing cells by a lentivirus vector. Science 272, 263-267.

Rosenberg, S. A., Aebersold, P., Cornetta, K., Kasid, A., Morgan, R. A., Moen, R., Karson, E. M., Lotze, M. T., Yang, J. C., Topalian, S. L., Merino, M. J., Culver, K., Miller, D., Blaese, M., and Anderson, W. F.(1990). Gene transfer into humans-immunotherapy of patients with advanced melanoma, using tumor-Infiltrating infiltrating lymphocytes modified by retroviral gene transduction. N. Engl. J. Med. 323, 570-578.

Shimada, T., Fujii, H., Mitsuya, H., and Nienhuis, A. W.(1991). Targeted and highly efficienct gene transfer into CD4+ cells by a recombinant human immunodeficiency virus retroviral vector. J. Clin. Invest. 88, 1043-1047.

Van Zeijl, M., Johann, S. V., Closs, E., Cunningham, J., Eddy, R., Shows, T. B., and O'hara, B.(1994). A human amphotropic retrovirus receptor is a second member of the gibbon ape leukemia virus receptor family. Proc. Natl. Acad. Sci. U. S. A. 91, 1615-1172.

Wattel, E., Vartanian, J. P., Pannetier, C., and Wain-Hobson, S.(1995). Clonal expansion of human T-cell leukemia virus type 1-infected cells in asymptomatic and symptomatic carriers without malignancy. J. Virol. 69, 2863-2868.

Enhancement of cellular productivity for a recombinant protein by a combination of promoter activation and gene amplification

X. -Y. DONG[¶], Y. KATAKURA, Y. ZHANG, T. MIURA, Y. DAIMON, H. OHASHI[*], AND S. SHIRAHATA
Graduate School of Genetic Resources Technology, Kyushu University
6-10-1 Hakozaki, Higashi-ku, Fukuoka 812-81, Japan
[*]*Pharmaceutical Laboratory, Kirin Brewery Co. Ltd.*
[¶]*Department of Chemical Engineering, Tianjin University*
Tianjin 300072, P. R. CHINA

ABSTRACT We have developed an oncogene activated production (OAP) system using amplified *ras* oncogene to activate the CMV promoter controlling the foreign gene in mammalian cells (Yano, T. *et al.*, 1994). CHO cells were demonstrated to be suited for the OAP system (Perry, S. *et al.*, 1996). Here we show that very high level production of recombinant protein can be achieved by combining the OAP system with the glutamine synthetase (GS) gene amplification system. The human interleukin 6 (hIL-6) gene regulated by the human CMV promoter was inserted into a GS-mini-gene expression plasmid pEE14. A highly productive host CHO cell line, *ras*-clone I containing amplified *ras* oncogene was further transfected with the plasmid expressing both hIL-6 and GS mini-gene, and selected with methionine sulphoximine (MSX) in G-DME medium supplemented with 10% dialyzed fetal bovine serum. We could establish a hIL-6 hyper-producing cell line DRGI-29 derived from *ras*-clone I, which exhibited a peak productivity value of about 40 μg hIL-6/10^6 cells/day through a combination of the OAP system and GS gene amplification system. The cellular productivity of DRGI-29 cells was about 13 times higher than the control hIL-6-producing cells derived from normal CHO cells whose hIL-6 gene was amplified by GS gene amplification system, and about 5 times higher than I13 cells established by the OAP system, which contains amplified *ras* oncogene and non-amplified hIL-6 gene (Perry, S. *et al.*, 1996). When DRGI-29 cells were cultivated for 30 days, the accumulation rate of about 80 μg hIL-6/ml/3days on 9th day. Whereas, the maximum productivity was achieved after long-term culture in the case of I13 cells (Perry, S. *et al.*, 1996). These results indicate that this OAP and GS hybrid system enables the efficient and rapid establishment of recombinant protein hyper-producing cell lines.

1. Introduction

To construct a highly productive cell line for a recombinant protein, we developed the oncogene activated production (OAP) system. This system uses the combination of a effector plasmid, which consists of an oncogene, and a reporter plasmid that carries the expression gene of interest under the control of a strong promoter. The product of the effector plasmid should stimulate the promoter of the reporter plasmid to enhance produciton of the foreign protein. The best combination of the oncogene and promoter for efficient protein production was shown to be ras and CMV (Yano, T. *et al.*, 1994). Furthermore, when E1A oncogene was introduced into a hIL-6 producing cell line that was already ras amplified, productivity was boosted 10 fold (Shirahata, S. *et al.*, 1995). This OAP system was shown to be more effective in CHO cells, where the peak productivity and the peak 3-day accumulation reached 13.6 μg/10^6 cells/day and 88.6 μg/ml/3 day, respectively (Perry, S. *et al.*, 1996).
In this study, we attempted to enhance the productivity in recombinant CHO cells generated with the OAP system by gene amplification. One of the most efficient gene amplification system is the glutamine synthetase (GS) system developed by Cockett *et al.* (1990). Here we demonstrate that very high level production of recombinant protein can be achieved by combining the OAP system with the GS gene amplification system.

2. Materials and Methods

2.1. VECTOR CONSTRUCTION

Vector used for production of hIL-6 with the GS gene amplification system was shown in Fig. 1. Human IL-6 cDNA was inserted downstream from the CMV promoter of pEE14 (LONZA Biologic plc, Slough, UK) (Cockett, M. I. *et al.*, 1990) and named pEE14-IL-6.

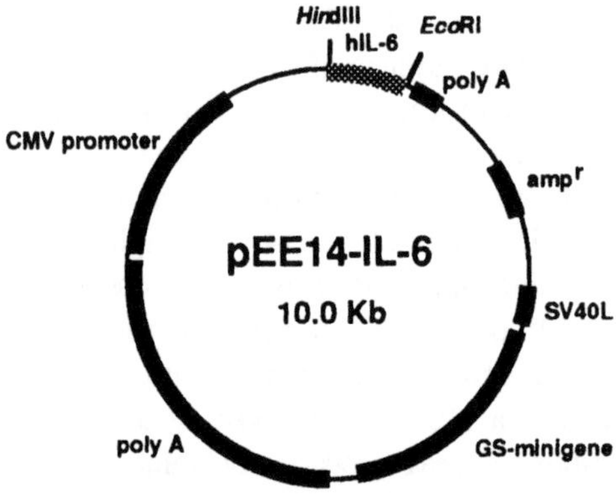

Fig. 1 Vector construction used for gene amplification

2.2. CULTURE MEDIUM

We prepared G-DME medium suited for the GS gene amplification system by adding 30 μM each of adenosine, guanosine, cytidine and uridine, 10 μM thymidine, 100 μM non-essential amino acids (Gibco BRL), 500 μM glutamic acid and 500 μM asparagine to Dulbecco's Modified Eagles Medium without *L*-glutamine (Bio WHITTAKER, Walkersville, MD, USA) supplemented with 10% fetal bovine serum

(IRVINE SCIENTIFIC, Santa Ana, CA, USA). Cells were cultured in 5% CO_2 at 37°C.

2.3. PRODUCTIVITY ENHANCEMENT OF RAS AMPLIFIED CHO CELL LINE BY GENE AMPLIFICATION

CHO cells and the *ras*-primed CHO cells (ras-clone I) which was established in our laboratory were used as the starting cell lines (Perry, S. et al., 1996). CHO cells and ras-clone I were seeded at the density of 6 x 10^4 cells/ml and transfected with the expression plasmids of pEE14-IL-6 and the selection marker of pSV2-bsr by the highly efficient calcium phosphate coprecipitation method (Teruya, K. et al., 1993). When confluency was reached, blasticidin S and methionine sulphoximine (MSX) were added for selection of the gene-amplified recombinant cells to the medium to the final concentrations of 5 μg/ml and 25 μM, respectively. The selected cultures from CHO cells and *ras*-clone I were named YCGI and DRGI, respectively. The individual clones (YCGIs and DRGIs) were isolated by a limiting dilutional cloning method.

2.4. ELISA

The amount of hIL-6 produced by the recombinant CHO cell lines was measured by the enzyme-linked immunosorbent assay (ELISA) as previously described (Perry, S. *et al.*, 1996). Briefly, after adding goat anti-hIL-6 antibody (R&D, Minneapolis, MN, USA) to the microtitration plate (Nunk, Roskilde, Denmark) precoated with rabbit anti-goat IgG (H&L) antibodies (Zymed, San Francisco, CA, USA), supernatant from hIL-6 producing cell lines, their appropriate dilutions, and the standard control samples were added. After adding anti-hIL-6 monoclonal antibody (Genzyme, Cambridge, MA, USA) which was previously biotinylated using N-hydroxysuccinimidobiotin (Sigma, St. Louis, MO, USA), hIL-6 was detected by incubating with streptavidin horseradish peroxidase conjugate (Amersham, Buckinghamshire, UK). and subsequently reacting with ABTS diammonium salt (WAKO, Osaka, Japan). Measurement at the wavelength 405 nm with a reference at 492 nm was taken.

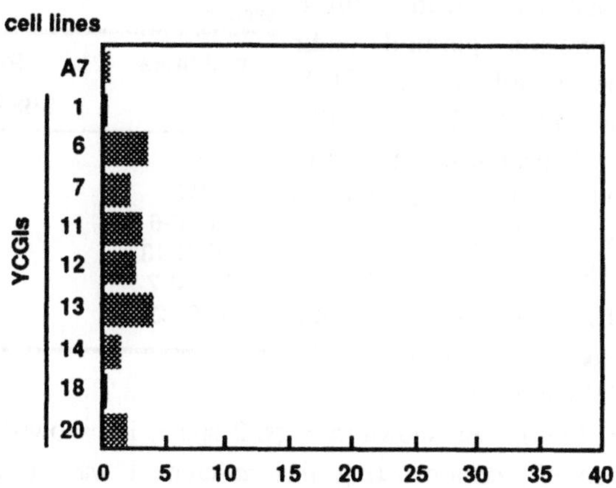

Fig. 2 Cellular productivity for A7 and YCGI clones

RESULTS AND DISCUSSION

We obtained 20 clones of YCGIs from the CHO cells transfected with pEE14-IL-6 by one round selection with 5 μg/ml blasticidin S and 25 μM MSX. As shown in Fig. 2, seven clones produced detectable levels of hIL-6. YCGI-13 showed the highest peak productivity value of 3.1 $\mu g/10^6$ cells/day among these clones. By transfecting *ras*-clone I with pEE14-IL-6, we obtained 43 clones of DRGIs by the one round of selection as shown in Fig. 3. The peak productivity values of 11 clones demonstrated to be more than 10 $\mu g/10^6$ cells/day, and the overall productivity of DRGIs were demonstrated to be much higher than YCGIs. DRGI-29 showed a peak productivity value of 38 $\mu g/10^6$ cells/day, which was over 13 times higher than YCGI-13.

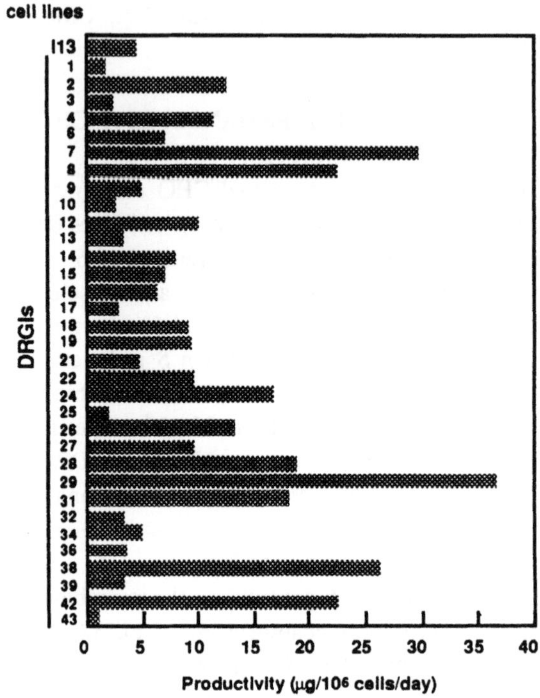

Fig. 3 Cellular productivity for I13 and DRGI clones

Perry *et al*. established hIL-6 producing CHO cell lines. A7 showed a highest peak productivity value of hIL-6 among the recombinant CHO cells transfected with hIL-6 expression plasmid under the control of the CMV promoter. Furthermore, hIL-6 hyper-producing CHO cells of I13 was established by transfecting *ras*-primed CHO cells (*ras*-clone I) with hIL-6 expression plasmid under the control of the CMV promoter (Perry, S. *et al*. 1996). As shown in Figs. 2 and 3, I13 showed much higher productivity of hIL-6 than A7, demonstrating that *ras*-clone I was suited for establishing a recombinant protein hyper producing cell line.

Table 1 Peak productivity values for recombinant CHO cells generated by conventional, OAP, GS, and OAP & GS hybrid system

cell lines	Productivity ($\mu g/10^6$ cells/day)	Accumulation (μg/ml/day)
A7	0.3	0.8
I13	7.3	6.6
YCGI-6	2.6	2.1
YCGI-13	3.1	2.4
DRGI-7	34.2	16.2
DRGI-29	39.5	14.0

As demonstrated in Table 1, the peak productivity value for YCGI-13 was 10 times greater than A7 and that for DRGI-29 was 5.4 times greater than I13, indicating that cellular productivity of hIL-6 markedly increased by gene amplification both in CHO cells and *ras*-clone I. When comparing the cellular productivity between YCGI-13 and DRGI-29, the peak productivity value of DRGI-29 was demonstrated to be 12.7 times greater than YCGI-13, indicating that *ras*-clone I is best suited host cell line for establishing a recombinant protein hyper producing cell line by the GS gene amplification system.

In this study, we described a novel system named the OAP & GS hybrid system to establish a recombinant protein hyper-producing cell lines through activation of the promoter controlling the recombinant gene expression by oncogene and gene amplification. The peak productivity value of the recombinant CHO cells generated with the OAP & GS hybrid system reached 39.5 $\mu g/10^6$ cells/day, which is enough to fulfill the commercial demand.

REFERENCE

1. Cockett, M. I., Bebbington, C. R., and Yarranton, G. T. (1990) "High level expression of tissue inhibitor of metalloproteinases in chinese hamster ovary cells using glutamine synthetase gene amplification." *Bio/technology*, **8**:662-667.
2. Perry, S., Katakura, Y., Zhang, Y., Inoue, Y., Teruya, K., Miura T., Ohashi, H. and Shirahata, S. (1996) "Productivity enhancement of recombinant proteins in chinese hamster ovary cells by oncogenes." *Animal Cell Technology: Basic & Applied Aspects*, **8**, Kluwer Academic Press, Netherland, (in press).
3. Teruya, K., Shirahata, S., Yano, T., Seki, K., Tachibana, H., Ohashi, H., and Murakami, H. (1993) "Secretory cell immunoscreening assay - A high sensitive screening method for secretory cells." *Anal. Biochem.*, **214**: 468-473.
4. Yano, T., Teruya, K., Shirahata, S., Watanabe, J., Osada, K., Tachibana, H., Ohashi, H., Kim, E. -H., and Murakami, H. (1994) "Ras oncogene enhances the production of a recombinant protein regulated by the cytomegalovirus promoter in BHK-21 cells." *Cytotechnology*, **16**: 167-178.

A TETRACYCLINE REGULATED RECOMBINANT PROTEIN PRODUCTION SYSTEM

YINGPEI ZHANG, YOSHINORI KATAKURA
AND SANETAKA SHIRAHATA
Graduate School of Genetic Resources Technology, Kyushu University
6-10-1 Hakozaki, Higashi-ku, Fukuoka 812-81, Japan

ABSTRACT We designed a bicistronic vector enabling high level, regulated production of recombinant protein in mammalian cells. In this vector, a tetracycline-responsive promoter PhCMV*-1 was used to control the transcription of a bicistronic unit composed of a gene of interest, an internal ribosomal entry site (IRES) and a chimeric transactivator (tTA). In the absence of tetracycline, the leaky expressed tTA transactivators bind to the PhCMV*-1 promoter and activate the expression of both the gene of interest and the tTA gene. The tTA transactivators translated from the bicistronic transcript will further activate the transcription of the bicistronic transcript, therefore, the expression system is autocatalytic. Addition of tetracycline prevents the binding of the tTA transactivators to the PhCMV*-1 promoter, therefore, represses the transcription of the bicistronic transcript. Using this vector, we established a recombinant cell line producing 980 ng β-lactoglobulin /10^6 cells/day with a regulation factor of several thousand folds.

1. Introduction

The productivity of a recombinant protein is to a great degree determined by the transcription efficiency of the gene. Since the promoter activity is determined by the interactions among various transcriptional factors, it is possible to achieve high level expression of the recombinant gene by specific recruitment of strong transactivators to a promoter in an artificial manner.

Gossen and Bujard have developed a stringently regulated expression system using activating domain of VP16 and the regulation elements derived from tetracycline operon of *Escherichia coli* transponson Tn10 [1]. The system is composed of two elements, one is a chimeric transactivator (tTA) consisting of the tetracycline repressor and the activating domain of VP16. The other is a tetracycline responsive

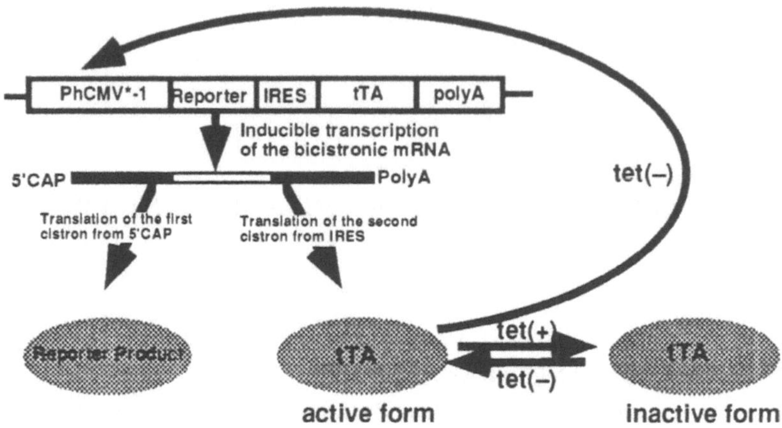

Figure 1. Principle of the autocatalytic expression system. The PhCMV*-1 promoter used to control the expression of the bicistronic unit composed of the reporter gene, the IRES and tTA gene is a tetracycline-responsive promoter. In the absence of tetracycline, the promoter is activated by the chimeric transactivator tTA, therefore, the expression is autocatalytic. In the presence of tetracycline, the tTA transactivators can not activate the PhCMV*-1 promoter, and the expression of both reporter and tTA gene is repressed.

promoter PhCMV*-1 consisting of the tetracycline operator sequences fused to a minimal promoter. In the absence of tetracycline, the tTA binds to PhCMV*-1 promoter and stimulates transcription. Addition of tetracycline blocks the binding of tTA to PhCMV*-1 promoter, therefore, abolishes the transactivating by tTA. In the original tetracycline inducible expression system, the tTA gene is controlled by the CMV promoter. Since the CMV promoter is regulated by various cellular signals, its activity varies in different cell types. In transgenic mice, it was observed that the reporter gene under the control of PhCMV*-1 promoter expressed poorly in some tissues where the CMV promoter does not possess strong transcriptional activity. This problem can be overcome by replacing the CMV promoter with PhCMV*-1 because the latter contains few cellular signal regulated *cis*-sequences. For this purpose, we designed a bicistronic expression vector in which the PhCMV*-1 promoter was used to control the transcription of a bicistronic unit composed of the gene of interest, an internal ribosomal entry site (IRES) [2] and the tTA gene. Insertion of internal IRES upstream of the tTA gene is to ensure the efficient translation of the second cistron through a CAP-independent translational initiation mechanism. As shown in Figure 1, this expression system is autocatalytic, therefore, should lead to high level expression of the reporter gene.

2. Materials and Methods

To construct the bicistronic vector, we replaced the CMV promoter in pUHD15-1 [1] with a 0.45 kb *Xho*I-*Eco*RI fragment containing the PhCMV*-1 promoter derived from pUHD10-3 , resulting pUTA-1. A 0.6 kb *Sal*I fragment of bovine β-lactoglobulin (β-LG) cDNA was inserted into the *Sal*I site upstream of poliovirus 5' UTR fragment functioning as IRES in pSBC1 [2]. The *Xho*I-*Eco*RI fragment of the resulting plasmid was then replaced by the above mentioned 0.45 kb *Xho*I-*Eco*RI fragment containing PhCMV*-1, resulting in pSTLG6. The pSTLG6 was then digested with *Not*I and *Xho*I, the 1.7 kb fragment containing PhCMV*-1, *β-LG* gene, and poliovirus 5' UTR was recovered and blunted, then inserted into the *Eco*RV site of pBluescript KS(+) (Stratagene), and the plasmid with the *β-LG* gene adjacent to the *Eco*RI site in the polylinker of the pBluescriptKS(+) was named pKSTLG3. The pKSTLG3 was then digested with *Eco*RI, and the 1.3 kb fragment containing *β-LG* gene and poliovirus 5'UTR was inserted into the *Eco*RI site of pUTA-1. The plasmid containing desired direction of *β-LG* gene, poliovirus 5' UTR and *tTA* was named pULTA18 (Figure 2A).

3. Results and Discussion

Transient expression assay was first carried out to examine whether the bicistronic vector can produce sufficient tTA to stimulate the transcription from the PhCMV*-1 promoter. Briefly, $3x10^5$ *dhfr*⁻ Chinese hamster ovary (CHO) cells maintained in MEM-α medium (GIBCO) supplemented with 10% fetal bovine serum (FBS) were transfected with 15 μg pULTA18 (Figure 2A) using calcium phosphate co-precipitation method. After washing twice with phosphate buffered saline (PBS) and changing fresh medium, tetracycline was added to one of the two paralleled dishes at the final concentration of 2 μg/ml. The cells were incubated for 48 hours, then the β–LG protein secreted into the supernatants was determined by enzyme-linked immunosorbent assay (ELISA). The cells in tetracycline-free medium produced 98 ng/ml β-LG protein, while the cells in tetracycline-supplemented medium produced about 9 ng/ml β-LG, probably synthesized from the mRNA before addition of tetracycline.

Having confirmed functioning of the bicistronic vector in transient expression assay, we further characterized the bicistronic expression in stable transfected cell lines. To this end, we co-transfected CHO cells with pULTA18 and pSV2/*bsr-dhfr*, a vector containing *bsr* and *dhfr* gene under the control of SV40 early promoter with calcium phosphate co-precipitation method. The cells were first selected by blasticidin S (5 μg/ml) in the absence of tetracycline, the drug-resistant cells were then cloned by limiting dilution method. 19 among 127 clones were found to produce β-LG protein. Five of them were expanded and their productivity was determined (Figure

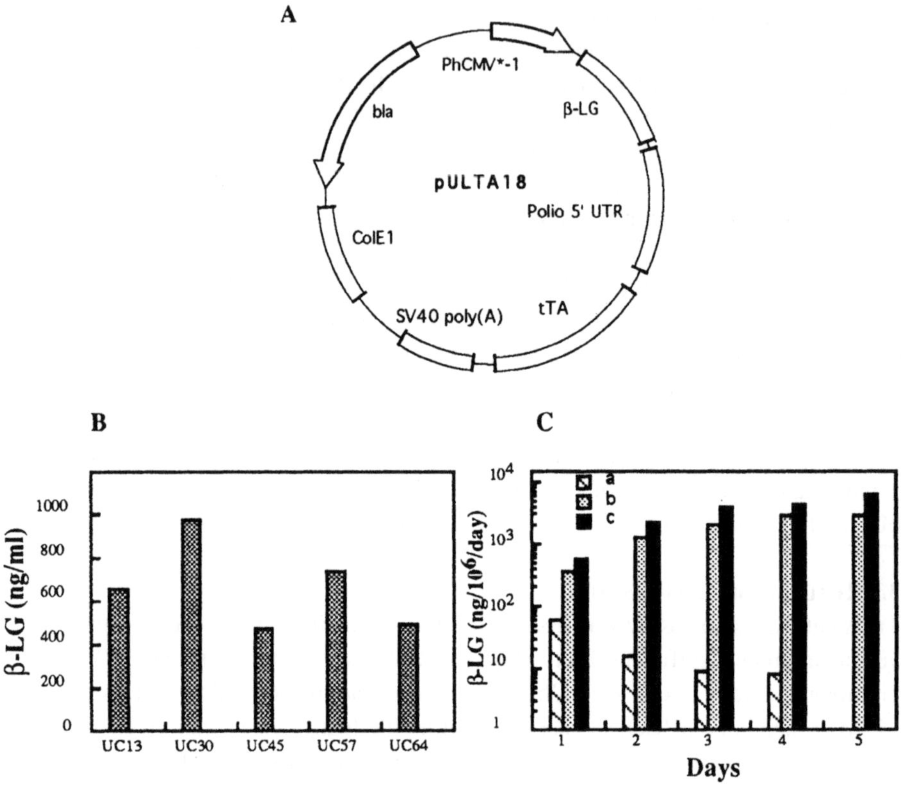

Figure 2. Expression of β-LG gene with the bicistronic vector. A. The bicistronic vector pULTA18. B. Productivity of the cell lines established by co-transfection of CHO cells with pULTA18 and pSV2/bsr-dhfr, selection with blasticidin S (5 μg/ml) followed by cloning with limiting dilution method. C. Regulation of β-LG production by tetracycline. UC30 cells (3×10^5) were plated into two 35 mm dishes and cultured in the presence (a) or absence (b) of tetracyclince (2 μg/ml). Every 24 hours, the supernatants were removed for β-LG quantitiation, and the cells were washed twice with PBS, then maintained in fresh medium either supplemented with tetracycline or not. (c): UC30 cells were plated into 60 mm dishes and allowed to grow overnight in tetracycline free medium. Every 24 hours, 50 μl of supernatants were removed for β-LG quantitiation.

2B). Clone UC30 shows highest productivity (980 ng/10^6 cells/day) and clone UC45 shows lowest productivity (460 ng/10^6/day) in tetracycline-free medium. These values are comparable to those which can be obtained using very strong promoter such as the CMV promoter in CHO cells. The UC30 cells were then further characterized. As shown in Figure 2C, the production of β-LG is repressed by tetracycline (a in Figure 2C), indicating that the expression of the reporter gene in the bicistronic system is dependent on the tTA. During a 5 day induction period, about 6.25 μg/ml β-LG was produced by fully induced cells in tetracycline-free medium (c in Figure 2C), while in the presence of 2 μg/ml tetracycline, no production of β-LG was detected in the control dishes inoculated with fully repressed UC30 cells maintained in the medium containing 2 μg/ml tetracycline. To examine the toxicity of the expression system, we compared the growth rate of UC30 with that of untransfected parent CHO cells, and did not observe significant growth inhibition (data not shown).

The bicistronic expression system has several advantages. It is better than conventional expression systems in that the expression level not only is very high, but can be stringently regulated also. It is better than the original tetracycline inducible system in that it is independent of the cellular regulation signals and has the potential to work efficiently in various types of cells.

4. References

1 Gossen, M. and Bujard, H. (1992) Tight control of gene expression in mammalian cells by tetracycline-responsive promoters, *Proc. Natl. Acad. Sci. USA.*, **89**, 5547-5551.

2. Dirks, W., Wirth, M. and Hauser H. (1993) Dicistronic transcription units for gene expression in mammalian cells, *Gene*, **128**, 247-249.

IMPROVEMENT OF TRANSIENT PROTEIN PRODUCTION BY COS-1 CELLS STABLY OVEREXPRESSING *bcl-2* GENE

TETSUO FUJITA , SATOSHI TERADA , HIROSHI UEDA , and EIJI SUZUKI
Department of Chemistry and Biotechnology, Faculty of Engineering, University of Tokyo
Bunkyo-ku, Tokyo 113, Japan

ABSTRACT : COS-1 cells were transfected with human *bcl-2* gene by using plasmid vector BCMGS-neo derived from bovine papilloma virus type-1; the plasmid can multiply in nucleus of mammalian cells. The transfectants overexpressed human Bcl-2 and showed remarkably improved survival when cultured at low (0.2%) serum concentration. This survival indicated that Bcl-2 protein functioned in COS-1 cells and then prevented the *bcl-2* transfectants from apoptosis caused by serum limitation. For examining efficiency of protein productivity, the *bcl-2* transfectants were further transfected with mouse lambda chain cDNA carried by an SV40-derived plasmid for transient expression. The culture supernatant of the *bcl-2* transfectants and control cells was sampled on day 4, 5, 6, 7, and 14 posttransfection. Control cells almost stopped producing lambda chain on day 6 posttransfection. On the other hand the *bcl-2* transfectants continued production after day 7, producing twofold amount of the lambda chain in total in comparison with control, presumably because Bcl-2 suppressed apoptosis in late culture. These COS-1 cells stably overexpressing Bcl-2 will be advantageous for transient expression of recombinant proteins.

KEY WORDS : Bcl-2, apoptosis, COS-1, transient expression

1. INTRODUCTION

The transient expression methods are well established and now used to express various recombinant proteins with mammalian cells for a rapid and simple preparation in many laboratories (1, 2), and recently it was reported to be applicable for large scale protein production (3). When SV40 origin-containing expression plasmids are transfected into nucleus of COS cells, plasmids are amplified up to 10^6 copies due to the SV40 large T antigen in the cells, and then recombinant proteins encoded on the plasmids can be easily and intensively expressed. But COS cells die after the transfected plasmids are highly amplified, to stop producing recombinant proteins. And the mechanism of this cell death and its type, apoptosis or necrosis, have been unknown yet.

Apoptosis is the suicidal cell death and Bcl-2 protein is functional to suppress apoptosis occurring on various conditions. *bcl-2* gene was first identified by Tsujimoto et al (4) and its function was by Vaux et al (5). Overexpression of Bcl-2 in many cells has been reported to result in suppression or delay of apoptosis induced by various causes, such as heat shock, ethanol, methotrexate, UV irradiation, serum or growth factor deprivation in culture medium, and viral infection. The death of COS cells during transient expression can be a case of apoptosis induced by viral infection since viral origin vectors are used and amplified too much in nucleus.

Itoh et al. reported that the transfection of hybridoma cells with *bcl-2* improved 4-fold the productivity of monoclonal antibody in a batch culture of cells in comparison with control (6, 7). In this present work, we enhanced the transient protein productivity successfully by generally engineering COS cells, one of most widely used cell lines in laboratories, to stably overexpress human *bcl-2* gene.

2. MATERIALS AND METHODS

Culturing COS-1 cells : COS-1 cells were cultured in Dulbecco's modified Eagle's (DME) medium (Nissui,Tokyo) supplemented with 10%(v/v) calf serum (CS), 20 mM N-(2 hydroxyethyl) piperazine-N'-2-ethanesulfonic acid (HEPES), 0.2% $NaHCO_3$, 2 mM glutamine and 0.06 mg/ml kanamycin. The cells were grown in ø100mm dish at 37 °C in humidified air containing 5% CO_2. To remove adherent cells from dish, culture medium was aspirated and 500 ml of 0.02%(w/v) trypsin and 0.05%(w/v) EDTA in PBS was poured and incubated at 37 °C for 5 min.

Construction of Bcl-2 expression plasmids : pCΔj-bcl-2 (4, 8, 9) includes 0.95 kbp *bcl-2* ORF between EcoRI sites. Its fragment was ligated into EcoRI site in pBluescript II KS+ (Stratagene). Properly oriented fragment of *bcl-2* cDNA between XhoI and NotI sites was inserted downstream of CMV promoter in BCMGS-neo (10, 11), yielding a plasmid termed BCMG-bcl2-neo (*Figure 1*).

Preparation of cell extracts and Western Blotting : The 10^5 of cells were harvested and lysed in 10 mM TrisHCl (pH 7.4), 1% Triton X-100, 0.15 mM NaCl, 50 µg/mL phenyl methanesulfonyl fluoride (PMSF), and 2 µg/ml aprotinin at 4°C for 30 min. After debris was deprived by centrifuge, the lysate was analyzed by Western blotting using mouse anti human *bcl-2* monoclonal antibody (Boehringer Manneheim) and goat anti mouse Ig polyclonal antibody conjugated with horse radish peroxidase (HRP) (Bio Source International, Inc .Tago Products). Human Bcl-2 was detected by ECL kit (Amersham).

Cultivation in Low serum condition : The 10^5 of cells were incubated in DMEM supplemented with 0.2% FBS in ø60mm dishes. After several incubation periods, cells were collected, washed twice with PBS to remove trypsin, stained with trypan blue, and counted with hemacytometer.

Examination of protein productivity : Mouse lambda chain cDNA was amplified by PCR and inserted between EcoRI and NotI sites on plasmid pcDNA3 (Novagen), yielding a plasmid termed pcDNA-λ . For expressing lambda chain transiently, 10 μg of the plasmid was transfected into 5x10[5] of the three types of COS-1 cells in ø100mm dishes with DEAE-dextran/chloroquine method (1). After transfection for transient expression, culture medium was sampled and the concentration of lambda protein in medium was measured by enzyme-linked immunosorbent assay (ELISA) using rabbit anti mouse lambda antisera (Organon Teknika Corp.) and goat anti mouse lambda antibodies conjugated with HRP (Southern Biotechnology Associates, Inc.).

3. RESULTS AND DISCUSSION

Preparing COS-1 cells overexpressing Bcl-2

BCMGS-neo is a plasmid derived from bovine papilloma virus type-1 (BPV-1) and can multiply 20-100 copies in nucleus. Hence cloned bcl-2 cDNA would be expressed intensively owing to the strong CMV promoter and the high copy number of plasmids in nucleus. BCMG-bcl2-neo and BCMGS-neo were transfected into COS-1 cells by electroporation, respectively, and then the obtained cells were termed COS-1/bcl-2 and COS-1/cont, respectively. As shown in *Figure 1*, Bcl-2 26kDa band was strongly seen in lysate of COS-1/bcl-2 but not COS-1/cont. The lysate of COS-1 cells showed the same pattern as that of COS-1/cont (data not shown).

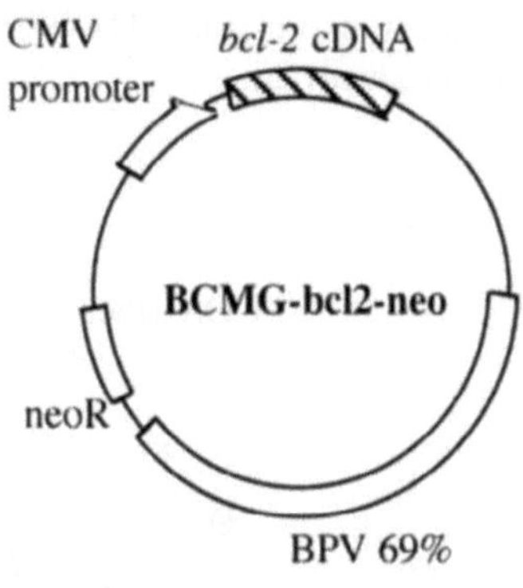

Figure 1. Construction of BCMG-bcl2-neo

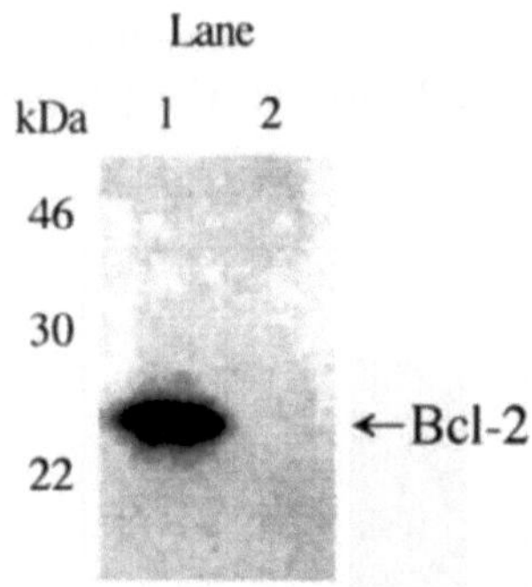

Figure 2. Detection of Bcl-2 protein overexpressed in COS-1 cells. Lane 1: lysate of COS-1/Bcl-2. Lane 2: lysate of COS-1/cont

Examination of apoptosis resistance

To examine whether COS-1 cells overexpressing human Bcl-2 acquired apoptosis resistance, we cultured COS-1/bcl-2 and COS-1/cont cells in serum reduced medium, and examined their viabilities. Reportedly culturing cells at a reduced serum concentration induces apoptosis and is commonly used to evaluate apoptosis-resistant cells (8, 9, 12). The 10[5] of COS-1/bcl-2 and COS-1/cont cells were incubated in DMEM with 0.2% FBS, respectively. After various incubation periods, cells were collected, stained with trypan blue, and counted with hemacytometer. The viable cell number of COS-1/cont gradually decreased and reached to half of the initial one on day 9. On the other hand that of COS-1/bcl-2 increased to 4-fold of the initial one on day 9 (*Figure 3*). This result indicated that human Bcl-2 expressed in COS-1 could suppress apoptosis, and induce COS-1/bcl-2 to proliferate much better than COS-1/cont.

Protein productivity

For comparing protein productivity of COS-1/bcl-2 and COS-1/cont, we chose mouse lambda protein as a model protein product. As this protein is secreted to culture medium when expressed in mammalian cells, daily protein production can be easily determined by measuring the concentration of the protein in culture supernatant. After transfection for transient expression, each cell was incubated in complete medium, and the culture supernatant was sampled on day 4, 5, 6, 7 and 14 posttransfection and the concentration of lambda chain in each sample was measured by ELISA. The product concentration in each culture was shown in *Figure 4* as the ratio to that of day 4 instead of the absolute quantity since the absolute concentration varied depending on the transfection efficiency of the lambda protein DNA for each run of the experiment that could not be stabilized. As shown in *Figure 4*, in comparison with COS-1/cont, COS-1/bcl-2 showed equal or less productivity by day 6, however, from day 7 onward, COS-1/bcl-2 continued production up to 2-fold at the end of the cultures, while COS-1/cont stopped production after day 7. The productivity of COS-1 cells showed almost the same time course as that of COS-1/cont (data not shown). This suggested that overexpression of Bcl-2 suppressed cell death at late culture. In conclusion, the overexpression of Bcl-2 in COS cells improved productivity of recombinant protein in transient expression. These more durable COS-1 cells will be widely applied for improving productivity.

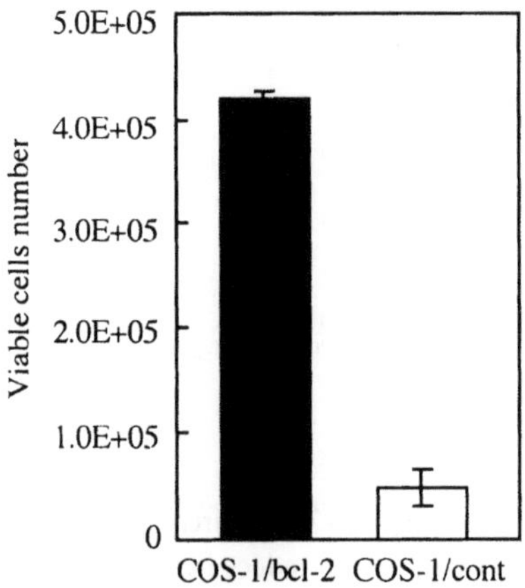

Figure 3. The number of viable cells in serum starvation

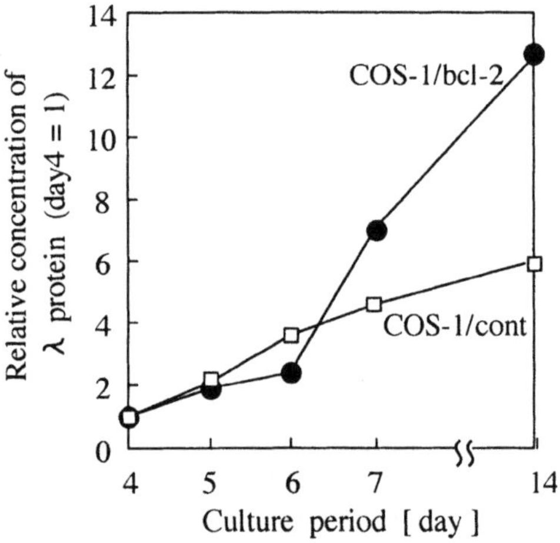

Figure 4. Measurement protein productivity by λ protein

ACKNOWLEDGEMENTS

We wish to thank Dr. Yoshihide Tsujimoto for providing us with *bcl-2* cDNA and Dr. Hajime Karasuyama for BCMGS-neo.

REFERENCES

1. Current Protocols in Molecular Biology, John Wieley & Sons, Inc.
2. Mellon, P., Parker, V., Gluzmam, Y., and Maniatis, T (1981) Identification of DNA sequences required for transcription of the human 1-globin gene in a new SV40 host-vector system, *Cell* **27**, 279-288.
3. Blasey, H. D., Aubry, J-P., Mazzei, G. J. and Bernard, A. R.(1996) Large scale transient expression with COS cells, *Cytotechnology* **16**, 183-192 .
4. Tsujimoto, Y., Cossman,J., Jaff, E., and Cruce, C. M.(1985) Involvement of the bcl-2 gene in human follicular lymphoma, *Science* **228**, 1440-1443.
5. Vaux, D. L., Cory, S., and Adams, J. M. (1988) Bcl-2 gene promotes haemopoietic cell survival and cooperates with c-myc to immortalize pre-B cells, *Nature* **335**, 440-442.
6. Itoh, Y., Ueda, H., and Suzuki, E. (1995) Overexpression of bcl-2, apoptosis suppressing gene: prolonged viable culture period of hybridoma and enhanced antibody production, *Cytotechnology* **48**, 118-122.
7. Suzuki, E., Terada, S., Ueda, H., Fujita, T., Komatsu, T., Takayama, S., and Reed, J. C. (in press) Establishing apoptosis resistant cell lines for improving protein productivity of cell culture, *Cytotechnology*
8. Tsujimoto, Y. (1989) Stress-resistance conferred by high level of bcl-2 protein in human B lymphoblastoid cell, *Oncogene* **4**, 1331-1336.
9. Tsujimoto, Y. (1989) Overexpression of the human BCL-2 gene product results in growth enhancement of Epstein-Barr virus-immortalized B cells, *Proc. Natl. Acad. Sci. USA* **86**, 1958-1962.
10. Karasuyama, H., and Melchers, F.(1988) Establishment of mouse cell lines which constitutively secrete large quantities of interleukin 2, 3, 4 or 5, using modified cDNA expression vectors, *Eur. J. Immunol.* **18**, 97-104.
11. Karasuyama, H., Kudo, A., and Melchers, F.(1990) The proteins encoded by the VpreB and 5 pre-B cell-specific genes can associate with each other and with μ heavy chain, *J. Exp. Med.* **172**, 969-972.
12. Henderson, S., Rowe, M., Gregory, C., Croom-Carter, D., Wang, F., Longnecker, R., Kieff, E., and Rickinson, A.(1991) Induction of bcl-2 expression by Epstein-Barr virus latent membrane protein 1 products infected B cells from programmed cell death *Cell* **65**, 1107-1115.

EFFECT OF CYTOKINES ON GROWTH OF A MOUSE HYBRIDOMA CELL LINE AND ITS ANTIBODY PRODUCTION

SATOSHI TERADA[1], FUSAO MAKISHIMA[2], HIROSHI UEDA[1] and EIJI SUZUKI[1]*

1 Department of Chemistry and Biotechnology, and

2 Department of Precision Machinery Engineering, Faculty of Engineering, The University of Tokyo, Hongo 7-3-1, Bunkyo-ku, Tokyo 113, Japan

Three cytokines, interleukin 6 (IL-6), leukemia inhibitory factor(LIF), and oncostatin M (OSM), that bind to composite receptors including a common signal transducer gp130 suppressed growth of a mouse B-cell hybridoma cell line 2E3-O cultured in serum-free medium, while they enhanced antibody production of the cells. The specific growth rate of the cells reduced from 1.0 day^{-1} for control to 0.6 day^{-1} for the cultures supplemented with 1ng/ml of IL-6, 4ng/ml of LIF, or 20ng/ml of OSM, respectively. The antibody productivity increased five-fold when the cells were cultured with 1ng/ml of IL-6, 25ng/ml of LIF, or 20ng/ml of OSM, respectively. Transforming growth factor β 1 (TGF-β 1) similarly suppressed growth of the cells at the concentration of 5 ng/ml, while it did not enhance the antibody production. As a whole, this work suggests that gp-130, which is commonly involved in each receptor for IL-6, LIF, and OSM, transduce signals for suppressing growth and possibly for enhancing antibody production in the hybridoma cells.

1. Introduction

IL-6 is known as a growth factor for B cell hybridoma / plasmacytoma cells[1], and a potent factor acting on hybridoma cells to regulate antibody production. We have investigated the effect of a recombinant human interleukin-6 (IL-6) on the growth of mouse hybridoma 2E3-O cell line in serum-free medium[2], and found that IL-6 suppressed growth of this cell line cultured in the medium. Furthermore, supplementation of IL-6 to the cultures of the hybridoma enhanced antibody production (daily antibody production per cell) up to five-fold[2].

There are similarities in gene structure and amino acid sequence among IL-6, leukemia inhibitory factor (LIF), oncostatin M (OSM), and granulocyte colony-stimulating factor (G-CSF). On the other hand, IL-6, LIF, OSM, interleukin-11 (IL-11), and ciliary

*Corresponding author

neurotrophic factor (CNTF) bind to composite receptors containing a common signal transducer gp130[3]. These factors show some *in vitro* biological activities in common such as induction of differentiation of the murine M1 myeloid leukemic cell line[4,5]. Such redundancy of cytokine function can be due to the common signal transducer gp130[6]. Such overlapping of function among these factors, however, is not always observed. For example, IL-6 but not LIF stimulated growth of the hybridoma line B9, while another cell line DA.1 showed growth-response to LIF but not to IL-6[7]. In the present work, we investigated whether some of these cytokines display activities overlapping with those of IL-6 on the hybridoma 2E3-O cells. We report that LIF, OSM, and IL-6 suppress growth and enhance antibody production of the hybridoma cells.

The suppression of growth and the enhancement of antibody production occurred almost simultaneously. We further examined the relation between the slowed growth and the enhanced antibody production by using transforming growth factor (TGF-β 1) as another growth suppressor that probably suppressed the growth of the cells through some mechanism different from that for IL-6, LIF, and OSM. TGF-β 1 does not involve gp130 in its signal transduction. TGF-β 1 inhibits growth of some types of leukemic cells and a subline of the mouse myeloid leukemia M1 cell line[8]. We also investigated the effect of TGF-β 1 on growth and antibody production of the hybridoma 2E3-O cells.

2. MATERIALS AND METHODS

Reagents

The cytokines used were purified recombinant human IL-6 (Wako Junyaku, Tokyo), murine LIF (Genzyme, Cambridge, MA), human OSM (Genzyme, Cambridge, MA), and human TGF-β 1 (Kurashiki Bouseki, Japan). A chemically defined serum-free medium, ASF103, was obtained from Ajinomoto (Tokyo).

Cell line and cell culture

Cell line 2E3-O is a P3U1 (P3X63 AG8U.1) derived mouse hybridoma[9]. The hybridoma is a high producer of an IgG1 specific to a trinitrophenyl(TNP)-hapten. The cells were cultured in a serum-free ASF-103 medium at 37℃ and at 5% CO_2. Cell number was counted with Coulter Counter (Coulter). Cell viability was measured by trypan blue dye exclusion test.

Determination of antibody concentration

The antibody concentration was measured using ELISA.

3. RESULTS

Leukemia inhibitory factor and oncostatin M displayed effects overlappingly with IL-6 on growth and antibody production.

We investigated the effects of LIF and OSM, which utilize a signal transducer gp130 as IL-6, on growth and antibody production of the 2E3-O cells cultured in a serum-free medium (ASF103). The exponentially growing hybridoma cells were washed, suspended in the serum-free medium and cultured in the presence of recombinant leukemia inhibitory factor (LIF) or oncostatin M (OSM) at various concentrations. Figure 1-A revealed that LIF dose-dependently inhibited growth of the 2E3-O cells. The specific growth rate (the daily increment of the total cell density divided by the viable cell density) reduced from 1.0 day^{-1} for control culture to 0.6 day^{-1} for the culture with 4 ng/ml of LIF. This anti-proliferative effect was comparable to that of IL-6 at 1 ng/ml. As with IL-6, the cell culture supplemented with LIF at 4 to 25 ng/ml was viably maintained at the cell viability over 80% for 10 days, a remarkably long period, and continuously produced the antibody throughout the period. The antibody productivity (pg of antibody / cell / day) shown in Fig.1-B was calculated for the 2E3-O cells cultured at various concentrations of LIF. The antibody productivity increased four to five times in comparison with the control.

OSM was also a growth suppressive factor for the 2E3-O cells. As shown in Fig.2, OSM suppressed the growth of the cells at 2 to 20 ng/ml as IL-6 did at 1 ng/ml. The productivity was nearly four-fold of the control. As the results, all the three cytokines, IL-6, LIF, and OSM, which involve a common signal transducer gp130, showed very similar effects on growth and antibody production of the 2E3-O cells.

TGF-β 1 suppressed cell growth of 2E3-O cells

We investigated the effect of recombinant TGF-β 1 on growth and antibody production of the 2E3-O cells. TGF-β 1 weakly inhibited growth of the 2E3-O cells at 0.05 ng/ml, and substantially did at 0.50 ng/ml as much as IL-6 at 1.0 ng/ml, reducing the growth rate to 0.50 day^{-1} from 0.97 day^{-1} for control. Its antiproliferative effect was almost saturated at above 0.5 ng/ml, while IL-6 further reduced the growth rate to 0.16 day^{-1} at 4 ng/ml. The cells in the presence of 5 to 50 ng/ml of TGF-β 1 still continuously grew at the slowed growth rate (as shown in Fig.3-A), maintaining high viability, for example, 77% and 72% at day five and eight, respectively.

TGF-β 1 did not enhance antibody production of 2E3-O cells

We determined the antibody concentration in the supernatant of the cultures. TGF-β 1 at all the concentrations tested had no effect on the antibody productivity (Fig.3-B), while IL-6 at 1 ng/ml enhanced antibody production 4.5-fold of the control as expected.

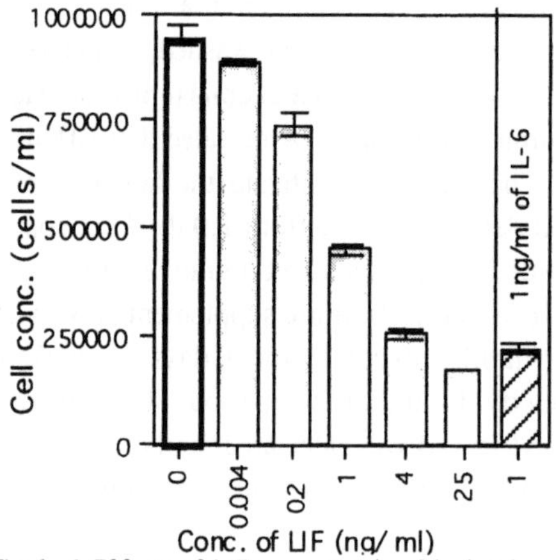

Fig.1-A Effect of LIF on growth of hybridoma.
Cells were cultured for 111 hours with LIF.

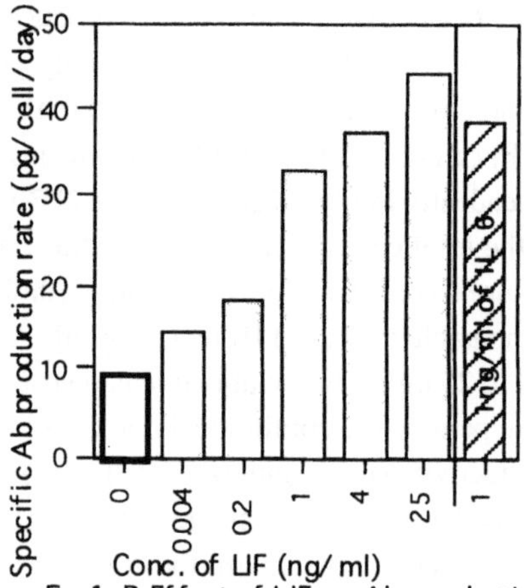

Fig.1-B Effect of LIF on Ab production

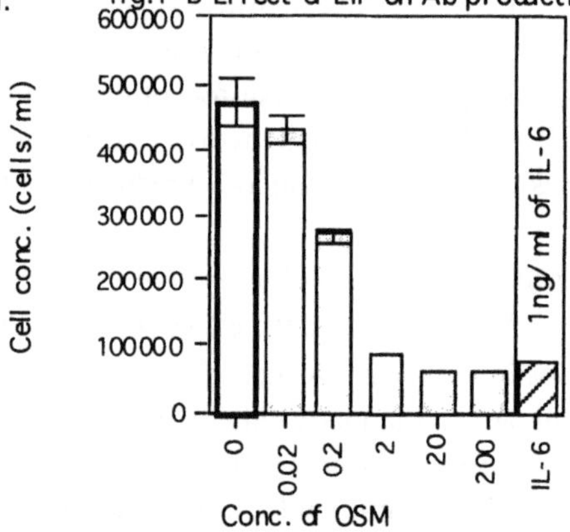

Fig.2 Effect of OSM on growth of hybridoma
Cells were cultured for 102 hours with OSM.

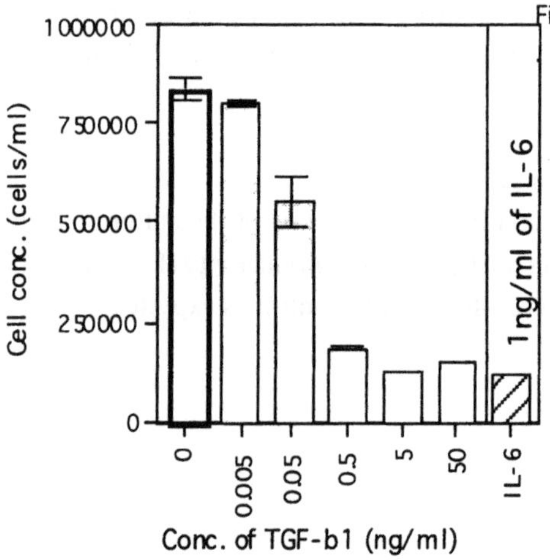

Fig.3-A. Effect of TGF on growth of hybridoma.
Cells were cultured for 93 hours with TGF.

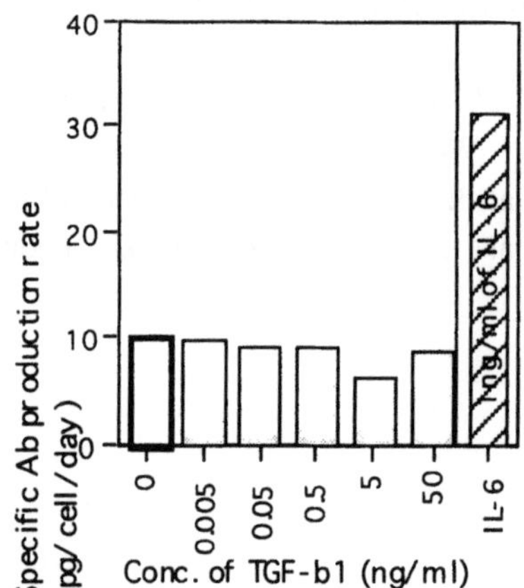

Fig.3-B. Effect of TGF on Ab production

4. DISCUSSION

Overlapping activities of the three cytokines

Among the four cytokines whose receptors involve gp130 in common, IL-6 is known to induce immunoglobulin secretion of B cells, while LIF and OSM are so far not to do so. The results of this study, however, showed that a hybridoma cell line derived from B cells responds to all the three cytokines when adapted to serum-free medium, being growth-suppressed and induced to enhance synthesis of antibody. A possible explanation of these phenomena is that acquisition of ability to grow in the serum-free medium accompanies selection of phenotypes expressing signal transducing cascades responsible for the phenomena. Another explanation is that the signals started by IL-6, LIF, or OSM which are countered with some serum activity are liberated to induce the phenomena when the cells are cultured in the serum-free medium.

Unenhanced antibody production during growth-suppression by TGF- β 1

When the four cytokines IL-6, LIF, OSM, and TGF- β 1 suppressed the cell growth, TGF- β 1 did not enhance antibody production while the others did. Three possible explanations of this difference are the following. First, the signal transduction associated with TGF- β 1 does not activate immunoglobulin gene expression, while those associated with the other cytokines do. Secondly, IL-6, LIF, and OSM more intensively than TGF- β 1 arrest the cells in G1 phase where the cells may synthesize antibody faster than in the other phases of the cell cycle as reported for some plasmacytoma and hybridomas[10, 11]. Thirdly, TGF- β 1 suppresses the cell growth, possibly by suppressing protein synthesis, resulting in unenhanced antibody production; this is supported by the reports that hybridoma cells maintained constant antibody production rate, when their growth was moderately suppressed by the culture conditions known to suppress protein synthesis such as the cycloheximide supplementation[12] and the amino acid limitation[12, 13].

There are several reports on effects of TGF- β 1 and IL-6 on growth and differentiation of the mouse M1 myeloid leukemia cells, which may be related to our results. For example, TGF- β 1 inhibited growth and dexamethason-induced differentiation of the M1 cells[9]; TGF- β 1 induced cell death by apoptosis in the M1 cells and this induction of apoptosis was inhibited by IL-6[14]; the M1 cells were induced to differentiate into macrophages in the presence of IL-6[4].

5. Acknowledgments

S.T. was supported by Fellowships of the Japan Society for the Promotion of Science for Junior Scientists. This study was supported in part by a grant for Biodesign Research Program from RIKEN to F.M.

6. REFERENCES

1. Van Damme J, Opdenakker G, Simpson RJ, Rubira MR, Cayphas S, Vink A, Billian A, Van Snick J (1987) Identification of the human 26-kD protein, interferon β 2, as a B-cell hybridoma / plasmacytoma growth factor induced by interleukin 1 and tumor necrosis factor. J Exp Med 165: 914-919.
2. Makishima F, Terada S, Mikami T, Suzuki E (1992) Interleukin-6 is antiproliferative to a mouse hybridoma cell line and promotive for its antibody productivity. Cytotechnol 10: 15-23.
3. Kishimoto T, Taga T, Akira S (1994) Cytokine signal transduction. Cell 76: 253-262.
4. Onozaki K, Akiyama Y, Okano A, Hirano T, Kishimoto T, Hashimoto T, Yoshizawa K, Taniyama T (1989) Synergistic regulatory effects of interleukin-6 and interleukin-1 on the growth and differentiation of human and mouse myeloid leukemic cell lines. Cancer Res 49: 3602-3607.
5. Tomida M, Yamamoto-Yamaguchi Y, Hozumi M (1984) Purification of a factor inducing differentiation of mouse myeloid leukemic M1 cells from conditioned medium of mouse fibroblast L929 cells. J Biol Chem 259: 10978-10982.
6. Hirano T, Matsuda T, Nakajima K (1994) Signal transduction through gp130 that is shared among the receptors for the interleukin 6 related cytokine subfamily. Stem Cells 12: 262-277
7. Barton BE, Jackson JV, Lee F, Wagner J (1994) Oncostatin M stimulates growth in B9 hybridoma cells: potential role of oncostatin M in plasmacytoma development. Cytokine 6: 147-153.
8. Okabe-Kado J, Honma Y, Hayashi M, Hozumi M (1989) Inhibitory action of transforming growth factor-beta on induction of differentiation of myeloid leukemia cells. Jpn J Cancer Res 80: 228-232.
9. Mikami T, Makishima F, Suzuki E (1991) Enhancing effect of mouse peritoneal exudate cells and their products on antibody productivity of hybridoma cells: Application of in vivo factors to in vitro culture. Cytotechnol. 7: 93-101
10. Garatun-Tjeldstf O, Pryme IF, Weltman JK, Dowben RM (1976) Synthesis and secretion of light-chain immunoglobulin in two successive cycles of synchronized plasmacytoma cells. J Cell Biol 68: 232-239.
11. Kromenaker SJ, Srienc F (1991) Cell cycle-dependent protein accumulation by producer and nonproducer murine hybridoma cell lines: A population analysis. Biotechnol. Bioeng. 38: 665-677.
12. Suzuki E, Ollis DF (1990) Enhanced antibody production at slowed growth rates: experimental demonstration and a simple structured model. Biotechnol Prog 6: 231-236.
13. Suzuki E, Takahashi K, Terada S, Makishima F (1993) Growth control of hybridoma cultures by adding or depriving interleukin-6, caffeine, amino acids, growth factors, or antisense c-myc. In Kaminogawa S, Ametani A, Hachimura S(eds) Animal Cell Technology: Basic & Applied Aspects, Kluwer Academic Publishers, Dordrencht, 5: pp237-242.
14. Lotem J, Sachs L (1992) Hematopoietic cytokines inhibit apoptosis induced by transforming growth factor beta1 and cancer chemotherapy compounds in myeloid leukemic cells. Blood 80: 1750-1757.

EFFECT OF HYPOOSMOTIC STRESS ON HYBRIDOMA CELL GROWTH AND ANTIBODY PRODUCTION

JOON SOO RYU & GYUN MIN LEE
Department of Biological Sciences, Korea Advanced Institute of Science and Technology, 373-1, Kusong-Dong, Yusong Gu, Taejon 305-701, Korea

1. Abstract

To investigate the response of hybridoma cells to hypoosmotic stress, S3H5/γ2bA2 and DB9G8 hybridomas were cultivated in the hypoosmolar medium (DMEM supplemented with 10% serum) resulting from sodium chloride subtraction. Both hybridomas showed similar responses to hypoosmotic stress in regard to cell growth and antibody production. The cell growth and antibody production at 276 mOsm/kg were comparable to those at 329 mOsm/kg (standard DMEM). Both cells grew well at 219 mOsm/kg, though their growth and antibody production were slightly decreased. When the osmolality was further decreased to 168 mOsm/kg, the cell growth did not occur. When subjected to hyperosmotic stress, both cells displayed significantly enhanced specific antibody productivity (q_{Ab}). However, the cells subjected to hypoosmotic stress did not display enhanced q_{Ab}. Taken together, both hyperosmotic and hypoosmotic stresses depressed the growth of S3H5/γ2bA2 and DB9G8 hybridomas. However, their response to hypoosmotic stress in regard to q_{Ab} was different from that to hyperosmotic stress.

2. Introduction

Osmolality is one of the most important physical properties of culture medium. Most cell culture media are designed to have the osmolality in the range of 270 to 330 mOsm/kg which is known to be quite acceptable for most cells [2].

Mammalian cells including hybridomas are sensitive to changes in medium osmolality. When subjected to hyperosmotic stress, hybridomas generally display enhanced specific antibody productivity (q_{Ab}) and depressed cell growth[1,3-7,9,10]. Hence, the enhanced q_{Ab} during batch culture does not result in a substantial increase in the final antibody concentration due to the depressed cell growth at higher osmolalities[3,6,7]. Physiological responses of hybridomas to hyperosmotic stress have been extensively studied because of the potential of commercial strategies based on hyperosmotic stress.

While much work has been performed on the response of hybridomas to hyperosmotic stress, there are, to date, little reported studies on the response of hybridomas to hypoosmotic stress. Hence, we study the effect of hypoosmotic stress resulting from NaCl subtraction on hybridoma cell growth and antibody production.

3. Materials and Methods

3.1 CELL LINES, MEDIUM, AND CULTURE MAINTENANCE

Two murine hybridoma cell lines, S3H5/γ2bA2 and DB9G8, were used in this study. The medium used for culture maintenance of both cell lines were Dulbecco's modified Eagle's medium (DMEM) supplemented with 10% (v/v) fetal bovine serum (FBS). No antibiotics were added to the media. The cells were maintained in 25 cm^2 T-flasks in a 5% CO_2/air mixture, humidified at 37°C. The DMEM with various osmolalities was prepared by adjusting NaCl concentration, while all other medium components being kept constant. The medium was supplemented with 10% FBS. Osmolalities were varied from 168 to 499 mOsm/kg. These values of osmolalities were measured after the initial pH of media was adjusted to 7.4.

3.2 CELL CULTURE

Identical experiments were carried out for S3H5/γ2bA2 and DB9G8 hybridomas. The cells exponentially growing in standard DMEM were inoculated into T-75cm^2 flasks containing 30mL of DMEM with various osmolalities. The initial cell concentration was approximately 1 x 10^5 cells/mL. The cultures were kept in a 5% CO_2/air mixture, humidified at 37°C. All the experiments were repeated at least three times.

3.3 ANALYTICAL METHODS

Cell growth and viability were monitored by counting cells with a hemocytometer by trypan blue dye exclusion method. Osmolality was measured using an Osmometer. Total secreted antibodies were quantified using an enzyme linked immunosorbent assay (ELISA).

4. Results and Discussion

To determine hybridoma response to hypoosmotic stress resulting from NaCl subtraction in a medium, S3H5/γ2bA2 and DB9G8 hybridomas were cultivated in the T-flasks containing hypoosmolar DMEM with various osmolalities. The response of these hypoosmotic stress can be compared with that to hyperosmotic stress.

4.1 S3H5/γ2bA2 CELLS

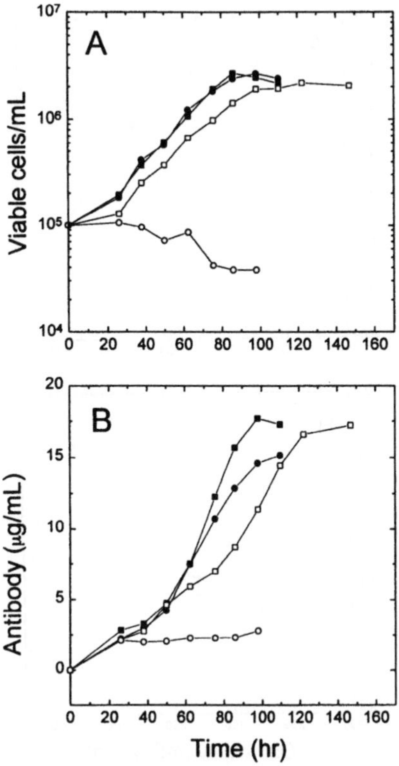

Figure 1. Cell growth (A) and antibody production (B) during a batch culture of S3H5/γ2bA2 hybridoma cells in hypoosmolar media. Data points: (○) 168, (□)219, (●) 276, (■) 329 mOsm/kg.,

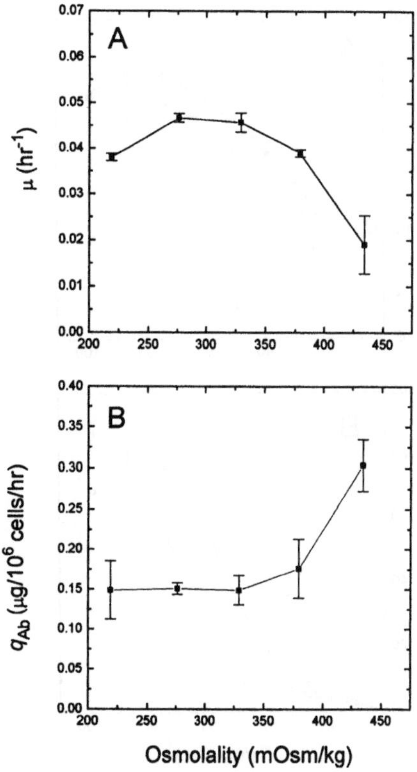

Figure 2. Effect of medium osmolality on S3H5/γ2bA2 hybridoma cells in regard to (A)μ, and(B)q_{Ab},

Figure 1 and Figure 3 shows typical cell growth and antibody production during a batch culture of S3H5/γ2bA2 and DB9G8 cells in hypoosmolar media, respectively. The cell growth and antibody production of both cell lines at 276 mOsm/kg were comparable to those at 329 mOsm/kg (standard DMEM), confirming that the osmolality in this range is quite acceptable for most mammalian cells. Both cell lines could still grow at 219 mOsm/kg, though their growth and antibody production were decreased. At this osmolality, cell viability remained over 90%. When the osmolality was further decreased to 168 mOsm/kg, both cell lines did not grow at all.

4.2 DB9G8 CELLS

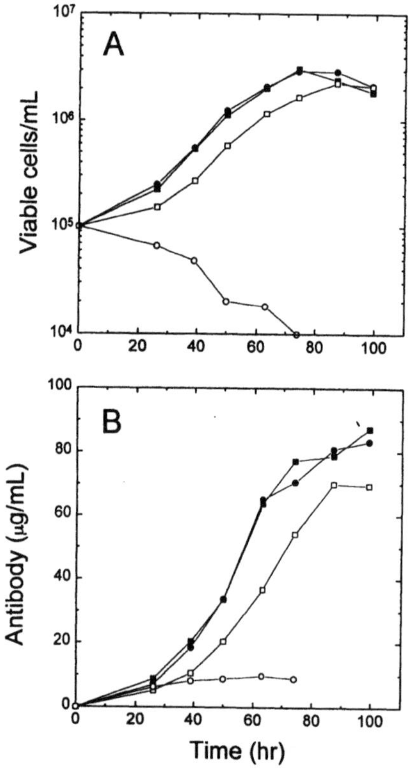

Figure 3. Cell growth (A) and antibody production (B) during a batch culture of DB9G8 hybridoma cells in hypoosmolar media. Data points: (O) 168, (□)219, (●) 276, (■) 329 mOsm/kg.,

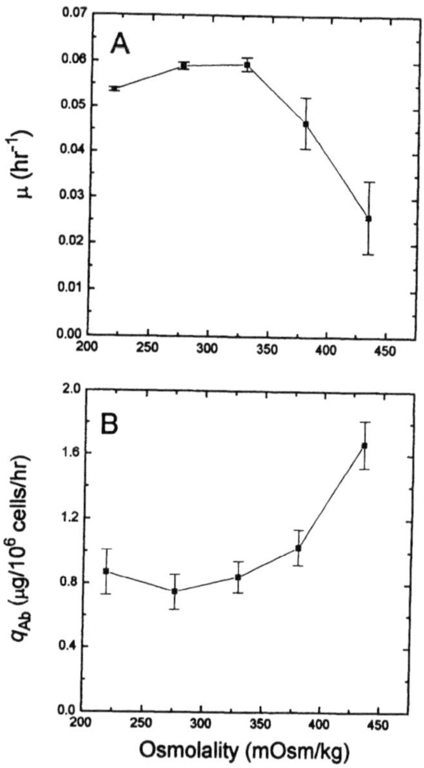

Figure 4. Effect of medium osmolality on DB9G8 hybricoma cells in regard to (A)μ, and(B)q_{Ab},

In order to compare the response of cells to hypoosmotic stress with that to hyperosmotic stress, specific growth and antibody production rates of S3H5/γ2bA2 cells at various osmolalities were plotted in Figure 2. When subjected to hyperosmotic stress, the cells displayed significantly enhanced q_{Ab}. On the other hand, when subjected to hypoosmotic stress, the cells did not display enhanced q_{Ab}.

Figure 4 shows specific growth and antibody and production rates of DB9G8 cells at various osmolalities. Enhanced q_{Ab}, which was observed at hyperosmolar condition, was not observed at hypoomotic condition.

5. Acknowledgements

The authors thank Dr. Mark Kaminski for providing S3H5/γ2bA2 hybridoma. This work was supported in part by the Ministry of Education, Korea.

6. References

1. Bibila, T. A., Ranucci, C. S., Glazomitsky, K., Buckland, B. C.,Aunins, J. G. 1994. Monoclonal antibody process development using medium concentrates. Biotechnol. Prog. **10**: 87- 96.

2. Freshney, R. I. 1987. Culture of animal cells. 2nd edition Alan R. Liss, New York.

3. Oh, S. K. W., Vig, P., Chua, F., Teo, W. K., Yap, M. G. S. 1993. Substantial overproduction of antibodies by applying osmotic pressure and sodium butyrate. Biotechnol. Bioeng. **42**: 601-610.

4. Øyaas, K., Ellingsen, T. E., Dyrset, N., Levine, D. W. 1994. Utilization of osmoprotective compounds by hybridoma cells exposed to hyperosmotic stress. Biotechnol. Bioeng. **43**: 77-89.

5. Øyaas, K., Ellingsen, T. E., Dyrset, N., Levine, D. W. 1994. Hyperosmotic hybridoma cell cultures : Increased monoclonal antibody production with addition of glycine betaine. Biotechnol. Bioeng. **44**: 991-998.

6. Ozturk, S. S., Palsson, B. O. 1991. Effect of medium osmolarity on hybridoma growth, metabolism, and antibody production. Biotechnol. Bioeng. **37**: 989-993.

7. Park, S. Y., Lee, G. M. 1995. Feasibility study on the use of hyperosmolar medium for improved antibody production of hybridoma cells in a long-term, repeated-fed batch culture. Bioprocess Engr. **13**: 79-86.

8. Reddy, S., Bauer, K. D., Miller, W. M. 1992. Determination of antibody content in live versus dead hybridoma cells : Analysis of antibody production in osmotically stressed cultures. Biotechnol. Bioeng. **40**: 947-964.

9. Reddy, S., Miller, W. M. 1994. Effects of abrupt and gradual osmotic stress on antibody production and content in hybridoma cells that differ in production kinetics. Biotechnol. Prog. **10**: 165 -173.

DEVELOPMENT OF HIGH OUTPUT EXPRESSION VECTORS FOR ANTIBODY PRODUCTION IN MAMMALIAN CELLS

CHRISTIAN GERVAIS[1,2], DENIS PAQUETTE[1,2], ANN BURNS-TARDIF[1,2], LUIS MARTIN[2] and BERNARD MASSIE[3]*
[1]*Biomira inc., 6100 Royalmount Ave., Montreal, QC, Canada H4P 2R2,*
[2]*Biomira inc., 9411 - 20 Ave., Edmonton, AB, Canada T6N 1H1,*
[3]*Biotechnology Research Institute/NRC, 6100 Royalmount Ave., Montreal, QC, Canada H4P 2R2.*

The design of efficient expression vectors is a major component in the development of a high level production system for recombinant antibodies. The BKV enhancer/Adenovirus major-late promoter (BKV/AdMLP) and human cytomegalovirus immediate-early enhancer/promoter (CMV) were evaluated. The Adenovirus tripartite leader containing a chimeric intron (AdTpl-SS), immunoglobulin leader intron and J/C chimeric intron were added and/or shifted. Cell lines COS-1, CV-1, 293S, CHO and NSO were used as hosts for transient assays. The secreted products were quantified by ELISA.
In all cell lines, the CMV enhancer/promoter yielded higher secretion levels than the BKV/AdMLP. AdTpl-SS and J/C intron each had a positive but not cumulative effect whereas splicing the immunoglobulin leader region only complicated the cloning and was without significant benefit. Positioning the J/C intron between the variable and the constant domain is necessary to obtain benefits except in CHO where the same intron inserted downstream of a cDNA yielded comparative results. Overall, adding splice site(s) to cDNA increased secretion 2 to 6 times. In transient expression, levels of 8 to 17 µg of Kappa chain per ml were measured respectively in COS-1 and in 293S cells using our best construct. Results obtained from transient assays led to the assembly of two cassettes in tandem to express both chains of B43.13 MAb. Stable expression levels in static batch cultures attain 80 µg/ml in 293S and over 350 µg/ml in NSO using serum-free medium.

1. Introduction

As part of the development of a high production system for recombinant antibodies, the design of an efficient expression vector plays a major role. Our main goals were to attain the maximum permissible level of expression in each particular cell line without compromising its survival and to maintain sufficient stability of expression levels through time to permit production in industrial quantities. Different approaches were developed and tested separately or in combination:

-Maximum high constitutive production level with genomic integration.

*Corresponding author

-Maximum high constitutive production level with episomal replication of expression vector.

-Low constitutive production level inducible through a switch or a transcription factor controlled by a switch.

Here we present the work leading to the assembly of an expression cassette for high level constitutive production of recombinant proteins in a variety of mammalian cells.

2. Methods

To measure the efficacy of the modifications implemented in expression vectors, we chose to use the Kappa Light Chain of chimeric MAbB43 as a marker protein since it will be a part of the final product and is well tolerated by the cells. Every modification made was tested separately or in combination in transient expression assays using several cell lines. Plasmid DNA's were purified on Cesium Chloride gradient. DNA concentration was determined by measuring O.D. at 260/280 nm and integrity verified on agarose gel. CV-1, COS-1, 293S and CHO cell lines were transfected using Lipofection; NSO cell line was transfected by electroporation. Transfections were repeated up to 4 times in duplicate each time. Supernatants were collected from 24 hours to 120 hours post-transfection. Systematic analysis were made on samples 48 hours and 120 hours post-transfection. In order to monitor secretion levels obtained with different constructs, fast and reliable tools were needed. Two immunoassays were developed:

-The first assay measures Human Kappa Light Chain in cell culture supernatants. Free and bound Kappa Chain can be quantified as well as any Human Antibody-Kappa when the proper standard is loaded. No cross-reaction with Lambda chains or with immunoglobulins from other species.

-The second assay measures Human IgG_3 in cell culture supernatants. No cross-reaction with immunoglobulin of other isotypes or species.

3. Results and Discussion

In transient expression assays, the kinetics of product secretion varied between cell lines. Unlike NSO and CHO, CV-1, COS-1 and 293S cell lines continued to secrete significant amounts of product after 48 hours (Fig. 1). All further results presented use data obtained 48 hours post-transfection. The trend after 120 hours in cells that continue to accumulate product after 48 hours is comparable.

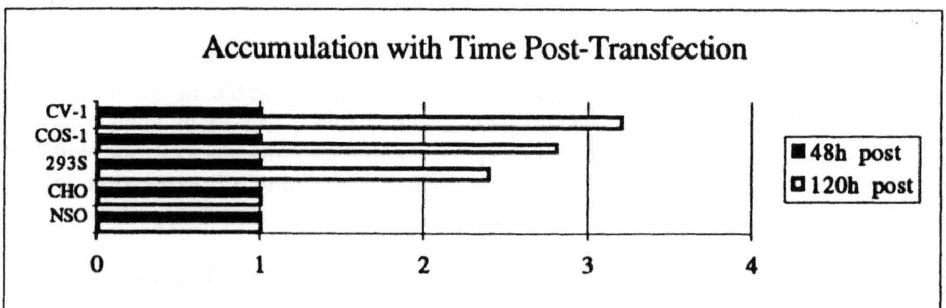

Figure 1. Relative levels of secreted human Kappa light chains with time post-transfection

For every cell line used for evaluation, levels of secreted Kappa Chain obtained with each construct were reported as relative values to each other within the same cell line. For each heading presented (Promoters, Splicing...), the less efficient was assigned the value of 1; the element under evaluation is then plotted as a multiple, quantifying the relative benefit. These data cannot be used to make comparisons between cell lines since their transfection efficiencies vary. However, a general positive trend for any element is a good indication of its potential usefulness.

3.1. PROMOTERS

We first assayed the relative strength of the BKV/AdMLP [1] and CMV IE enhancer/promoter [2](Fig. 2). In the five cell lines tested, with promoters either driving a cDNA or a gene (genomic construct), CMV yielded higher secretion levels than BKV/AdMLP. The advantage of CMV ranges from twice as good in CHO to over 10 times better in 293S and NSO.

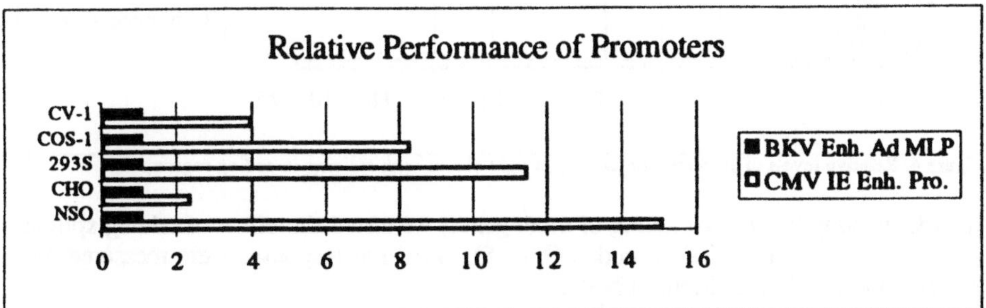

Figure 2. Relative levels of secreted human Kappa light chains with different promoters

3.2. SPLICING

Although it is widely recognized that using a gene yields better results than a cDNA for protein expression [3], it can be difficult to fit most genes into a reasonable sized vector. Evenmore so if two gene products are to be expressed by the same vector. Our goal was to find the best combination of splice site number and position without oversizing our vector.

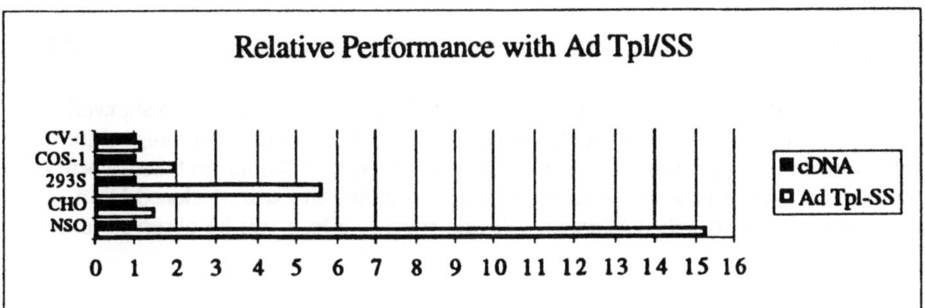

Figure 3. Relative levels of secreted human Kappa light chains with addition of Ad Tpl and splicing sites at the 5' end

Using a CMV driven expression cassette with a cDNA as base line, splice site(s) upstream, downstream or within the coding region were added and/or shifted. Splicing in 5' of the AUG start codon was tested concomitantly with the addition of Ad tripartite leader (Tpl) [1, 4, 5]. A beneficial effect was measured in every cell line tested except CV-1, more significant in CHO, COS-1, 293S and NSO (Fig. 3). It can be as good as five fold in 293S and up to fifteen fold in NSO.

In general, adding splicing sites within the coding region had a slightly better effect than splicing at the 5' end with the Ad Tpl (Fig. 4). Again, no major effect in CV-1 but a more significant effect in CHO cells.

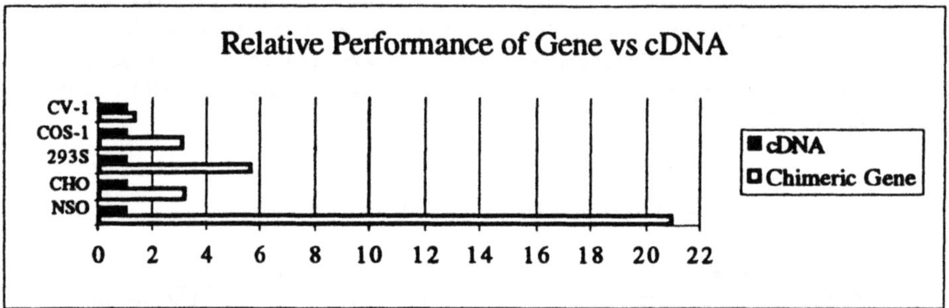

Figure 4. Relative levels of secreted human Kappa light chains with splicing in coding region

In order to simplify future cloning of other genes, we tested the effect of adding splicing sites downstream of the STOP codon (Fig. 5). Positive responses were measured only in CHO and to a lesser extent in NSO.

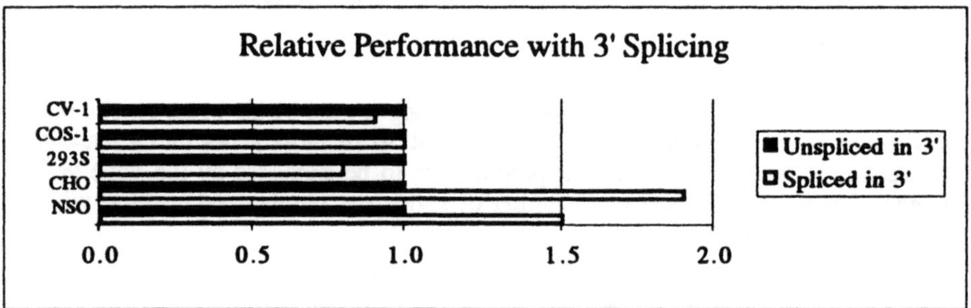

Figure 5. Relative levels of secreted human Kappa light chains with splicing downstream of STOP codon

The analysis of these results led us to assemble and test what we designated as a judicious combination of splicing sites at optimal positions yielding significant improvement over a simple cDNA (Fig. 6). However, care should be taken when sites are removed from a genomic sequence in order to reduce the size of cloned DNA. All splice sites are not equal; deleterious cell line specific effects can be observed (data not shown).

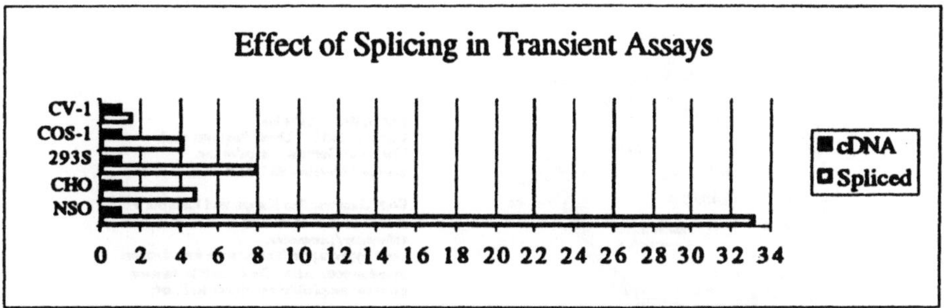

Figure 6. Relative levels of secreted human Kappa light chains with 'judicious' splicing

4. Conclusion

Our results show that optimization of the expression cassette is required to obtain maximum expression levels. Simply cloning a cDNA downstream of a strong promoter is not good enough. With our optimized expression cassette, the actual levels of transient expression of Kappa light chain obtained with different cell lines can be fairly high as shown in Figure 7. Although transfection protocols used might not have been optimal for each cell line, the quantity of product obtained after five days of culture could fulfill most R & D needs without having to generate stable clones.

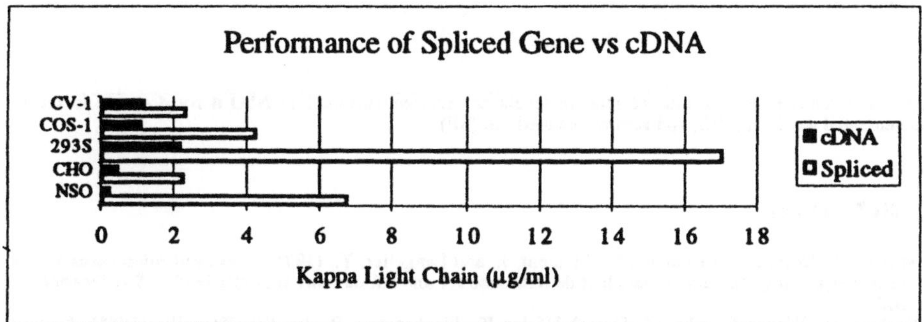

Figure 7. Actual levels of transient expression of Kappa Light Chain obtained in five cell lines

Both chains of chimeric B43.13 MAb were also successfully expressed in transient transfection. The Gamma 3 Heavy Chain gene was modified as the Kappa Chain. Levels of up to 8 µg of antibody per ml were obtained with 293S cell line using two small shuttle vectors expressing respectively the light and the heavy chain under the control of our optimized cassette. These results led to the assembly of 'pMPGB43' expressing both chains of chimeric B43.13 MAb (Fig. 8). Stable clones were generated in 293S and NSO cell lines. Levels in static batch cultures attain 80 µg/ml in 293S and over 350 µg/ml in NSO using serum-free medium. Moreover, adapting our clones to serum-free culture conditions not only improved secretion levels but also significantly enhanced long term stability of secretion, even without selective pressure (Fig. 9).

Other MAbs were cloned in this vector. Preliminary results obtained with chimeric B67.4 MAb (Gamma 1 and Gamma 3 Heavy Chains) are promising, leading us to predict yields similar to B43.13 MAb.

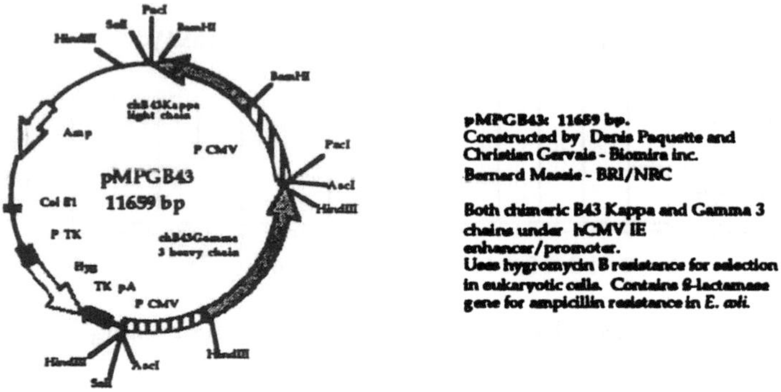

Figure 8. Schematic view of expression vector for chimeric antibody

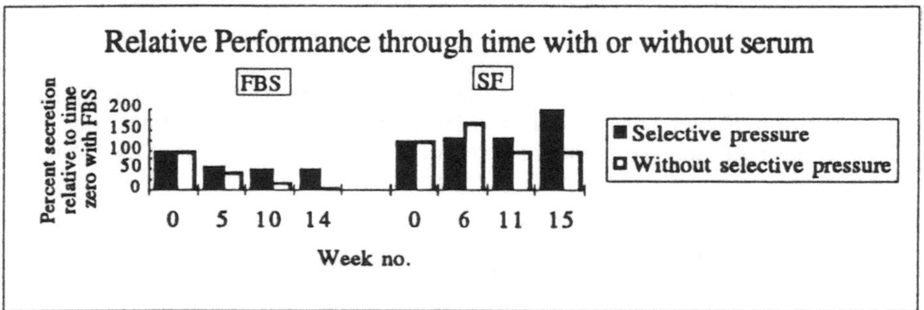

Figure 9. Evolution through time of relative levels of secreted chB43.13 in NSO derived clones in serum supplemented medium (FBS) and serum-free medium (SF)

4. References

1. Massie, B., Dionne, J., Lamarche, N., Fleurent, J., and Langelier, Y. (1995) Improved Adenovirus vector provides Herpes simplex virus ribonucleotide reductase R1 and R2 subunits very efficiently. *Bio/Technol.*, 13: 602-608.

2. Boshart, M., Weber, F., Jahn, G., Dorsch-Häsler, K., Fleckenstein, B., and Schaffner, W. (1985) A very strong enhancer is located upstream of an immediate early gene of human cytomegalovirus. *Cell*, 41: 521-530.

3. Kriegler, M. (1990) Assembly of enhancers, promoters, and splicing signals to control expression of transferred genes. *Methods in enzymology*, 185: 512-527.

4. Logan, J., Shenk, T. (1984) Adenovirus tripartite leader sequence enhances translation of mRNA's late after infection. *Proc. Natl. Acad. Sci. USA*, 81: 3655-3659.

5. Sheay, W., Nelson, S., Martinez, I., Chu, T.-H.T., Bhatia, S., and Dornburg, R. (1993) Downstream insertion of the adenovirus tripartite leader sequence enhances expression in universal eukaryotic vectors. *BioTechniques*, 15: 856-862.

Acknowledgements

We wish to thank Sylvain Mercille, Mark Johnson, Pierre Jolicoeur, Johanne Côté, Antoine W. Caron, Dick D. Mosser and Amine Kamen for their contributions. This work was supported by a contribution grant from the National Research Council of Canada to Biomira Inc.

MAGNETOLIPOSOMES AS INTRACELLULAR HEAT STRESS MEDIATORS FOR THE TEMPERATURE SENSITIVE EXPRESSION OF CYTOKINE

ISABELLE A. BOUHON, MASASHIGE SHINKAI, HIROYUKI HONDA and TAKESHI KOBAYASHI*.
Department of Biotechnology, Graduate School of Engineering, Nagoya University,
*Furo-cho, Chikusa-ku, Nagoya 464-01, Japan. (*Author for Correspondence)*

ABSTRACT. A dual way to turn temperature increase into account to improve hyperthermia was investigated. Application of intracellular heating by magnetoliposomes, enabled us to enhance expression of interferon-β gene regulated by the MMTV promoter. The expression obtained in an 39°C environment was higher than that in extracellular heating at 43°C. Cell killing effect was also raised by intracellular heating although this effect was not directly proportional to the cellular expression of IFN-β.

1. Introduction

The heating properties of magnetic fluids by hysteresis loss in a magnetic field have launched the concept of intracellular incorporation of thermoinducible material (Viroonchatapan,1996; Jordan, 1993) to meet the necessity of specific heating in hyperthermia. We have investigated the heating properties of cationic magnetoliposomes (Shinkai *et al.*, 1996), and their ability to be adsorbed and accumulated in cells place them in advantageous position compared to classical hyperthermia devices. However, the uniform distribution of these magnetite particles remains a critical aspect to realize *in vivo* and consequently, part of the tissue may not reach the fatal temperature. However, this deficiency can be advantageously exploited if the sublethal temperature augmentation triggers locally the expression of formerly transfected gene encoding cytokines.

In the present paper, we report on the enhancement of expression of interferon-β gene related to heat shock generation by magnetoliposomes. The interferon-β was chosen for the high cytotoxicity of its endogenous expression in glioma cells as demonstrated by Mizuno *et al.* (1990). We chose the MMTV (Mouse Mammary Tumor Virus) promoter which is induced by a non toxic substance and of which the induction is enhanced in the range of temperature afforded by mild hyperthermia (Sanchez *et al.*, 1994). The enhancement of interferon-β expression driven by the promoter and the correlated killing effect were evaluated under several heat stress conditions with intracellular (magnetoliposomes incorporation and magnetic induction) or extracellular heating (water bath).

2. Materials and Methods

2.1. MAGNETOLIPOSOMES AND PLASMID ENTRAPPED LIPOSOMES PREPARATION

Cationic magnetoliposomes were prepared by evaporation of chloroform solution with the molar ratio of 1:2:2 as TMAG:DLPC:DOPE ((*N*-(α-trimethylammonioacetyl)-didodecyl-D-glutamate chloride, Dilauroyl phosphatidylcholine and Dioleoyl phosphatidylethanolamine). After formation of a lipidic membrane in the flask, an aqueous colloidal solution of magnetite particles was addded. The lipids were dispersed by vortexing and the liposomes were formed by sonication.

The plasmid liposomes were prepared by the same method as above but the pSVMTVIFN-β plasmid (Mizuno *et al.*, 1990) solution was added instead of the magnetite particles and the liposomes were not sonicated for avoiding impairment of the plasmid.

2.2. GENE TRANSFECTION AND MAGNETITE INCORPORATION

The U87 MG glioma cells were seeded in 10 cmφ diameter dishes and grown until close to confluence. The cells were cultivated in Eagle's minimal essential medium (Life Technologies, Grand Island, NY) supplemented with 10 % fetal calf serum, 5 mM non-essential amino acids, 0.1 mg ml^{-1} streptomycin and 100 U ml^{-1} penicillin. They were transfected with pSVMTVIFN-β at the concentration of 0.2 mg ml^{-1} in DNA and 10 nM in lipids. A minimum of 8 hours later, the magnetoliposomes were added to reach 20 mg ml^{-1} in the medium.

2.3. HEAT STRESS TREATMENT AND INDUCER ADDITION

Extracellular heating (EH): The dishes were placed in an air tight vessel in contact with hot water (43°C) during 1 hour.
Intracellular heating (IH): 5 hours after magnetite incorporation, the cells were collected in pellet in an microcentrifugal tube. The tube was placed inside the coil of a high frequency electromagnetic field generator (118 kHz, Dai-ichi High Frequency Co., Tokyo, Japan). The temperature was controlled by insertion of an optical fiber probe at the center of the pellet which was maintained at 39°C in a thermostatic agarose gel (Shinkai *et al.*, 1996). The cells were reseeded in other dishes to count the viability.

Dexamethasone (Dex) was added to the dishes 2 hours after the heat treatment at the final concentration of 1 μM.

2.4. INTERFERON-β CONCENTRATION ASSAY AND GROWTH INHIBITORY EFFECT

The medium was replaced with the fresh one every day. IFN-β concentration in the medium was determined by using an enzyme immunoassay kit (Toray Industries Inc., Tokyo, Japan) which can detect human interferon-β until 2.5 IU ml^{-1} (International Unit)

The viable cell number was estimated by trypan blue exclusion method with a hemacytometer

3. Results and Discussion

3.1. ENHANCEMENT OF EXPRESSION

In the present study, the basal expression of IFN-β was 5 mIU $cell^{-1}$, because analogs of Dex were contained in the serum supplemented culture medium. In the case of EH for 60 min at 43°C, the consecutive enhancement was not stronger than 25% either with or without addition of Dex (1.1 without Dex, 1.3 with Dex) (Fig. 1). On the contrary, in the case of IH at 39°C, the IFN-β expression by heat stress alone reached 180 % of the basal expression in absence of the inducer and that by heat stress with Dex reached 160 % of the expression in presence of Dex.

Moreover, the enhancement appeared to depend on the heating interval duration, the optimum being 20 min in presence of Dex while 40 min heating gave the maximum expression without Dex. In average, the expression level in presence of the inducer was three times higher than that in its absence.

The difference of optimum heating period for maximizing the expression either in presence or in absence of the inducer can not be explained by a different mechanism of potentiation via glucocorticoid receptor activation according to the explanations provided by Sanchez *et al.* (1994). However, the change in pH (Kiang *et al.*, 1991) and pO_2 (Bicher *et al.*, 1980) generated by the hyperthermia may counteract the efficacy of the dexamethasone at one stage or other. This period of unfavorable intracellular environment is assumed to last proportionally with the heating period.

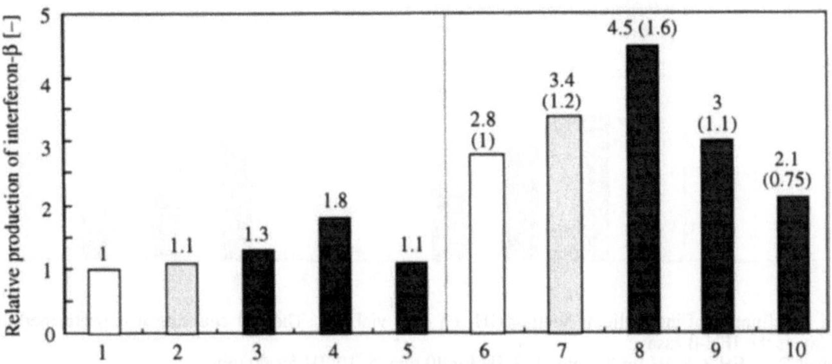

Figure. 1. Effect of heat stress on IFN–β expression either by extracellular (EH, by water bath) or intracellular (IH, by magnetic induction) heating. The basal expression for the control (5 mIU $cell^{-1}$) is taken as 1.0. The figure in parenthesis is relative value to the expression in column 6.
1; control, 2, 7; EH 3, 8; IH for 20 min, 4, 9; IH for 40 min, 5, 10; IH for 60 min.
Dexamethasone: 1–5; no addition, 6–10; addition.

3.2. CYTOTOXICITY

The considerable toxicity (68 % cell viability) was shown by expression of the plasmid pSVMTVIFN-β in cells (Fig. 2), which is correlated with the high sensitivity of the cell line to endogenous cytokine. In the case of the inducer addition, the viable cell number decreased to 47%, but the influence by EH on cell killing potential was not significant (45 % cell viability). These observations were correlated with the IFN-β production.

By intracellular heat shock, the hyperthermic effect was added to IFN-β cytotoxicity. The maximum killing effect was obtained with the longer duration of the magnetic field application, i.e. 60 min, which reduced the cell viability until 21 %. However, the influence of interferon-β level and the effect of the inducer addition were fading. Actually, even if the expression of cytokine was higher with addition of Dex, the difference in the viable cell number was comparable with that in absence of the inducer, especially when the duration of heating was extended.

The higher effect of the hyperthermia than the cytokine in cell killing with the extended duration of the stress may be explained from the work of Arancia *et al.* (1994) who demonstrated that the lipid toxicity increased with the temperature augmentation. Though we did not apply temperature above 39°C, we can assume that the intensity of the stress is comparably deepened by extension of the duration and by augmentation of the temperature. It can be deduced from these results that the transfected cells are more sensitive to hyperthermia than nontransfected cells. Thereby, the proportion of transfected cells which were heated insufficiently decreased with the time and the production of IFN-β weakened by loss of expressing cells. The observation that 60 min during heating impaired the protein expression supported this explanation.

In addition, Yoshida *et al.*, (1992) demonstrated by pSVMTVIFN-β and pSV2neo cotransfectants. The neomycin-selected cells were actually disappeared. Thus, at least the remaining 20 % cells in the case of 60 min IH seems to be nontransfected cells. Therefore, if the transfection efficiency is improved, all cells will be killed in 60 min IH.

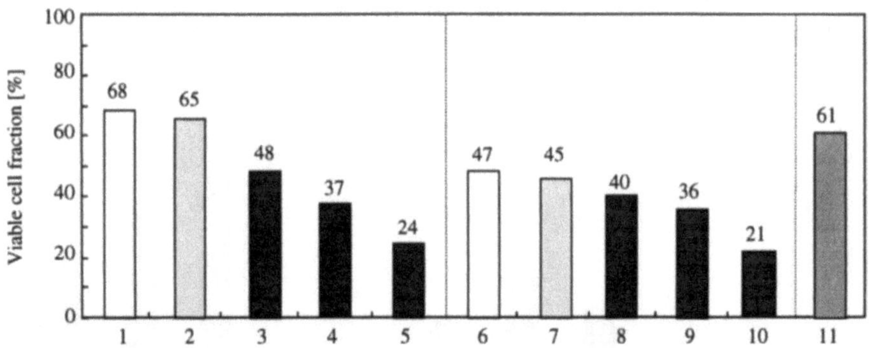

Figure 2. Influence of intracellular heating (IH) on cells viability. The cell counting was performed on the same day as the IFN-β assay.
1; control, 2, 7; EH 3, 8; IH for 20 min, 4, 9; IH for 40 min, 5, 10; IH for 60 min.
Dexamethasone: 1–5; no addition, 6–10; addition, 11: IH for 60 min for non-transfected cells.

4. Conclusions

The results obtained from the transfection of cancer cells confirmed that the MMTV promoter expression activity was sufficient to trigger the death of the transformed cells although the cell killing was not directly proportional to the cellular expression of IFN-β.

The feasibility to use the magnetoliposomes as satisfying hyperthermia mediator in order to increase the expression driven by a thermoinducible promoter was investigated in this study and we could conclude satisfactory results. In fact, the magnetoliposomes proved to be an efficient heat shock operator as the intracellular heating triggered high cellular expression in an 39°C environment, which was slightly higher than normal body temperature.

5. Acknoledgements
We thank Dr. Masaaki Mizuno and Prof. Jun Yoshida, School of Medicine, Nagoya University, for supply of the plasmid pSVMTVIFN-β.

6. References

Arancia, G., Calcabrini, A., Matarrese, P., Marcocci, L., Pietrangeli, P. and Mondovi, B. (1994) Effects of incubation with liposomes at different temperatures on cultured melanoma cells. *Int. J. Hyperthermia* **10**, 101–114.

Bicher, H.I., Hetzel, F.W. Snadhu, T.S. Frinka, S., Vaupel, P., O'Hara, M.D. and O'Brien, T. (1980) Effect of hyperthermia on normal and tumor microenvironment, *Radiology* **137**, 523–530

Jordan, A., Wust, P., Fähling, H., John, W., Hinz, A. and Felix, R. (1993) Inductive heating of ferrimagnetic particles and magnetic fluids: physical evaluation of their potential for hyperthermia, *Int. J. Hyperthermia* **9**, 51–68

Kiang, J.G., Mc Kinney, L.C. and Gallin, E.K. (1990) Heat induces intracellular acidification in human A-431 cells: role of Na^+-H^+ exchange and metabolism, *Am. J. Physiol.* **C28**, 727–737.

Mizuno, M., Yoshida, J., Sugita, K., Inoue, I., Seo, H., Hayashi, Y., Koshizaka, T. and Yagi, K. (1990) Growth Inhibition of glioma cells transfected with the human β-interferon gene by liposomes coupled with a monoclonal antibody, *Cancer Res.* **50**, 7826–7829.

Sanchez, E.R., Hu, J.L., Zhong, S., Shen, P., Greene, M.J. and Housley, P.R. (1994) Potentiation of glucorcorticoid receptor-mediated gene expression by heat and chemical shock, *Mol. Endo.* **8**, 408–421.

Shinkai, M., Yanase, M., Honda, H., Wakabayashi, T., Yoshida, J. and Kobayashi, T. (1996) Intracellular hyperthermia for cancer using magnetite cationic liposomes: in vitro study,. *Jpn. J. Cancer Res.* **87**, 1179–1183.

Viroonchatapan, E., Sato, H., Ueno, M., Adachi, I., Tazawa, K. and Horikoshi, I. (1996) Magnetic targeting of thermosensitive liposomes to mouse livers in an situ on-line perfusion system, *Life Sci.* **58**, 2251–2261.

Yoshida, J., Mizuno, M. and Yagi. K. (1992) Cytotoxicity of human β-interferon produced in human glioma cells transfected with its gene by means of liposomes, *Biochem. Int.* **28**, 1055–1061.

EFFECT OF AMMONIUM ION ON GLYCOFORM OF ERYTHROPOIETIN PRODUCED BY RECOMBINANT CHO CELLS USING LECTIN-BLOTTING TECHNIQUE

KERN HEE CHANG, KYUNG SOO KIM AND JUNG HOE KIM
Department of Biological Sciences
Korea Advanced Institute of Science and Technology
373-1, Ku sung dong, Yu sung gu, Taejon, Korea

Abstract The glycosylation pattern of Erythropoietin (EPO), produced by recombinant CHO cell, was studied using rapid and simple method of 'Lectin-blotting'. We used three different lectins in this experiment; MAA (Maackia amurensis agglutinine), RCA (Ricinus communis agglutinine), DSA (Datura stramonium agglutinine) which binds to terminal sialic acid, galactose and N-acetyllactosamine chain respectively. The effect of ammonium ion on glycosylation of EPO was examined since it accumulates in the medium as a byproduct mainly resulted from the glutamine metabolism. Consequently, treatment of ammonium chloride from 8mM to above substantially inhibited the sialylation of terminal galactose residue.

1. Introduction

Due to the recent recognition of environmental factors that affect glycosylation during cell culture process, the structural integrity of these oligosaccharides has attracted much attention for industrial scientists and engineers. Glucose starvation, serum concentration and change of extracellular pH were reported to be common factors that can affect on glycosylation [1,2,3]. Ammonium ion, a metabolic byproduct which accumulates in the medium, has been also reported to give disruptive effects on oligosaccharide processing [4]. The growing interest on the potential factors that can affect the glycosylation made it necessary for manufacturer to monitor glycosylation during the production of recombinant glycoprotein to ensure the structural integrity of oligosaccharides in the aspect of quality control. Thereafter, glycosylation analysis technique has become an important aspect in the structural characterization of glycoprotein.

In this experiment we chose three-lectin combination for characterizing the carbohydrate residue of Erythropoietin (EPO). The ammonium chloride was employed in order to examine the effect of ammonium ion on glycosylation of EPO, since it accumulates in the medium as a byproduct mainly resulted from glutamine metabolism.

2. Materials and Methods

2.1. CELLS AND THEIR MAINTENANCE

The culture was performed using recombinant Chinese hamster ovary (CHO) cell expressing human EPO. The human EPO gene in recombinant CHO cells was amplified with DHFR by methotraxate selection thereafter maintained in the medium without methotraxate. This cell line was kindly shared by Dr. H. J. Hong of Korea Research Institute of Bioscience and Biotechnology (KRIBB). Cells were grown in Minimal eagle's medium alpha (MEMα) (Life technologies, NY, USA) supplemented with 2g/L D-glucose, and 10 % fetal bovine serum (FBS) (Gibco laboratories, NY, USA). For serum-free production medium, 10% fetal bovine serum was replaced by ITS (insulin 5μg/mL, transferrin 5μg/mL, sodium selenite 20nM). Cells were cultured on 75cm^2 T-flask (Falcon, NJ, USA) and grown in an incubator at 37°C in an atmosphere of 5% CO_2 and 95% humidity.

2.2. CELL CULTURE

The culture started with 2 x 10^6 cells in T-75 flask with 20mL of growth medium and incubated for 3 days until confluence in order to ensure that overgrowth of the cell monolayer had not occurred and that all experiments used similar cell numbers. The medium was removed once when the T-flasks were confluent and each flasks were rinsed twice with phosphate buffered saline (potassium phosphate monobasic; 0.2g/L, potassium chloride; 0.2g/L, sodium chloride; 8.0g/L and anhydrous sodium phosphate dibasic; 1.15g/L). The flasks were then switched to 20mL of serum-free production medium supplemented with either 0, 4, 8, 12mM of ammonium chloride (NH_4Cl) and cultured for 12 hours maintaining the viability above 90%. After the production phase, medium was withdrawn and stored in -20°C freezer. The cultures were all duplicated.

2.3. SDS-PAGE AND LECTIN, IMMUNOBLOTTING

Supernatant samples were concentrated using 10,000 NMWL ultrafree-CL ultrafiltration membranes (Millipore, Bedford, MA) to 40-fold then separated by SDS-PAGE(12% acrylamide gel). Once the sample has been separated, the protein bands were transferred to 0.45 micron nitocellulose membrane (Hybond C, Amersham, UK) in Transfer buffer (Tris; 15.6mM and Glycine; 120mM). After blocking the unoccupied protein binding sites on the nitrocellulose sheets with 1% bovine serum albumine (BSA) in TBS, the nitocellulose sheets were incubated respectively with 3 types of digoxigenin (DIG) conjugated lectins (Table 1.)[5,6]. The sheets were washed 3 times in TBST buffer and subsequently incubated with alkaline phosphatase conjugated anti-DIG antibody (Behringer Mannheim, Mannheim, Germany). Finally, sheets were washed 3 times in TBST buffer then incubated in 5-bromo-4-chloro-3-indol phosphate/4-nitroblue tetrazolium chloride (BCIP/NTP) reagent (Promega, WI, USA). For the immunoblotting of samples, DIG conjugated lectin and anti-DIG antibody and alkaline phosphatase conjugated anti-DIG antibody should be replaced with monoclonal anti-EPO antibody and alkaline phosphatase conjugated anti-mouse IgG antibody (Promega, WI, USA) , respectively.

TABLE 1. Lectins used in blotting experiment and their specificity

Lectin	Source	Specificity	Specificity on EPO
DSA	*Datura stramonium*	Galβ(1-4)GlcNAc	N-linked chain: N-acetyl-lactosamine chain
		GlcNAc-Ser / Thr	O-linked chain: N-acetyl-glucosamine chain
MAA	*Maackia amurensis*	NeuNAcα(2-3)Gal	Terminal sialic acid residue
RCA	*Ricinus communis*	βGal > αGal	Galactose residue (rather than sialic acid) exposed at chain end (asialyo-chain)

3. Results and Discussion

3.1 THE INTERPRETATION OF BLOTS

The blot patterns of samples can provide information on glycan microheterogeneity (Fig 1.) Since DSA(Datura stramonium agglutinine) binds mainly to N-acetyllactosamine chain (Galβ1-4GlcNAc), the structural integrity of N-acetyllactosamine residue of the sample can be verified by the appearance of blot. MAA(Maackia amurensis agglutinin) recognizes glycan when it has sialic acid at the end of glycan (NeuAcα2-3Gal) whereas RCA(Ricius communis agglutinin) recognizes sialic acid missing glycan end with galactose residue exposed. The combination of using these two lectins, MAA and RCA, in the analysis enables one to clearly judge the existence of terminal sialic acid. Considering the fact that there is a close relationship between sialic acid and biological activity , it is highly necessary to evaluate the sialylation of EPO. In this way, it is possible to deduce the structural information of glycan of EPO

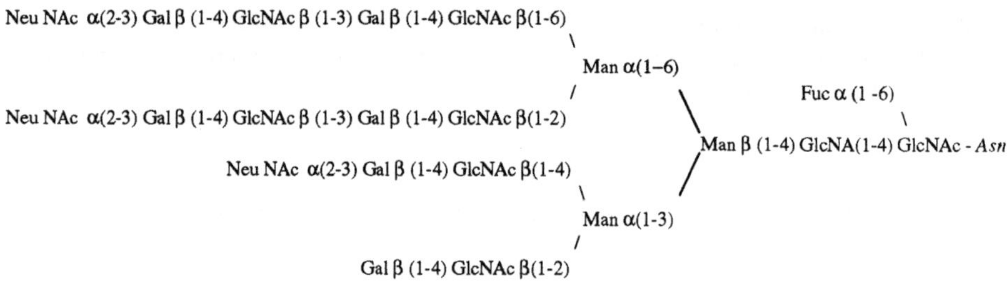

Figure 1. Structure of major N-linked carbohydrate chains of EPO

3.2. EFFECT OF AMMONIUM CHLORIDE ON EPO GLYCOSYLATION

Supernatant of the culture supplemented with 0, 4, 8, 12mM of ammonium chloride were collected and analyzed using immunoblotting and lectin blotting. The immunoblots and the lectin-blots by DSA did not show any significant change of the pattern throughout all concentrations of ammonium chloride (Fig 2A). However, the results of the lectin-blots using RCA and MAA were different. The blot samples which probed with MAA showed a remarkably decreased density from 8mM to above and in samples with RCA, clear blots appeared from 8mM to above (Fig 2C, 2D lane 6 to 9). These results indicate that ammonium chloride did not affect on synthesis of N-acetyllactosamine but strongly inhibited the terminal sialylation from the concentration of 8mM to above. Considering the fact that, in case of perfusion culture system, the accumulation of ammonium ion often reached as much as high concentration that can affect on the sialylation as we can see in this result, it is quite necessary to monitor effect of culture conditions on the glycoform of the proteins during the culture process. This study also showed that the lectin-blotting method using the combination of DSA, MAA, RCA could be useful for the characterization of glycoform of EPO owing to its simplicity and rapidity.

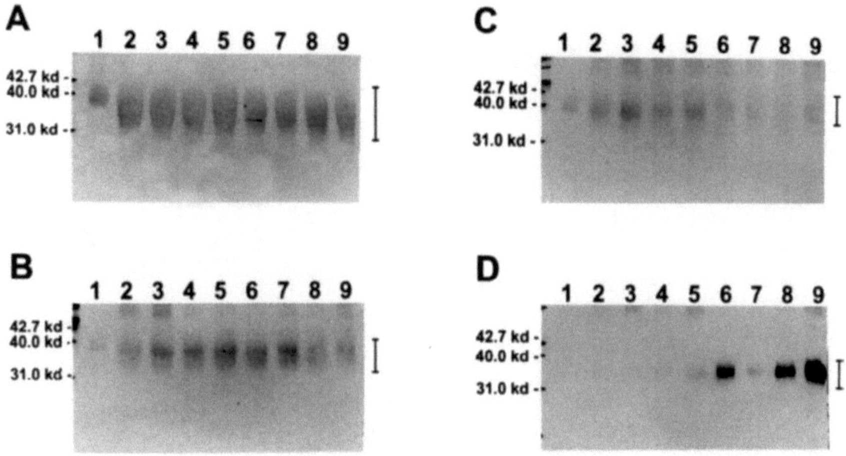

Figure 2. Blot patterns of EPO produced in media supplemented with various concentration of NH_4Cl (A) Immunoblots of anti-EPO antiblody. (B) Lectin-blots of DSA. (C) Lectin-blots of MAA. (D) Lectin-blots of RCA. Lane 1: Authentic sample of EPO, lane 2,3: culture sample without NH_4Cl, lane 4,5: 4mM NH_4Cl, lane 6,7: 8mM NH_4Cl, lane 8,9: 12mM NH_4Cl.

5. References

1. Hayter, P. M., Curling, E. M.A., Bains, A. J.., Jenkins, N., Salmon, I., Strange, P. G., Tong, J. M., Bull, A. T. (1992) Glucose-limited chemostat culture of Chinese hamster ovary cells producing recombinant human interferon-γ, *Biotechnol. Bioeng.*, **39**, 327-335
2. Borys, M., Linzer, D. I. H., and Papoutsakis E. T.(1993) Culture pH affects expression rates and glycosylation of recombinant mouse placental lactogen proteins by Chinese hamster ovary cells (CHO), *Bio/Technol.*, **11**, 720-724
3. Moellering, B. J.., Tedesco, J. L., Scott, R., Townsend, R., Hardy, M. R. and Prior, C. P. (1989) Electrophoretic differences in a Mab expressed in three media, Biopharm., **3**, 30-38
4. Borys, M., Linzer, D. I. H., and Papoutsakis E. T.(1994) Ammoium affects the glycosylation patterns of recombinant mouse placental lactogen-I by Chinese hamster ovary cells in a pH-dependent manner, *Biotechnol. Bioeng.*, **43**, 505-514
5. Haselbeck, A., Schickaneder, E., Eltz, H. and Hosel, W. (1990) Structural characterization of glycoprotein carbohydrate chain by using Digoxigenin-labeled lectins on blots, *Anal. Biochem.* **191**, 25-30
6. Haselbeck, A. and Hosel, W. (1993) 'Immunological detection of glycoproteins on blots base on labeling with Digoxigenin', in Hounsell, E. F.(eds.) *Glycoprotein analysis in biomedicine*, Humana press inc., New Jersey, USA, pp.161-173

Effect of Cloned Gene Dosage on the Expression Level and Stability of Chimeric Antibody Producing CHO Cells

SANG JICK KIM, NO SOO KIM, HYO JEONG HONG[1], GYUN MIN LEE
Department of Biological Sciences, KAIST, Taejon 305-701, Korea
[1]Antibody Engineering Research Group, KRIBB, Taejon 305-606, Korea

1. Abstract

Recombinant Chinese hamster ovary (CHO) cells expressing high level of chimeric antibody against S surface antigen of hepatitis B virus (HBV) were obtained using stepwise methotrexate (MTX)-dihydrofolate reductase (dhfr) mediated gene amplification procedure and subsequent cell cloning by limiting dilution method. Specific growth rate of amplified clone was inversely proportional to the MTX level used in gene amplification. Specific antibody productivity (q_{Ab}) of amplified clones rapidly increased with MTX level up to 0.32 μM and thereafter, q_{Ab} became saturated at 20 $\mu g/10^6$cells/day. The heavy chain gene copy number of amplified clones directly correlated with q_{Ab}. The q_{Ab} of all tested clones gradually decreased during the long-term culture in the absence of MTX, which resulted from the loss of antibody cDNA. Despite the decreased q_{Ab}, the 1 μM amplified clone could maintain high volumetric antibody productivity over three months because cell growth improved during long-term culture in the absence of MTX.

2. Introduction

For high level expression of recombinant antibodies, the most widely used mammalian expression system is probably the gene amplification procedure offered by the use of dihydrofolate reductase (DHFR) deficient CHO cells. When subjected to successive rounds of selection in medium containing stepwise increments in methotrexate (MTX), recombinant CHO (rCHO) cells with specific antibody productivity (q_{Ab}) as high as 100 $\mu g/10^6$cells/day were obtained by Page and Sydenham [4].

For large-scale production of recombinant antibodies for *in vivo* therapy, not only the expression level but the stability of cell lines used is critical. Changes in rCHO populations after extended culture in the presence and absence of selective pressure have been reported [1, 6, 9]. However, there are few reports assessing the quantitative effect of increasing the number of cloned gene copies on the stability of CHO cells regarding q_{Ab} [3].

In the present study, the effect of cloned heavy and light chain gene copies on the expression level as well as the stability of CHO cells was examined systematically during a long-term culture in the presence and absence of selective pressure.

3. Materials and Methods

Parental CHO cells expressing a chimeric antibody with human constant regions and mouse variable regions against the S surface antigen of HBV were made by cotransfecting the heavy and light chain expression plasmids into DHFR-deficient CHO cells (DG44), and selection in α-MEM supplemented with 10% dialyzed fetal bovine serum (d-FBS) and G418 (550 μg/mL). Positive parental cells were subjected to increasing levels of MTX (0.02, 0.08, 0.32, 1.0., and 4.0 μ*M*). Clones with the highest antibody productivity from each amplified pool were isolated by limiting dilution method. All the amplified clones at various MTX levels were grown in α-MEM with 10% d-FBS and the corresponding level of MTX in a 5% CO_2/air mixture, humidified at 37°C. For long-term cultures, clones amplified at various levels of MTX were cultivated as monolayer cultures in 25 cm^2 T-flasks containing 5 mL of α-MEM with 5% d-FBS in the absence as well as in the presence of the corresponding level of MTX.

Viable cell concentration and viability were determined by the trypan blue dye exclusion method using a hemocytometer. The secreted chimeric antibody demonstrating anti-S specificity was quantified using an ELISA as described previously [2]. The intracellular antibody was quantified using a flow cytometry. Fluorescein-isothiocyanate (FITC)-labeled goat anti-human IgG (heavy chain specific, Southern Biotechnology Associates, Birmingham, AL) was used for detection of intracellular antibody. For estimation of gene copies per cell, Southern blot analysis was conducted as described in the DIG user's guide (Boehringer Mannheim).

4. Results and Discussion

4.1 CHARACTERIZATION OF AMPLIFIED CLONES

Fig. 1 shows that μ of amplified clones was inversely related to the concentration of MTX. The μ of parental clone was 1.01 day^{-1} whereas the μ of 4 μ*M* amplified clone was 0.42 day^{-1}. As expected, q_{Ab} rapidly increased with MTX level up to 0.32 μ*M* and thereafter, q_{Ab} became saturated at 20 μg/10^6 cells/day.

The relative gene copy numbers of amplified clones, which were determined by comparison of band intensities on Southern blot, were plotted as a function of MTX concentration in Fig. 2. Both light and heavy chain gene copy numbers rapidly increased with MTX level up to 0.08 μ*M*. Thereafter, the light chain gene copy number continued to increase gradually with increasing MTX level. On the other hand, the heavy chain gene copy number became saturated. This result suggests that q_{Ab} of amplified clones was related with their heavy chain gene copy number rather than light chain gene copy number.

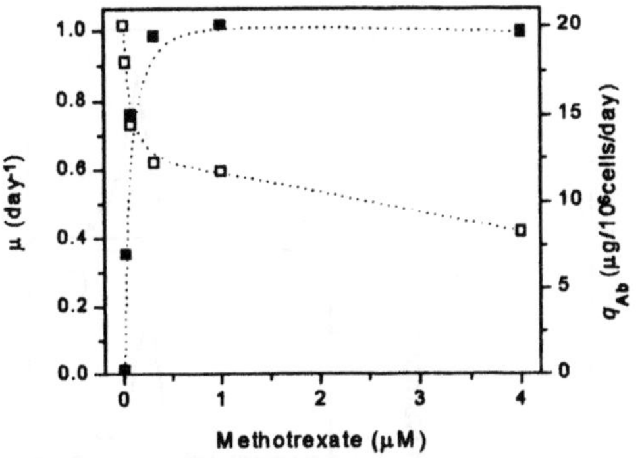

Figure 1. Specific growth rate (μ) and antibody production rate (q_{Ab}) of amplified clones selected from various amplified pools:(□) μ and (■) q_{Ab}.

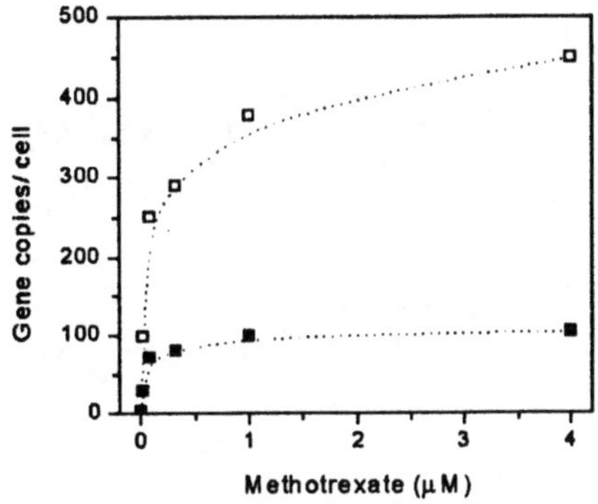

Figure 2. Light (□) and heavy (■) chain gene copy numbers of amplified clones.

Flow cytometric analysis also suggested that the expression level of heavy chain in amplified clone limited the enhancement of q_{Ab} (data not shown). The intracellular antibody content increased with MTX level up to 0.32 μ*M* and thereafter, no further increase in intracellular antibody content was observed. Since FITC-labeled goat anti-human IgG used in the quantification of intracellular antibody content was heavy chain specific, the fluorescent intensity of cells actually account for the heavy chain content. On the other hand, the intracellular DHFR, like the light chain gene copies, continued to increase with MTX level up to 4 μM. Since the dhfr gene was placed only in the light chain expression plasmid, the dhfr gene was coamplified with the light chain gene. Thus, the expression level of dhfr is likely to correlate with that of the light chain. This disproportionate amplification and/or expression of Ig chains may be a drawbacks in using two vector systems with each Ig chain on a separate plasmid [5].

4.2 STABILITY OF AMPLIFIED CLONES

Fig. 3 shows q_{Ab} of amplified clones during the culture. In the presence of MTX, the q_{Ab} of amplified clones slightly decreased during the culture. In the absence of MTX, the decrease in q_{Ab} of amplified clones was significant. Furthermore, the relative extent of decrease in q_{Ab} varied among the amplified clones. The 1 μ*M* amplified clone was most stable and retained 55% of the initial q_{Ab} after the 21st passage (approximately 60 generations). The 0.08 μ*M* amplified clone, on the other hand, was most unstable and retained only 24% of the initial q_{Ab} after the 21st passage. Thus, the antibody production stability of amplified clones was not related to the MTX concentrations where the cells were amplified, and hence to extent of elevated gene dosage.

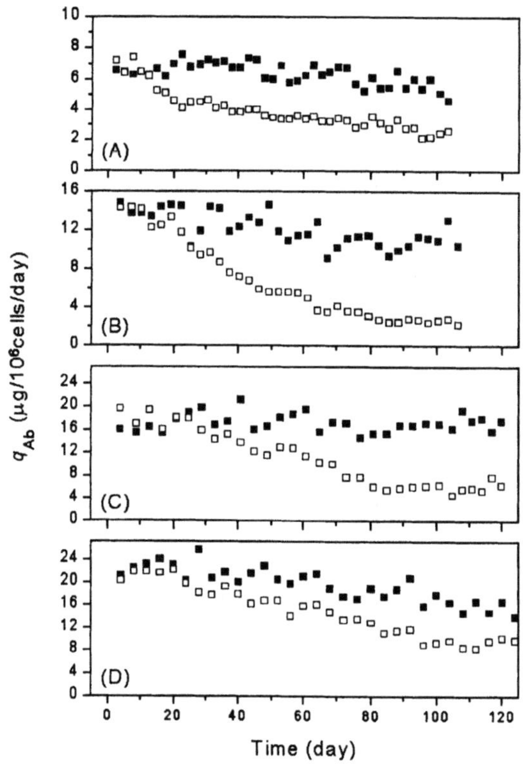

Figure 3. Antibody production characteristics of amplified clones during long-term culture in the presence (■) and absence (□) of MTX: (A) 0.02 μ*M*, (B) 0.08 μ*M*, (C) 0.32 μ*M*, (D) 1 μ*M* clone.

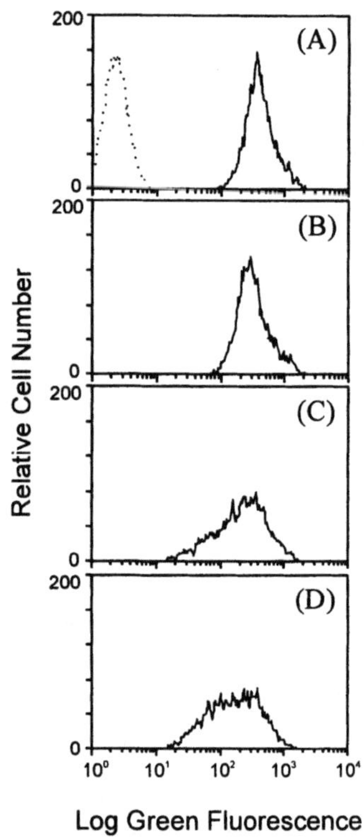

Figure 4. Flow cytometric analysis of intracellular antibody content of 1 μ*M* amplified clone during long-term culture ((A) 6th, (B) 11st, (C) 21st, (D) 31st passage) in the absence of MTX: Dotted line represents DG44 used as negative control.

As observed in q_{Ab} of amplified clones during long-term culture, the 0.08 μ*M* amplified clone in the absence of MTX displayed the most significant loss of gene copies. It retained only 14% of initial heavy chain copies and 10% of light chain gene copies. On the other hand, the 1 μ*M* amplified clone, which was most stable, could retain 59% of initial heavy chain copies and 65% of light chain gene copies. The 1 μ*M* amplified clone displayed the highest volumetric productivity. Furthermore, this high volumetric antibody could maintain over three months even in the absence of MTX. This result could be achieved because the cell growth rate in the absence of MTX was higher than that in the presence of MTX and it was improving during long-term culture.

Like q_{Ab}, the intracellular antibody content of amplified clones gradually decreased in the absence of MTX. There is a clear trend that the heterogeneity of cell population with respect to intracellular antibody content increased during long-term culture (Fig. 4). Fluorescence in situ hybridization (FISH) study of CHO amplicon by Wurm et al. showed meaningful insight to the stability of amplified sequence in chromosomes of cells grown for extended period in the presence or absence of MTX [7, 8]. According to their explanation, the 1 μM amplified clone in this study might have increased stable amplified structure known as master integations by further round of MTX-mediated gene amplification and thereby was relatively stable in the absence of MTX.

In conclusion, rCHO cells expressing high level of chimeric antibody against S surface antigen of HBV were obtained by cotransfection of heavy and light chain cDNA expression vectors into dhfr-deficient CHO cells and subsequent gene amplification followed by limiting dilution. The proportionate amplification of Ig chain genes may further improve the expression level of these rCHO cells. The rCHO cells amplified at 1 μM MTX could maintain their high volumetric productivity over three months in the absence of MTX.

5. Acknowledgments

The authors would like to thank H.G. Lee for help with the flow cytometry. This work was supported in part by Ministry of Science and Technology, Korea.

6. References

1. Kaufman, R.J., Wasley, L. C., Spiliotes, A. J., Gossels, S. D., Latt, S. A., Larsen, G. R., Kay, R. M. 1985. Coamplification and coexpression of human tissue-type plasminogen activator and murine dihydrofolate reductase sequences in Chinese hamster ovary cells. J. Mol. Biol. **5**: 1750-1759.
2. Kim, J.H., Bae, S.W., Hong, H.J. Lee, G.M. 1996. Decreased chimeric antibody productivity of KR12H-1 transfectoma during long-term culture results from decreased antibody gene copy number. Biotechnol. Bioeng. **51**: 479-487.
3. Michel, M. L., Sobczak, E., Malpi ce, Y., Tiollais, P., Streeck, R. E. 1985. Expression of amplified hepatitis B virus surface antigen genes in Chinese hamster ovary cells. Bio/Technology 3: 561-566.
4. Page, M.J., Sydenhaim, M.A. 1991. High level expression of the humanized monoclonal antibody Campath-1H in Chinese hamster ovary cells. Bio/Technology **9**: 64-68.
5. Trill, J. J., Shatzman, A. R., Ganguly, S. 1995. Production of monoclonal antibodies in COS and CHO cells. Curr. Opin. Biotechnol. 6: 553-560.
6. Weidle, U. A., Buckel, P., Wienberg, J. 1988. Amplified expression constructs for human tissue-type plasminogen activator in Chinese hamster ovary cells: instability in the absence of selective pressure. Gene **66**: 193-203.
7. Wurm, F. M., Pallavicini, M. G., Arathoon, R. 1992. Integration and stability of CHO amplicons containing plasmid sequences. Devlop. Biol. Standard. **76**: 69-82.
8. Wurm, F. M., Petropoulos, C. J. 1994. Plasmid integration, amplification and cytogenetics in CHO cells: questions and comments. Biologicals **22**: 95-102.
9. Zettlmeissl, G., Ragg, H., Karges, H. 1987. Expression of biologically active human antithrombin III in Chinese hamster ovary cells. Bio/Technology **5**: 720-725.

A NOVEL APPROACH TO PREPARE APPROPRIATE ANTIBODIES FOR IMMUNOAFFINITY CHROMATOGRAPHY

ATSUSHI ENOMOTO, NAOSHI KAMATA, JUN-ICHI KURISAKI*, KOZO NAKAMURA† and NAOHIRO SHIRAGAMI
Department of Biological and Chemical Engineering, Faculty of Engineering, Gunma University, 1-5-1 Tenjin, Kiryu, Gunma 376, Japan
**National Institute of Animal Industry, Tsukubanorindanchi, Ibaraki 305, Japan*
†*Present address: Graduate School of Agriculture and Agricultural Life Sciences, The University of Tokyo, Yayoi, Bunkyo, Tokyo 113, Japan*

1. Introduction

Immunoaffinity chromatography, which is based on the fine specificity of antigen-antibody interactions, has been frequently employed for the spectacular one-step purification of proteins and enzymes from such diverse and complex mixtures as cell extracts, tissue culture fluids and microbial fermentation broths [1]. The extremely high affinity between antigens and antibodies, however, may often require hostile conditions (usually low pH or high concentration of chaotropic agents) for the elution step, resulting in denaturation and inactivation of both the purified products and the immunoadsorbents. In this respect, an appropriate antibody for this method can be defined as one effectively binding the target molecule in the adsorption step and subsequently dissociating the bound antigen with mild elution conditions [1-5]. A rational strategy for the screening and selection of such desired antibodies will become a critical point to make immunoaffinity chromatography more facilitative and attractive, particularly for the purification of unstable materials.

To develop and establish an effective method for preparing such appropriate antibodies, we here selected bovine ß-lactoglobulin (ß-Lg) as a model protein antigen, and three kinds of murine antisera raised after the primary, secondary and tertiary antigenic challenges were examined and compared for their ability to bind the antigen under differing pH and ionic strength conditions by an enzyme-linked immunosorbent assay (ELISA). In addition, seven kinds of anti-peptide antisera with distinct epitope specificity and four anti-peptide 42-56 monoclonal antibodies (mAbs) with highly similar specificity were also tested and compared for their reactivity to ß-Lg under various conditions of ionic strength.

2. Materials and Methods

2.1. Antigens

Bovine ß-lactoglobulin (genetic variant A), one of the major milk whey proteins, was prepared from fresh raw skim milk and purified by anion exchange chromatography on DEAE-Sephacel, as described elsewhere [6]. The peptide fragments of ß-Lg, 22-36, 32-46, 42-56, 62-76, 72-86, 119-133 and 149-162, were prepared by a peptide synthesizer and coupled with keyhole limpet hemocyanin (KLH, a carrier protein), according to the protocol of Mariani *et al.* [7].

2.2. Antisera

Groups of 4 or 5 BALB/c mice (female, 6 week-old) were injected intraperitoneally (i.p.) with 100 μg of ß-Lg or each peptide-KLH conjugate in complete Freund's adjuvant [4]. Two and four weeks later, the animals were boosted i.p. with a similar dose of the same antigen in incomplete Freund's adjuvant and were bled one week after each immunization.

2.3. Monoclonal antibodies

Four distinct mAbs used in this study were produced by cell fusions of spleen cells derived from BALB/c mice, which had been hyperimmunized with the peptide 42-56-KLH conjugate, with the myeloma cell line X-63-Ag.8-6.5.3. [5]. According to the manufacture's instruction, these mAbs were purified from the mouse ascites fluids by an Affi-Gel Protein A MAPS-II Kit.

2.4. Enzyme-linked immunosorbent assay

ELISA was carried out as described previously [4, 5]. Microtiter plates were first coated with a 0.001 % ß-Lg solution by incubating them overnight at 4 °C. Next, antisera or mAbs diluted in a phosphate, citrate or glycine buffer containing 0.05 % Tween 20 to an appropriate concentration giving about 50 % saturation was placed in the wells for further incubation. Before starting the experiments, we had confirmed that the antigen remained to be adsorbed to the wells in the range of pH and ionic strength that was examined [5]. The bound antibodies were then reacted with alkaline phosphatase-conjugated goat anti-mouse Ig(G+M+A) antiserum. The plates were finally incubated with a 0.1 % *p*-nitrophenylphosphate buffer solution at 37 °C for 30 min, and the resulting color development at 405 nm was measured. All experiments were done in triplicate, and the background level for non-specific binding was subtracted from the mean of individual values.

3. Results and Discussion

3.1. Antibodies elicited by the primary response

Four BALB/c mice were immunized with ß-Lg 3 times at intervals of 2 weeks, and the antisera elicited by the primary and secondary responses were tested for their ability to bind the antigen under various conditions of pH and ionic strength. As shown in Fig. 1, the binding activity of antibodies produced by the primary response was usually found to be highly sensitive to the pH value, especially in the acidic region, as compared with those following the second and third antigenic stimulations. In the case of the 3 mice except for no. 2, the relative binding activity of antibodies from the primary response was markedly less below pH 6 and became quite small at pH 4.5. On the other hand, both antibodies raised after the second and third immunizations could effectively react with ß-Lg over a wider pH range, and more than 50 % of the maximum binding activity remained at pH 4.5.

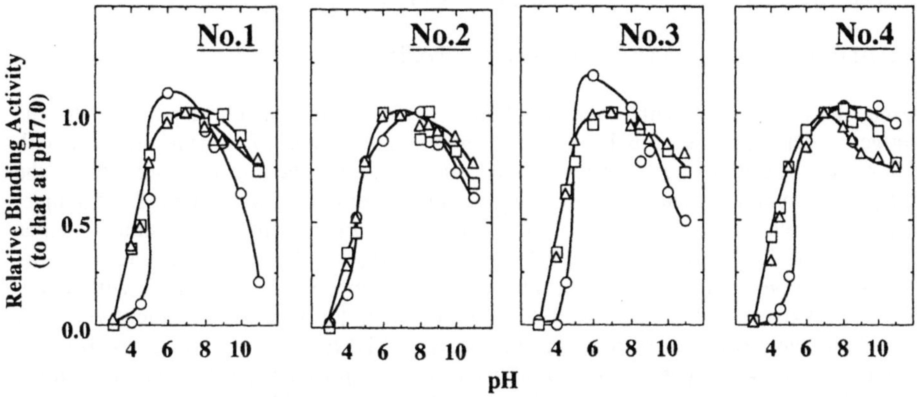

Figure 1. Effects of pH on the relative binding activity of mouse antisera raised after the first (○), second (□) and third (△) immunizations with ß-Lg.
The antisera were diluted in a phosphate, citrate or glycine buffer at the indicated pH value and with a constant ionic strength (0.02 mol/kg), and each binding activity relative to that at pH 7.0 was determined by ELISA.

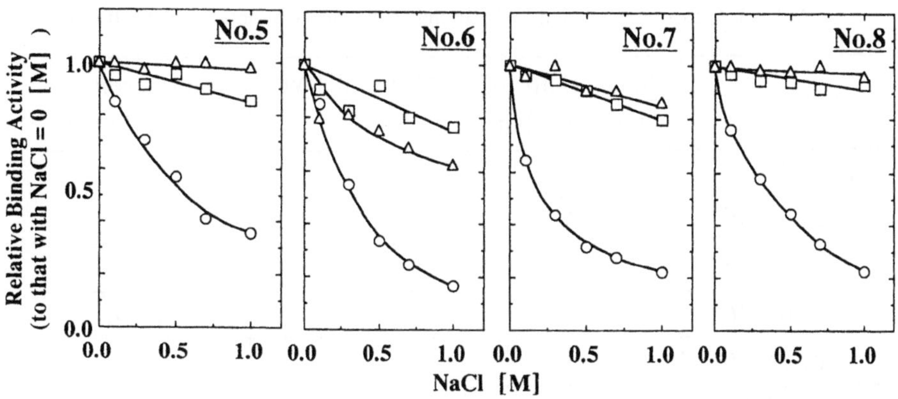

Figure 2. Effects of ionic strength on the relative binding activity of murine antisera raised after the first (○), second (□) and third (△) immunizations with ß-Lg.
Each antiserum was diluted in 20 mM phosphate buffer containing the indicated molar concentration of NaCl at a constant pH value (7.0), and the binding activity relative to that with the phosphate buffer (NaCl = 0 M) was determined by ELISA.

A similar phenomenon was also observed when determining the effects of ionic strength on the antigen-antibody interactions (Fig. 2). The binding activity of antisera from the primary response sharply decreased with increasing ionic strength, while that from the secondary response could form a stable antigen-antibody complex over an extremely wide range of ionic strength. These results from Figs. 1 and 2 suggest that antibodies raised during the early stage of immune response may be useful as ligands for immunoaffinity chromatography. Although the antibody titer from the primary response was generally shown to be extensively low [4], the application of hybridoma and gene technology will facilitate a sufficient supply of the optimum antibody from the repertoire of the primary response.

3.2. Anti-peptide antisera and monoclonal antibodies

Using KLH coupled with each synthetic peptide fragment corresponding to the antigenic determinants of ß-Lg as an immunogen, 7 kinds of murine antisera with distinct epitope specificity were prepared and tested for their cross-reactivity to the native protein and susceptibility to changing ionic strength. Among them, for example, antibodies specific to the peptide 42-56 including 6 charged amino acid residues such as Glu, Asp and Lys were capable of effectively binding to ß-Lg, and their reactivity was highly sensitive to ionic strength, although the other peptide fragments with many charged residues, 62-76, 72-86 and 119-133, could not induce the sufficient levels of peptide-native protein cross-reactive antibodies (TABLE 1). Furthermore, the binding activity of 3 in 4 anti-peptide 42-56 mAbs against ß-Lg also strongly depended on the slight change in ionic strength; the relative binding activity of LG5.2, LG5.3 and LG5.7 dramatically decreased with increasing ionic strength, and was almost lost at the condition of phosphate buffer including 0.5 M NaCl (Fig. 3). From these results, these mAbs may be mainly directed to the charged residues, thereby resulting in their high sensitivity to the ionic strength.

TABLE 1. Cross-reactivity and ionic strength dependence of anti-peptide antisera

Antisera*	Charged amino acid number	Cross-reactivity to β-Lg	Ionic strength dependence
Anti 22-36	2	○	×
Anti 32-46	4	○	○
Anti 42-56	6	○	○
Anti 62-76	7	×	N. D.
Anti 72-86	5	△	△
Anti 119-133	5	△	N. D.
Anti 149-162	3	○	○

O, highly reacted (depended); Δ, slightly reacted (depended); X, not reacted (depended); N. D., not determined

*Antisera raised after the third antigenic stimulation were used in this study.

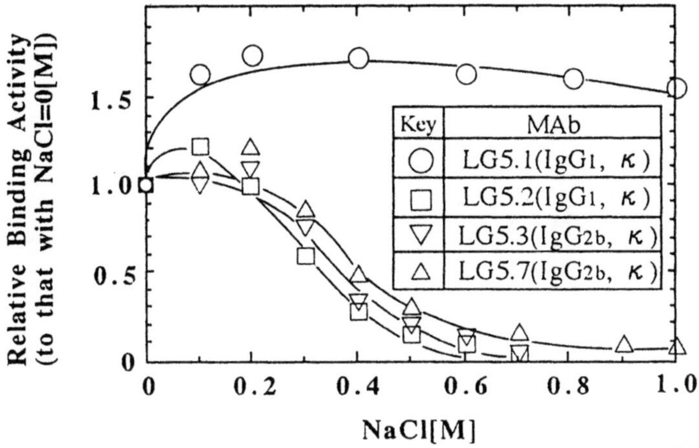

Figure 3. Effects of ionic strength on the relative binding activity of anti-peptide 42-56 mAbs against ß-Lg.
Each mAb was diluted in 20 mM phosphate buffer containing the indicated molar concentration of NaCl at a constant pH value (7.0), and the relative binding activity against ß-Lg was determined by ELISA.

4. Acknowledgments

We thank Mr. Shun-ichi Kinoshita, Mr. Satoshi Ishida, Mr. Tomoaki Takayanagi and Mr. Yuji Masuzawa of Gunma University for their experimental works.

5. References

1. Calton, G. J.: Immunosorbent separations, *Methods Enzymol.* **104** (1984), 381-387.
2. Katoh, S.: Applications of immunoaffinity to separation and analysis, *Hakkokogaku* (in Japanese) **69** (1991), 495-505.
3. Yarmush, M. L., Antonsen, K. P., Sundaram, S., and Yarmush, D. M.: Immunoadsorption: strategies for antigen elution and production of reusable adsorbents, *Biotechnol. Prog.* **8** (1992), 168-178.
4. Enomoto, A., Kamata, N., and Nakamura, K.: Antibodies elicited by the primary response are useful as ligands for immunoaffinity chromatography, *Biochem. Biophys. Res. Commun.* **201** (1994), 1008-1013.
5. Kamata, N., Enomoto, A., Ishida, S., Nakamura, K., Kurisaki, J., and Kaminogawa, S.: Comparison of pH and ionic strength dependence of interactions between monoclonal antibodies and bovine β-lactoglobulin, *Biosci. Biotech. Biochem.* **60** (1996), 25-29.
6. Kaminogawa, S., Shimizu, M., Ametani, A., Hattori, M., Ando, O., Hachimura, S., Nakamura, Y., Totsuka, M., and Yamauchi, K.: Monoclonal antibodies as probes for monitoring the denaturation process of bovine β-lactoglobulin, *Biochim. Biophys. Acta* **998** (1989), 50-56.
7. Mariani, M., Bracci, L., Presentini, R., Nucci, D., Neri, P., and Antoni, G.: Immunogenicity of a free synthetic peptide: carrier-conjugation enhances antibody affinity for the native protein, *Mol. Immunol.* **24** (1987), 297-303.

CHARACTERIZATION OF MONOCLONAL ANTIBODIES PRODUCED BY HYBRIDOMA CELL LINES PREPARED AGAINST HEPATITIS B VIRUS X PROTEIN

MI-KYUNG CHUNG, HEE GU LEE[1], EUN SUK OH[1], SUNG-SHIK MIN, HAE KYUNG LEE, TAE GYU LEE[2], JONG SOON LIM[2], CHANG MIN KIM[3] and SUE NIE PARK*

Virus Oncology Research Unit, [1]Cellular Oncology Research Unit, Molecular and Cell Biology Research Division, Korea Research Institute of Bioscience and Biotechnology, P.O.Box 115, Yusong, Taejon 305-600,
[2]LG Biotech Research Center, LG Chem Research Park, Yusong, Taejon,305-380,
[3]Korea Cancer Center Hospital, 215-4Kongneung-dong, Nowon-Ku 139-240, Seoul, Korea

The X gene of HBV encodes HBx protein which is required for virus replication and has been implicated in virus mediated liver oncogenesis. However, the molecular mechanism of HBx activity and its possible influence on cell proliferation have to be solved further in detail. To understand more about the immune responses against HBV and to make tools useful for preparing diagnostic and therapeutic agents, we have produced monoclonal antibodies against HBx protein. Eight independent hybridoma cell lines including X-01A were isolated for this purified HBx protein. The isotype of the antibody produced from X-01A hybridoma cell line was IgG1. Some HBs-positive sera tested contained antibodies that specifically recognized the HBx protein. We conclude that the HBx protein is antigenic in some patients.

1. Introduction

Hepatitis B Virus (HBV) is one of the causative agents for both acute and chronic hepatitis. The X gene of HBV encodes HBx protein which is required for virus replication. This protein has been also implicated in virus-mediated oncogenesis of hepatocytes. The molecular mechanism of HBx activity related to the hepatocytes proliferation and the progression to hepatocarcinoma have been extensively studied. Specifically, its association to tumor suppressor p53 and its transcription activator functions have been elucidated[1]. However, there are still controversial reports on its mechanism related to its oncogenicity [2,3]. Thus, more studies should be carried out to identify the roles of HBx

Keywords: Hepatitis B virus, HBx protein, monoclonal antibody, serological response

*Corresponding author

in the hepatocarcinogenesis. While studies on the molecular mechanism of HBx functions have been quite extensively carried out[4-6], the correlation of humoral and cellular immune responses against HBx protein and the progression to hepatocarcinoma has not been studied extensively[7-10]. Based on the immunological studies, we might get more insights on its role in hepatocarcinogenesis and we could make more proper tools useful for preparing diagnostic and therapeutic agents against this hepatocarcinogenesis-associated antigen. Therefore, in our laboratory, we have been trying to analyze humoral and cellular immune responses against HBV X antigen in human infected with HBV as following.

2. Materials and Methods

2.1. BACTERIAL STAIN

E.coli strain DH5 α (recA1 supE44 endA1 hsdR17 gyrA96 relA1 thi-1 Δ lacU169 lacZ Δ M15) was used for the expression of recombinant plasmid.

2.2. CONSTRUCTION OF PLASMID FOR EXPRESSION OF GST-X FUSION PROTEIN

To construct a hybrid protein that would encode most of the sequences of the X ORF, the procedure depicted in Fig.1 was followed. The 441bp fragment containing most of X ORF was isolated by PCR from plasmid pHBV315 carrying the HBV genome, which is a HBV adr subtype isolated from a Korean hepatitis patient. This fragment was inserted into the pGEX-3X vector to create the pGEX-X plasmid.

2.3. EXPRESSION OF GST-X FUSION PROTEINS IN *E.COLI*

E.coli transformed with pGEX-X was grown overnight in LB containing 100 μg/ml ampicillin. The overnight culture was added to LB (1:50 dilution) and grown for 90 min. Bacteria were induced with 0.5 mM IPTG for 3h at 37°C. Cells were pelleted, washed once with ice-cold STE (10mM Tris, pH 8.0, 150mM NaCl, 1mM EDTA), resuspended in STE containing 100 μg/ml of lysozyme, and incubated on ice for 15 min. Dithiothreitol (DTT) was then added to a final concentration of 5mM. Bacteria were lysed by the addition of N-laurylsarcosine (sarkosyl) to the final concentration 1.5% from a 10% stock in STE. After 1min of sonication on ice, the lysate was clarified by centrifuging at 10,000g (5min, 4°C). Supernatants were adjusted to 2 or 4% triton X-100, vortexed, incubated with glutathione agarose beads, and washed 6-8 times with ice-cold PBS (8.4mM Na_2HPO_4, 1.9mM NaH_2PO_4, pH 7.4, 150mM NaCl) by repeated centrifugation and resuspended in storage buffer (50mM HEPES, pH 7.4, 150mM NaCl, 5mM DTT, 10% v/v glycerol).

2.4. IMMUNOBLOT ANALYSIS

For immunoblotting with recombinant proteins, a protein extract was prepared by boiling the bacteria in sodium dodecyl sulfate (SDS) sample buffer. The proteins of 10 ul of these extracts were separated by SDS-polyacrylamide gel electrophoresis (PAGE) and immunoblotted.

2.5. HUMAN SERA

Sera of HBV-infected patients were obtained from Korea Cancer Center Hospital, Seoul, Korea. Blood samples negative for HBV markers were obtained form the Taejon Chungnam Red Cross Blood Center, Korea National Red Cross, Korea. Classical serological markers of HBV infection were determined by commercial assays (Behringwerke AG, Marburg/Lahn, Germany)

3. Results and Discussion

To elucidate the presence of any correlation between immune responses against HBx protein and hepatitis pathogenicity, the prevalence of antibodies against HBx protein in HBV-associated patients was determined. Diagnostic tools were prepared as following.

3.1. Production of HBx recombinant protein

HBV X gene was cloned into pET or pGEX plasmid vector for the production of recombinant proteins in *E. coli* (Fig.1). HBx protein was produced GST-fused form in *E. coli*. This protein was produced in large quantity of more than 60% of total proteins (Fig.2). As HBx recombinant GST-fusion protein was produced in inclusion bodies, this protein was solubilized by sarkosyl. The solubilized GST-fusion protein was purified by glutathione sepharose affinity column and gel filtration column chromatography (Fig.3). The purified HBx fusion protein was good immunogen eliciting humoral immune responses in mice.

3.2. Establishment of Hybridoma Cell Lines Producing Monoclonal Antibodies against HBx

We have produced monoclonal antibodies against HBx protein. Eight independent hybridoma cell lines including X-01A were isolated for this purified HBx protein and characterized further. The isotype of the antibody produced from X-01A hybridoma cell line was IgG. This antibody reacted well with recombinant antigen and Hep3B X antigen in ELISA, western blot, and immunoprecipitation(Fig.4). We investigated whether these monoclonal antibodies can recognize the HBx antigen in the sera from patients with the hepatocellular carcinoma. So far, this antibody could rarely detect X antigens in the

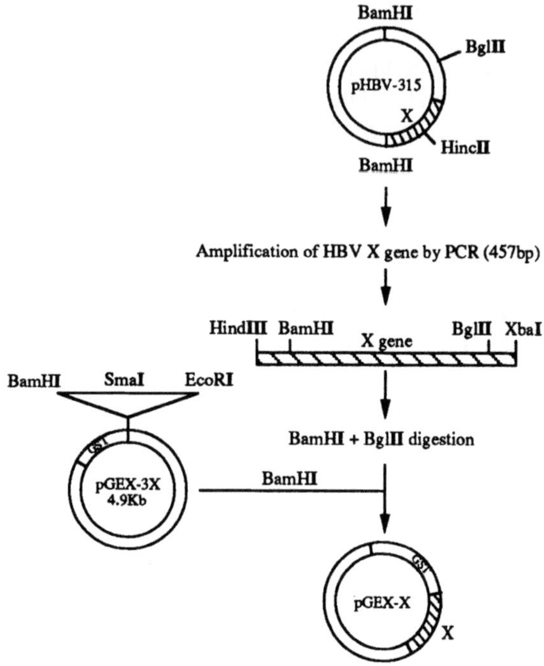

Figure 1. Construction scheme of pGEX-X expression vector

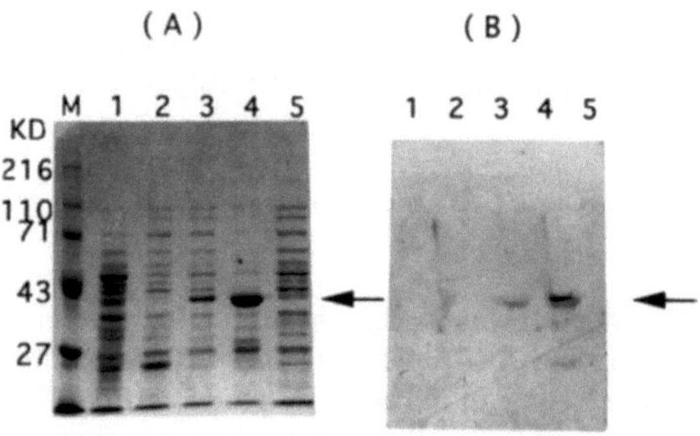

Figure 2. Expression of GST-X fusion protein (A) and western blot analysis with anti-X antibody(B). Lanes 1, total proteins of uninduced control clone control clone containing pGEX-3X vector; 2, total proteins of induced control clone; 3, total proteins of induced pGEX-X recombinant clone; 4, insoluble pellet of induced pGEX-X recombinant clone; 5, soluble supernatant of induced pGEX-X recombinant clone; M, molecular weight standard markers. The position of GST-X is indicated by the arrows.

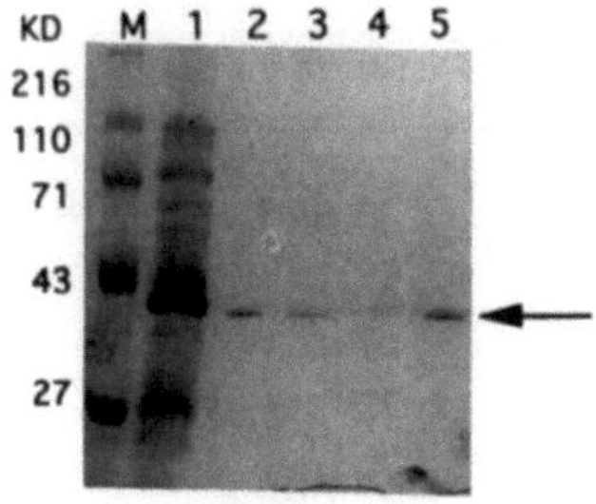

Figure 3. Purification of GST-X fusion protein using glutathione agarose beads. Glutathione agarose beads bound with GST-X protein were subjected to sequential elution as described under Materials and methods. Lanes 1, total lysate of pGEX-X recombinant clone; 2-4, eluted GST-X fusion protein by 10 mM reduced glutathione; 5, glutathione agarose bead suspension (50% v/v); M, molecular weight standard markers. The position of GST-X is indicated by the arrow.

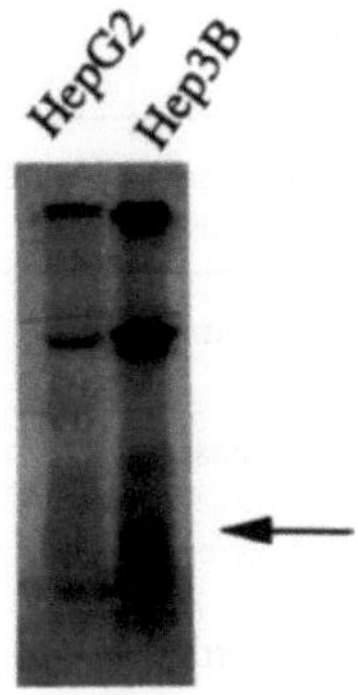

Figure 4. Identification of HBx protein by immunoprecipitation of [^{35}S] methionine-labeled extracts of hepatocellular carcinoma cell lines HepG2 and Hep3B with anti-HBx monoclonal antibody.

human sera. Currently we are trying to establish more hybridoma cell lines that could produce monoclonal antibodies which might be more efficiently reactive to X antigen in sera. Using the X-01A monoclonal antibody, we are now examining the expression of X protein in chronic active hepatitis and hepatocarcinoma tissues associated with HBV to elucidate HBx role in hepatocarcinogenesis and to see whether it could serve as a tumor antigen or not.

3.3. Determination of Serological Responses against HBV X Antigen

With purified X recombinant protein, the presence of antibodies against X protein was determined in the sera obtained from various stages of hepatitis and hepatocarcinoma patients associated with HBV infections. While no anti-X antibody positivity was observed in sera from acute hepatitis patients and chronic healthy carriers, more than 60% of sera from liver cirrhosis and hepatocellular carcinoma were positive for anti-X antibody (Table 1). This result suggests that the presence of anti-X antibody in the liver cirrhosis could be used as a prognostic marker for developing HCC. In addition, this also implies that the X antigen might have a role in the progression to HCC. Currently, we are testing more samples to get stronger evidence of this correlation.

Table 1. Correlation between the presence of anti-X antibody in patient sera, other serological markers and clinical diagnoses

Clinical diagnosis	No. of patients	No. of patients positive for anti-X antibody	Serological markers
Normal	12	0	4 $HBsAg^-$ 8 HBsAg-/anti-HBs^+
Acute hepatitis B	2	0	2 $HBsAg^+$
Chronic carrier	15	0	15 $HBsAg^+$
Liver cirrhosis	4	3	3 $HBsAg^+$ 1 $HBsAg^-$/anti-HBs^+
Hepatocellular carcinoma	19	12	17 $HBsAg^+$ 2 $HBsAg^-$/anti-HBs^+

4. References

1. Wang, X.W., Forrester, K, Yeh, H., Feitelson, M.A , Gu, J.-R., and Harris, C.C. (1994) Hepatitis B virus X protein inhibits p53 sequence-specific DNA binding, transcriptional activity, and association with transcription factor ERCC3. *Proc. Natl. Acad, Sci. USA* **91**, 2230-2234.
2. Henkler, F., Waseem, N., Golding, C., Alison, M.R., and Koshy, R. (1995) Mutant p53 but not Hepatitis B virus X protein is present in Hepatitis B virus-related human hepatocellular carcinoma. *Cancer Research* **55**, 6084-6091.
3. Ueda, H., Ullrich, S., Gangemi, J., Kappel, C.A, Ngo, L., Feitelson, M.A., and Jay, G. (1995) Functional inactivation but not structural mutation of p53 causes liver cancer. *Nature Genetics* **9** , 41-47.
4. Chirillo, P., Falco, M., Puri, P.L., Artini, M., Balsano, C., Levrero, M., and Natoli, G. (1996) Hepatitis B virus pX activates NF-kB-dependent transcription through a Raf-independent pathway. *J. Virol.* **70**, 641-646.
5. Takada, S., Kaneniwa, N., Tsuchida, N., and Koike, K. (1996) Hepatitis B virus X gene expression is activated by X protein but repressed by p53 tumor suppressor gene product in the transient expression system. *Virology* **216**, 80-89.
6. Doria, M., Klein, N., Lucito, R., and Schneider, R.J. (1995) The hepatitis B virus HBx protein is a dual specificity cytoplasmic activator of Ras and nuclear activator of transcription factors. *EMBO J.* **14**, 4747-4757.
7. Guilhot ,S., Fowler, P., Portillo, G., Margolskee, R.G., Ferrari, C., Bertoletti, A. , Chisari, F.V. (1992) Hepatitis B virus(HBV)-specific cytotoxic T-cell response in humans: production of target cells by stable expression of HBV-encoded proteins in immortalized human B-cell lines. *J. Virol.* **66**, 2670-2678.
8. Rehermann, B., Pasquinelli, C., Mosier, S.M., and Chisari, F.V. (1995) Hepatitis B virus(HBV) sequence variation in cytotoxic T lymphocyte epitopes is not common in patients with chronic HBV infection. *J Clin. Invest.* **96**, 1527-1534.
9. Elfassi , E., Haseltine, W.A., and Dienstag, J.L. (1986) Detection of hepatitis B virus X product using an open reading frame *Escherichia coli* expression vector. *Proc. Natl. Acad. Sci. USA* **83**, 2219-2222.
10. Pfaff , E., Salfeld, J., Gmelin, K., Schaller, H., and Theilmann, L. (1987) Synthesis of the X-protein of hepatitis B virus in vitro and detection of anti-X antibodies in human sera. *Virology* **158**, 456-460.

CULTURED-ANIMAL-CELL ASSAY FOR COLLAGEN-SYNTHESIS-ACTIVATION FACTOR PRODUCED BY CALLUS OF WOODY PLANT *EUCOMMIA ULMOIDES*

AKINORI ONO, XIN-HUI XING and HAJIME UNNO *
Department of Bioengineering, Tokyo Institute of Technology
4259 Nagatsuta-cho, Midori-ku, Yokohama 226 , Japan

In order to quantify the undefined factor enhancing collagen synthesis of animal cells (FECS) which is contained in a medicinal woody plant, *Eucommia ulmoides* and its cultured callus, an assay system using human fibroblast cell (HS-K) capable of synthesizing collagen was established. The HS-K cells were cultured in DMEM medium containing 10% FBS. Extracts of dried leaves and cultured calli of *E.ulmoides,* and broth of the callus culture were added into the HS-K culture. After 120-h incubation, collagen amount accumulated in the HS-K cells and in the medium was determined by measuring the amount of hydroxyproline obtained from the hydrolysis of the collagen. Intensity of the FECS was evaluated by the relative ratio of the synthesized collagen amount per HS-K cell under the presence of the additives to that of the control. The amount of collagen synthesized by the HS-K cells was found to depend on the amount of the added extracts. The relative intensity of FECS in the leaf extract, callus culture broth and callus extract were 2.5, 2.3 and 1.7, respectively.

1. Introduction

Eucommia ulmoides (Tochu) has been known as a medicinal woody plant with its hypotensive effect, diuretic effect, etc., from ancient time in China and Japan. About 36 kinds of medicinal compounds have been detected in its barks and leaves[1]. In addition, a recent research demonstrated that leaves of *E.ulmoides* include a factor which enhances collagen synthesis of animal cells (FECS)[2]. The research indicated that the muscle of eel

* corresponding author

cultured with feed supplemented with *E.ulmoides* leaf powder contained much collagen compared with that of control, which resulted in 1.8 times increase in the yield stress of its muscle. A mass production system of FECS by plant cell culture technology will be beneficial to food, cosmetic and medicinal industries, even though the details of FECS are still remained unclear.

The object of the present research is to develop an effective FECS production process by culture of *E.ulmoides* callus. For this purpose an appropriate FECS's assay method should be established. A bioassay system using animal cell culture of human fibroblast cell (HS-K) was thus proposed in the present research.

2. Materials and Methods

Animal cell culture

HS-K cell was obtained from Riken Cell Bank of Japan. The HS-K cells were routinely subcultured in a $0.0025m^2$ tissue culture flask using Dulbecco's modified Eagle's medium (DMEM) supplemented with $1.2kg/m^3$ sodium bicarbonate and 10% fetal bovine serum (FBS). The culture was carried out in a 5% CO_2 incubator at 310K.

Preparation of analytical samples

E.ulmoides calli were induced from infant plants and leaves. The calli fragmented into different ranges in size (<0.6mm, 0.6~1.0mm) were prepared to examining the effect of their size on the growth. After 4-week flask culture, the content of the calli was extracted by these different media, i.e., fresh culture medium, methanol and pure water. Both the callus extract and the callus culture broth diluted by 100 times with PBS were added into HS-K cell culture. This was based on a preliminary experiment that the samples diluted by over 100 times showed no inhibition on the growth of HS-K cells.

Extract of dried leaf of *E.ulmoides* was prepared by boiling 5g dried leaves in 50ml pure water for 3 hours. The extract was also diluted by 100 times with PBS and assayed.

The samples for assaying FECS were summarized in **Table 1**.

Assay of FECS using HS-K cell culture

The HS-K cells were inoculated at 3.0×10^5cells/ml in tissue culture flasks containing 5.0ml DMEM and 1.0ml of additives of calli culture broths, calli extracts and dried leaf extract. PBS of the same volume was used as a control. After 120 hours of incubation, collagen amounts accumulated in the HS-K cells and in the medium were determined by measuring the amount of hydroxyproline obtained from hydrolysis of the collagen. The procedure of collagen amount determination was followed the method of Inayama et al.[3].

The intensity of the FECS was evaluated by the relative ratio of the synthesized collagen amount per HS-K cell under the presence of the additives to that of the control.

3. Results and Discussion

All the FECS intensities extracted by the different solvents/solutions of fresh medium or the callus culture, methanol and pure water showed almost the same values, and these solvents/solutions showed no effect on collagen synthesis of HS-K cells.

In order to examine the relationship between FECS intensity and FECS concentration, HS-K cell culture under the addition of *E.ulmoides* dried leaf extracts diluted by different ratios were carried out. **Fig.1** shows the effect of the dilution ratio on FECS intensity. This result confirmed that synthesized collagen amount per HS-K cell depended on FECS concentration of the additives.

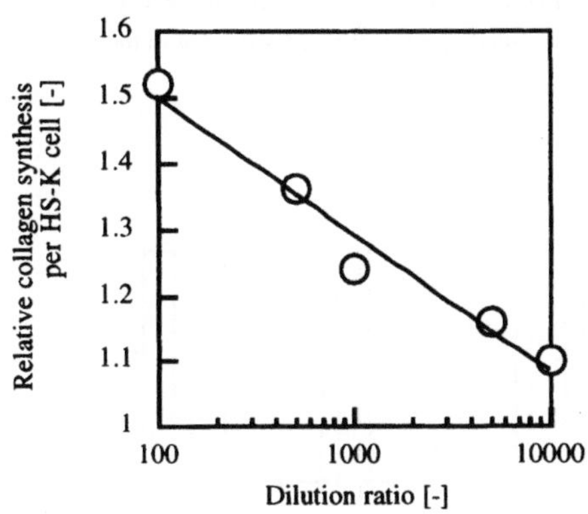

Fig.1 Effect of dilution ratio on amount of synthesized collagen

Fig.2 shows cell numbers of HS-K in the tissue culture flask after 5 days of incubation under the addition of different samples shown in Table 1. This result indicates that the additives showed no effect on the HS-K cell growth.

Fig.3 shows the FECS intensities under the addition of the above different samples. The FECS intensity under the addition of *E.ulmoides* dried leaf extract was 2.5. The FECS intensities under the addition of culture broth and extract of calli cultured under fragmented condition (0.6~1.0mm) which had been induced from leaves and infant plants, were almost the same (1.7~1.8). These were higher than those of calli with no fragmentation. These results show that the explants used for the callus induction did not affect the FECS production. As growth rate of the calli fragmented into 0.6~1.0mm was higher than that of the control, in both cases of the callus induced from the leaves and

Table 1 Samples for FECS assay

Number	Sample	Explants for callus induction
1.	dried leaf extract	
2. 3. 4. 5.	calli extract (control)* 〃 (0.6~1.0mm)** culture broth (control)* 〃 (0.6~1.0mm)**	leaves
6. 7. 8. 9. 10. 11.	calli extract (control)* 〃 (0.6~1.0mm)** 〃 (<0.6mm)*** culture broth (control)* 〃 (0.6~1.0mm)** 〃 (<0.6mm)***	infant plants
12.	PBS	

* : calli were not fragmented prior to the culture
** : calli were fragmented into 0.6~1.0mm prior to the culture
*** : calli were fragmented into less than 0.6mm prior to the culture

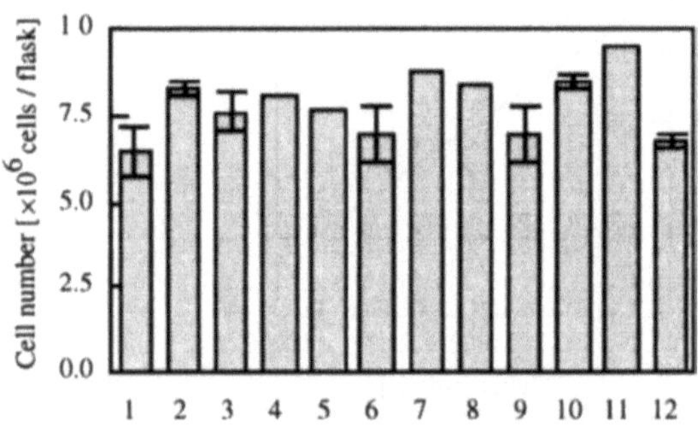

Fig.2 Effect of addives on HS-K cell growth (n=3)

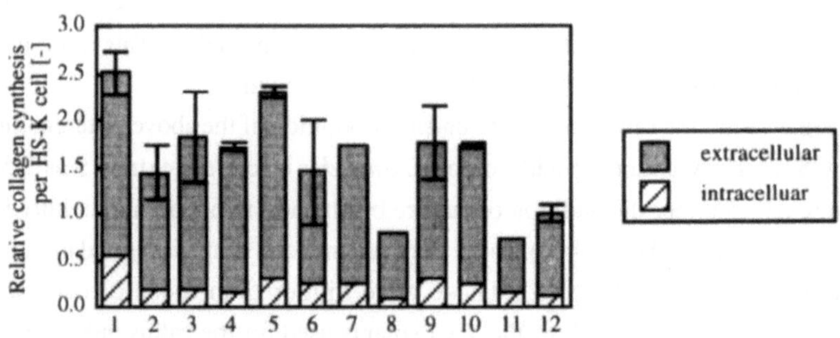

Fig.3 Effect of callus fragmentation on FECS production (n=3)

infant plants, fragmentation also promotes the FECS production. The FECS intensities under the addition of culture broth and extract of calli cultured by having been fragmented into smaller than 0.6mm were relatively small. The growth rate of calli fragmented into smaller than 0.6mm was found to be smaller than that of the control, which was probably due to the callus damage by the too much fragmentation.

In order to examine FECS production kinetics in callus culture, culture of calli fragmented into 0.6~1.0mm were carried out. **Fig.4** shows the time course of FECS intensity in the callus culture. This result shows that FECS was produced rapidly during the early time of the culture, and the FECS production was associated with the callus growth.

The above results show that HS-K cell culture was available for the assay of FECS produced by the callus culture.

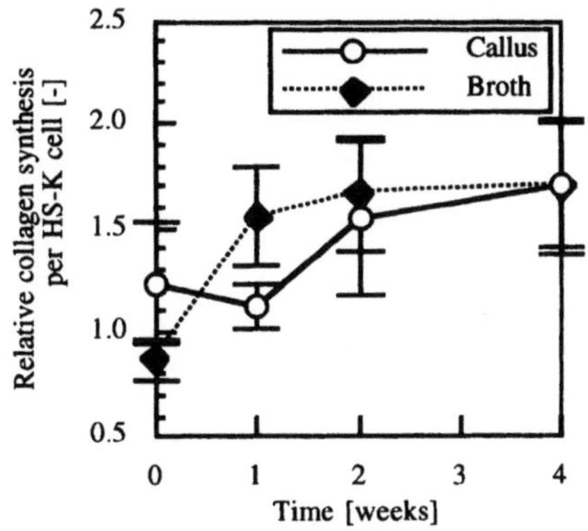

Fig.4 Time course of FECS productivity (n=3)

4. Conclusion

HS-K cell culture was available for the assay of FECS produced by calli culture of *E.ulmoides*. The FECS intensity was dependent on the FECS amount, and callus culture condition. With this assay system, the production of FECS by *E.ulmoides* calli culture has become to be characterized .

Acknowledgment

This research was supported by a Grant-in Aid (No.08650937) for Scientific Research from the ministry of Education, Science, Sports and Culture of Japan.

The authors thank Dr. Deyama T. of Yomeishu Central Research Laboratories for kindly supplying the seeds.

References

1) T.Deyama, (1994) , *Private communication*

2) S.Tanimoto, K.Ikuma and S.Takahashi, (1993), Improvement in Raw Meat Texture of Cultured Eel by Feeling of Tochu Leaf Powder, *Biosci. Biotech. Biochem.*, **57**(2), 205-208.

3) S.Inayama, T.Shibata, J.Ohtsuki and S.Saito, (1978) , *Keio J. Med.*, **27**, 43-46.

Author index

Subject index

MIX
Papier aus verantwortungsvollen Quellen
Paper from responsible sources
FSC® C105338
FSC
www.fsc.org

If you have any concerns about our products,
you can contact us on
ProductSafety@springernature.com

In case Publisher is established outside the EU,
the EU authorized representative is:
Springer Nature Customer Service Center GmbH
Europaplatz 3, 69115 Heidelberg, Germany

Printed by Libri Plureos GmbH
in Hamburg, Germany